Lecture Notes in Civil Engineering

Volume 608

Lecture Notes in Civil Engineering (LNCE) publishes the latest developments in Civil Engineering—quickly, informally and in top quality. Though original research reported in proceedings and post-proceedings represents the core of LNCE, edited volumes of exceptionally high quality and interest may also be considered for publication. Volumes published in LNCE embrace all aspects and subfields of, as well as new challenges in, Civil Engineering. Topics in the series include:

- Construction and Structural Mechanics
- Building Materials
- Concrete, Steel and Timber Structures
- Geotechnical Engineering
- Earthquake Engineering
- Coastal Engineering
- Ocean and Offshore Engineering; Ships and Floating Structures
- Hydraulics, Hydrology and Water Resources Engineering
- Environmental Engineering and Sustainability
- Structural Health and Monitoring
- Surveying and Geographical Information Systems
- Indoor Environments
- Transportation and Traffic
- Risk Analysis
- Safety and Security

To submit a proposal or request further information, please contact the appropriate Springer Editor:

- Pierpaolo Riva at pierpaolo.riva@springer.com (Europe and Americas);
- Swati Meherishi at swati.meherishi@springer.com (Asia—except China, Australia, and New Zealand);
- Wayne Hu at wayne.hu@springer.com (China).

All books in the series now indexed by Scopus and EI Compendex database!

Weiqiang Wang · Chengzhi Wang · Yang Lu
Editors

Hydraulic Structure
and Hydrodynamics

 Springer

Editors
Weiqiang Wang
Hohai University
Nanjing, China

Chengzhi Wang
Chongqing Jiaotong University
Chongqing, China

Yang Lu
Hohai University
Nanjing, China

ISSN 2366-2557   ISSN 2366-2565 (electronic)
Lecture Notes in Civil Engineering
ISBN 978-981-97-7250-6   ISBN 978-981-97-7251-3 (eBook)
https://doi.org/10.1007/978-981-97-7251-3

This work was supported by Guangzhou KEO Info Technology Co., Ltd.

This Springer imprint is published by the registered company Springer Nature Singapore Pte Ltd.
The registered company address is: 152 Beach Road, #21-01/04 Gateway East, Singapore 189721, Singapore

If disposing of this product, please recycle the paper.

Preface

Entitled *Hydraulic Structure and Hydrodynamics*, this book collects a bunch of selected papers from the academic conference 2023 9th International Conference on Hydraulic and Civil Engineering (ICHCE 2023), contributions recommended by the organizing committees, invited speakers and research institutes, and manuscripts from other sources.

This book delves into discussions central to hydraulic structures and research in the realm of hydrodynamics. Hydraulic structures stand as pivotal components within civil engineering and construction, playing a safeguarding role for structures vital to human development. Monitoring the safety and ensuring the structural stability of hydraulic structures have long remained a focal point within hydraulic engineering. Thus, factors affecting the safety of hydraulic structures, water pressure, and loading demand meticulous attention.

The book aims to furnish global civil engineers with cutting-edge research and engineering examples pertaining to the safety and hydrodynamics of hydraulic structures, with a particular emphasis on dam safety and inspection. It endeavors to inspire novel insights and research avenues for the readers and provide some experiences and results for researches in this field.

The book's content spans a wide spectrum of cutting-edge topics and emerging trends of hydrostructures and hydrodynamics, encompassing specific topics such as structural safety and testing of dams, hydraulic soil stability and seepage effects, and hydrodynamics and rheology. Each chapter is the result of collaboration by experts from various domains and provides a detailed overview of the potential that new technologies in the field of hydraulic engineering currently offer or will offer in the near future. We believe that the book will inspire new ideas and facilitate collaboration among the different disciplines involved.

The readers would benefit from enhancing their knowledge and skills in the domain areas. Also, this book may help the readers in developing new and innovative ideas. The book can be a valuable reference for beginners, researchers, and professionals interested in researches related to hydraulic engineering.

January 2024   The Editor in Chief

Contents

Study of Hydraulic Soil Stability and Seepage Effects

Structural Safety and Intelligent Monitoring of Dams

.

Experimental Study on Shear Characteristics of Gravelly Soil of Core Wall of Lianghekou Dam Under Freeze–Thaw Conditions

Bo Zheng and Enlong Liu

Abstract In order to study the mechanical properties of the upper loosely-paved gravelly soil used to protect the compacted dam material under the freeze–thaw cycle during winter construction on the engineering site, the gravelly soil with the soil-rock ratio of 7:3 was mixed evenly and sealed under the corresponding dry density and moisture content, and then subjected to freeze–thaw cycles. After reaching the corresponding freeze–thaw times, the compaction samples were taken out. The direct shear tests were conducted on gravelly soil after freeze–thaw cycles under different vertical pressures and the change pattern of mechanical properties were analyzed. The experimental results show that freeze–thaw cycles have no significant effect on the shear strength of soil samples under the lower vertical pressure. Under the higher vertical pressure, the shear strength of soil samples decreases with the number of freeze–thaw cycles and then stabilizes. The effect of freeze–thaw cycles on the shear strength parameters of the gravelly soil is more obvious. As the number of freeze–thaw cycles increases, the cohesive force tends to increase, while the internal friction angle decreases, tending to stabilization after 10 freeze–thaw cycles.

Keywords Gravelly soil · Freeze–thaw cycle · Mechanical properties · Direct shear test

1 Introduction

Lianghekou Hydropower Station is at the position of about 2 km downstream from the gathering mouth of Xianshui River, the main stream of Yalong River [1]. The watershed area controlled by the station is about 65,700 km^2, and the average flow

B. Zheng · E. Liu (✉)
State Key Laboratory of Hydraulics and Mountain River Engineering, College of Water Resource and Hydropower, Sichuan University, Chengdu, Sichuan, China
e-mail: liuenlong@scu.edu.cn

B. Zheng
e-mail: zhengbo@stu.scu.edu.cn

© The Author(s) 2025

W. Wang et al. (eds.), *Hydraulic Structure and Hydrodynamics*, Lecture Notes in Civil Engineering 608, https://doi.org/10.1007/978-981-97-7251-3_1

rate of the river is about 666 m^3/s for many years at the dam [2, 3]. The dam of Lianghekou Hydropower Station is a gravelly soil core-wall rockfill dam with a maximum height of 295 m, which is the second highest earth and rock dam in China and the third highest in the world [4–7]. Since the dam is as high as 300 m, the performance of the core wall impermeable soil material is very important for the safety of the dam body, so the requirements for the impermeable soil material are very high. However, the coarse grain content in the soil is low, which makes the mechanical index low and the compressibility high, and it is not suitable for the core wall impermeable material of 300 m level core-wall rockfill dam [8].

Rocks and soils that are at or below 0 °C and are cemented by ice are called frozen soil. Lianghekou Hydropower Station is located on the western Sichuan Plateau, with an average altitude of 3000 m, and the lowest temperature can reach − 15.9 °C. During the winter filling process of dams, the core wall materials will be affected by freezing and thawing, and the related characteristics will be changed, which is very unfavorable to the construction of dams. Regarding the effects of freeze–thaw action on hydroelectric dams, studies have focused on concrete-face rockfill dams and roller-compacted concrete dams to explore the effects of freeze–thaw action during winter construction, and a large amount of practical experience has been accumulated in this area at home and abroad for engineering reference [9]. There is currently limited research on the impact of freezing and thawing on the core wall materials of rockfill dams during winter construction, and there is limited practical data and research literature available for Ref. [10]. Lianghekou Hydropower Station Dam is a 300 m gravelly core-wall rockfill dam. The safety of the dam is related to the safety of multiple counties and reservoirs downstream. Studying the changes in mechanical properties of the gravelly soil core wall material of Lianghekou Hydropower Station under freeze–thaw conditions is of great significance and is very urgent. In this paper, a direct shear test is conducted on the gravelly soil after freeze–thaw cycles for the gravelly soil core wall material of Lianghekou Hydropower Station to analyze the changes of the strength properties of the gravelly soil under different number of freeze–thaw cycles and normal pressure.

2   Specimen Preparation Methods

The gravelly soil is prepared according to the determined sample dry density, moisture content, and specimen size. The gravelly soil with the soil-rock ratio of 7:3 is prepared with dry density $\rho_d = 1.97$ g/cm^3, moisture content $w = 11.8\%$, and the specimen size is: $L \times W \times H = 150 \times 150 \times 100$ mm. The gravelly soil is mixed and stewed at the corresponding moisture content for 24 h and then sealed in a sealed bag (as each specimen is compacted in three layers, it is divided into three parts), and then place it into the freeze–thaw testing machine for freeze–thaw cycle test under closed conditions (without external water compensation). Based on the situation at the engineering site, the freezing temperature is set to − 15 °C, the melting temperature is 20 °C, and the freeze–thaw cycle is one day. The experiment sets up five sets of freeze–thaw cycle times, including 0, 1, 5, 10, and 20.

After reaching the set number of freeze–thaw cycles, remove the specimen and let it stand at room temperature. Direct shear tests use DZJ-15 strain-controlled direct shear apparatus in the vertical pressure of 50, 200, 600, 1200 kPa. The shear rate of this experiment is taken as 1.5 mm/min and the horizontal displacement, horizontal shear stress, and vertical displacement are recorded every 0.1 mm. The experiment ends when the horizontal displacement reaches 18 mm.

3   Test Result and Analysis

Referring to Test Regulations, direct shear tests are carried out on gravelly soil, and the shear stress–shear displacement relationship curves and vertical displacement–shear displacement relationship curves are obtained.

The shear stress–shear displacement relationship curves are shown in Fig. 1, and the vertical displacement–shear displacement relationship curves are shown in Fig. 2, where the vertical displacement is positive in compression.

As seen from the shear stress–shear displacement relationship curves, it can be seen that for the same number of freeze–thaw cycles, there are no significant peaks in the relationship curve under vertical pressures of 50, 200, 600, and 1200 kPa. The overall performance is strain hardening, and the greater the vertical pressure hardening more obvious. Under the action of vertical pressure, the specimen is compacted, and the particles on the shear surface are rearranged during the shear process, and the small particles are filled into the pores composed of large particles, which makes the shear stress continuously increase with the increase of shear displacement, without obvious peak. Under the same number of freeze–thaw cycles, the greater the vertical pressure, the steeper the trend of the relationship curve, and the greater the shear stress corresponding to the same shear displacement. The increase in vertical pressure reduces the pore size of the specimen and increases the constraint, resulting in a significant increase in the strength and stiffness of the specimen; Under the condition of vertical pressure 1200 kPa, the shear stress–shear displacement relationship curve will appear obvious fluctuations. It is considered that this is due to particle fragmentation and reorientation arrangement. The higher vertical pressure makes the specimen denser, the particles are subjected to stronger constraints. During the shear process, the higher the shear stress makes the particles broken, and the shear stress suddenly decreases. With continuing to shear, the broken particles are rearranged and filled, and the shear stress will gradually increase.

From the vertical displacement–shear displacement relationship curve, it can be seen that under the same number of freeze–thaw cycles, the gravelly soil exhibits slight shear shrinkage first and then significant shear expansion at the vertical pressure of 50 kPa. When the vertical pressure is 200 kPa and above, it exhibits shear shrinkage, and the greater the pressure, the more obvious the shear shrinkage; The edges of soil particles are serrated and arranged in a staggered arrangement. The particles bite

Fig. 1 Shear stress–shear
displacement curves under
different number of
freeze–thaw cycles

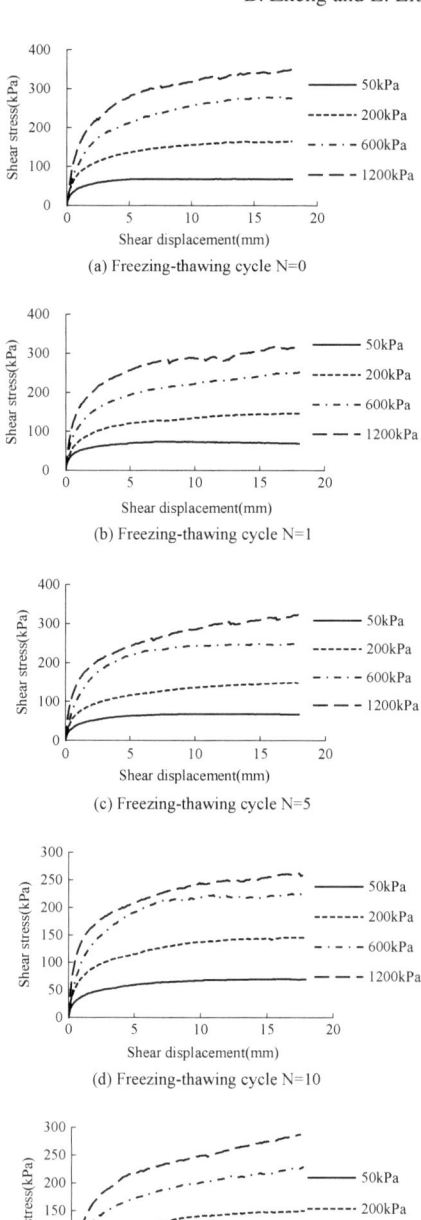

(a) Freezing-thawing cycle N=0

(b) Freezing-thawing cycle N=1

(c) Freezing-thawing cycle N=5

(d) Freezing-thawing cycle N=10

(e) Freezing-thawing cycle N=20

Fig. 2 Vertical displacement–shear displacement curves under different number of freeze–thaw cycles

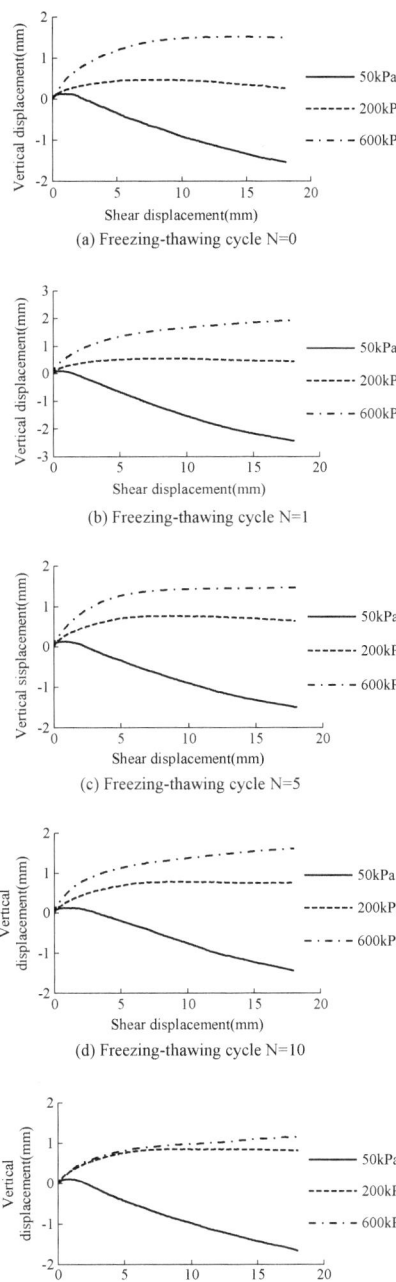

(a) Freezing-thawing cycle N=0

(b) Freezing-thawing cycle N=1

(c) Freezing-thawing cycle N=5

(d) Freezing-thawing cycle N=10

(e) Freezing-thawing cycle N=20

each other to hinder relative movement. During the shear process, the particles first rotate and stand upright, then across the adjacent particles before moving, which will generally increase the volume, shear expansion. The smaller the vertical pressure, the smaller the constraint, and the more obvious the shear expansion. However, when the vertical pressure is high, strong constraints can be generated, and the phenomenon of shear expansion no longer occurs.

The strength of soil is one of the important mechanical properties of soil. Soil is a collection of scattered particles, and the connection strength between soil particles is much smaller than the strength of the soil particles themselves. Therefore, under external forces, soil failure is mainly due to the relative dislocation between soil particles, that is, the strength of soil mainly depends on the interaction force between particles, rather than directly determined by the strength of the soil particles themselves. The strength of soil is actually the shear strength of soil, which refers to the ultimate ability of the soil to resist shear failure. The shear strength of soil is mainly composed of cohesion and internal friction, which is expressed using the Mohr Coulomb theory as:

$$\tau = c + \sigma \tan \varphi \tag{1}$$

where τ is the shear strength of the soil; σ is the normal stress on the shear plane; c is the cohesive force of the soil; φ is the internal friction angle of the soil.

Due to the lack of significant peaks in the shear stress–shear displacement relationship curve of gravelly soil under different freeze–thaw cycles and vertical pressures, so the shear stress of 10% strain is taken as the shear strength. The shear strength of gravelly soil under different freeze–thaw cycles and different vertical pressures is summarized in Table 1.

According to the data in Table 1, the variation curve of the shear strength of gravelly soil with the number of freeze–thaw cycles is shown in Fig. 3.

From Fig. 3, it can be concluded that under the same number of freeze–thaw cycles, the shear strength of gravelly soil increases with the increase of vertical pressure. When the vertical pressure is 50 kPa, the freeze–thaw cycle has no significant effect on the shear strength of the soil sample; At a vertical pressure of 200 kPa, the shear strength of the soil sample decreases by 11.52% after one freeze–thaw cycle, and then stabilizes with the increase of freeze–thaw cycles; At a vertical pressures of

Table 1 Summary of shear strength of gravelly soil

Vertical pressures (kPa)	Shear strength (kPa)				
	$N = 0$	$N = 1$	$N = 5$	$N = 10$	$N = 20$
50	67.44	70.77	67.44	69.44	66.10
200	162.25	143.55	144.89	143.55	146.89
600	275.76	241.04	247.04	220.34	216.36
1200	341.86	306.47	309.81	260.39	268.75

Fig. 3 Changes in shear strength under different number of freeze–thaw cycles

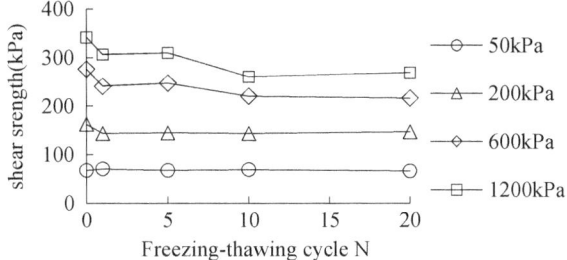

Table 2 Summary of shear strength parameter

Shear strength parameters	$N = 0$	$N = 1$	$N = 5$	$N = 10$	$N = 20$
Cohesion c (kPa)	66.71	67.92	69.55	71.52	71.52
Internal friction angle φ (°)	19.75	16.49	16.75	14.41	14.14

600 and 1200 kPa, the shear strength generally shows a decreasing trend with the increase of freeze–thaw cycles, and tends to stabilize after 10 freeze–thaw cycles.

According to the Mohr Coulomb strength theory, the vertical pressures of 50, 200, 600 kPa, and corresponding shear strength of gravelly soil under the same number of freeze–thaw cycles are fitted in a straight line to obtain the cohesion c and internal friction angle φ.

The shear strength parameters of gravelly soil under different freeze–thaw cycles are summarized in Table 2.

With the increases of freeze–thaw cycles, the cohesion of gravelly soil increases and tends to stabilize after 10 freeze–thaw cycles. From 10 to 20 freeze–thaw cycles, the change in cohesion is less than 1%; The internal friction angle decreases and tends to stabilize after 10 freeze–thaw cycles. From 10 to 20 freeze–thaw cycles, the change of internal friction angle is less than 2%. The freeze–thaw cycle makes the coarser particles split into lower particle sizes, the sticky particles agglomerate to higher particle size, while the particles move and rearrange. The increase in fines content leads to a weakening of the intergranular occlusion and a decrease in roughness, resulting in a decrease in the internal friction angle. The changes of gravelly soil structure caused by freeze–thaw cycles gradually tend to dynamic equilibrium as the number of freeze–thaw cycles increases, and the internal friction angle remains in a stable state.

4 Conclusions

(1) Under different freeze–thaw cycles, the shear stress–shear displacement curves exhibit strain hardening, and the greater the vertical pressure, the more obvious the hardening. Under the condition of the vertical pressure of 1200 kPa, the

shear stress–shear displacement relationship curves will exhibit significant fluctuations.

(2) For the vertical displacement shear-displacement relationship curve it can be found that at the vertical pressure of 50 kPa it first has slight shear shrinkage after obvious shear expansion. When the vertical pressure is 200 kPa and above, it exhibits shear shrinkage, and the greater the pressure shear contraction is more obvious.

(3) When the vertical pressure of gravelly soil is 50 kPa, the freeze–thaw cycle has no significant effect on the shear strength of the soil samples; At a vertical pressure of 200 kPa, the shear strength of the soil sample decreases after one freeze–thaw cycle, and then the shear strength stabilizes with the increase of the number of freeze–thaw cycles; at the vertical pressures of 600 and 1200 kPa, the shear strength generally shows a decreasing trend with the increase of freeze–thaw cycles, and tends to stabilize after 10 freeze–thaw cycles.

(4) As the number of freeze–thaw cycles increases, the cohesion of gravelly soil increases, and the internal friction angle decreases, and stabilizes after 10 freeze–thaw cycles.

References

1. Z. Yuan, Excavation and blasting technology of tailrace surge chamber of Lianghekou hydropower station on Yalong River in SiChuan, no. 01 (2022), pp. 64–67+77
2. X.J. Wu, Z.Y. Shi, Technology and roller compaction test of core gravelly earth material of the Lianghekou hydropower project dam. Yunnan Water Power **32**(03), 33–38 (2016)
3. T. Gu, J.H. Shen, Y. Liu et al., The stability of the left abutment slope in Lianghekou hydro-electric power station of the Yalong River. J. Geol. Hazards Environ. Preserv. **20**(04), 71–75 (2009)
4. Y.H. Mu, X.Y. Zhu, P. Yue et al., Monitoring investigation on winter freezing-thawing of dam core wall soils in cold regions. J. Glaciol. Geocryol. **40**(04), 756–763 (2018)
5. X. Jiang, N.W. Xu, Z. Zhou et al., Failure mechanism of surrounding rock of bus-bar tunnels at Lianghekou hydropower station subjected to excavation. Rock Soil Mech. **40**(01), 305–314 (2019)
6. Z.J. Zhou, X.W. Zhu, Y. Han et al., Influence of different rockfill zoning schemes in dam body stress deformation at Lianghekou core rockfill dam. Des. Hydroelectr. Power Station **37**(03), 6–7+32 (2021)
7. Q.H. Li, Slope stability analysis and reinforcement of super river-crossing bridge in reservoir area of Lianghekou hydropower station. Yangtze River **54**(02), 159–164 (2023)
8. P. Gao, S.Y. Wu, An experimental study of gravelly impervious materials for core wall in Lianghekou hydropower station. Adv. Sci. Technol. Water Resour. **32**(05), 64–66 (2012)
9. S. Leroueil, J. Tardif, M. Roy et al., Effects of frost on the mechanical behaviour of Champlain Sea clays. Can. Geotech. J. **28**(5), 690–697 (1991)
10. Q. Wang, Y.W. Zhou, Management and technology of the dam construction for the Jilintai I hydropower station in Xinjiang Uygur Autonomous Region. Water Resour. Hydropower Eng. (06), 1–4+7 (2004)

Study on Risk Assessment Method of Cascade Reservoirs Based on Hidden Danger Investigation

Yuwei Xie, Pubucireng, Yuanwu Wan, Xuehui Peng, and Peiran Jing

Abstract There are numerous cascade reservoirs in China's small and medium-sized river basins. Once the cascade reservoirs break, the disastrous consequences will result in extreme economic, social, and ecological losses. According to the hidden danger investigation results of cascade reservoirs, a risk assessment method and standard for cascade reservoirs were proposed and established based on the risk matrix method, which provides a foundation for the risk assessment of cascade reservoirs. Taking the D, C, and H cascade reservoirs as the case study, according to the hidden danger investigation results and dam break consequences of the reservoirs, it was concluded that the risk levels of the three reservoirs were in order of extremely high risk, high risk, and high risk. Therefore, it is suggested that engineering and non-engineering measures should be taken to reduce the cascade reservoir's risk to ensure the high-quality and safe operation of the cascade reservoir group in China.

Y. Xie (✉) · Y. Wan · X. Peng · P. Jing
Nanjing Hydraulic Research Institute, Nanjing, Jiangsu, China
e-mail: hhu_xyw@163.com

Y. Wan
e-mail: 2018100067@nit.edu.cn

X. Peng
e-mail: xhpeng@nhri.cn

P. Jing
e-mail: 286910843@qq.com

Y. Xie
College of Water Conservancy and Hydropower Engineering, Hohai University, Nanjing, Jiangsu, China

Pubucireng
Shigatse Everest Urban Investment and Development Group Co., Ltd., Shigatse, China
e-mail: puci.212@163.com

P. Jing
State Key Laboratory of Water Resources Engineering and Management, Wuhan University, Hubei, China

W. Wang et al. (eds.), *Hydraulic Structure and Hydrodynamics*, Lecture Notes in Civil Engineering 608, https://doi.org/10.1007/978-981-97-7251-3_2

Keywords Cascade reservoirs · Hidden danger investigation · Risk assessment · Risk criteria

1 Introduction

Due to the frequent extreme climate events and aggravated human activities, the complexity of safety and potential risks of cascade reservoirs in river basins are significantly increased. Once a specific reservoir in the cascade reservoir group breaks due to flash floods, strong earthquakes, and significant engineering hazards, it is probable to form a "domino effect" and bring catastrophic consequences to the safety of the entire cascade reservoir group. Therefore, it is necessary to strengthen the theoretical research and engineering practice on the risk evolution mechanism and risk assessment of river basin cascade reservoirs, which has profound and significant implications for effectively ensuring the safety of river basin cascade reservoirs.

Many risk assessment methods have been conducted for individual reservoir dams. Meng et al. employed the entropy-weighted fuzzy comprehensive evaluation method to assess the risk of reservoir flood operation [1]; Peng et al. proposed dam risk standards for life, economy, society, and environment in China based on domestic and international risk standards, providing a reference basis for evaluating the consequences of the dam break [2]. Research on risk assessment of the cascade reservoir groups is still under development. Li et al. discussed the cascade reservoir groups risk studies in China from the perspectives of accident risk probability, risk loss, risk standards, and cascade failure risk [3]; Zhou et al. proposed risk standards for ultra-high dams and their cascade reservoirs based on domestic and international risk standards [4, 5].

Attributable to the relatively small number of cascade reservoirs failure cases at home and abroad [6, 7], previous studies are mainly based on numerical methods for simulating dam break floods [8, 9]. The evolution algorithms and models have many advantages and disadvantages, and the objects are concentrated on cascade reservoirs on large rivers. Further in-depth research is needed for cascade reservoirs in small and medium-sized watersheds. Owing to the complexity of risk transmission in cascade reservoirs, it is difficult to calculate the probability of dam break and assess the risk. This study proposes a novel approach by investigating potential safety hazards in reservoirs and dams and adopts a semi-quantitative analysis method to conduct risk assessment research on cascade reservoirs.

2 The Investigation of Potential Safety Hazards

2.1 The Contents of Potential Safety Hazards

The potential safety hazards of dams mainly include engineering and management hazards, which are primarily investigated through data consulting, on-site inspection, informal exchange, calculation, and analysis.

The actual contents of the potential danger investigation include flood control capacity, water retaining structure, water release structure, water conveyance structure, metal structure, electrical equipment, and auxiliary management facilities. During the on-site inspection, the general part is first followed by the local part, highlighting the essential parts and observing the critical phenomena. Vital parts primarily include the dam crest, upstream and downstream dam surface, dam foot and adjacent areas, rock (soil) mass on both sides of the spillway, inlet–outlet of water conveyance structures, and adjacent upstream reservoirs. The dangerous signs, such as leakage, collapse pits, cracks, landslides, and flow soil or piping at the dam foundation, should be particularly mentioned. Management facilities include flood control roads, monitoring facilities, communications, lighting facilities, flood control emergency and rescue facilities, and management rooms.

The potential management hazards mainly include the implementation of system and management, embodied in the following aspects: the safety responsibility system, daily patrol of essential links (e.g., rainwater regime monitoring and reporting ability, dispatching application plan, and emergency plan), and maintenance and comprehensive management (e.g., registration, safety identification, risk removal, and reinforcement, dispatching and operation, management organization, management system, management fees, and archive management).

2.2 The Hazard Severity Classification

According to China's *Criteria for Determining Potential Hazards of Major Accidents in Production Safety of Hydraulic Projects* (No. 344, Hydraulic Safety Supervisor [2017]), production safety accidents of hydraulic projects are divided into general potential and significant potential hazards. Moreover, in the *Manual of Supervision and Inspection for Safe Operation and Management of Small Reservoirs* issued by the Department of Transportation and Management of the Ministry of Water Resources, hidden dangers of reservoir engineering are divided into general, heavy, and serious ones. Referring to the grading standard of the Department of Transportation and Management of the Ministry of Water Resources, the hidden danger severity of reservoir dam can be divided into three levels as follows:

(1) Generally hidden dangers: minor hazards. The engineering hidden dangers or management problems that can be eliminated immediately through rectification;

(2) Heavy hidden dangers: there are specific hazards that may aggravate the engineering hidden risks or management problems if not corrected in time;

(3) Serious hidden dangers: significant hazards, great impact on dam safety, and possible consequences. It is easier to ensure the safe operation of the reservoir project by taking corrective measures such as risk removal and reinforcement.

3   The Risk Assessment Method

This study employed the risk matrix method to assess the risk of cascade reservoir groups. The risk matrix method combines two factors, the possibility of a dam break of the cascade reservoirs and the severity of the dam break consequence. It is an evaluation method that combines qualitative and quantitative analysis in which the possibility and the consequence of a dam break are assessed accordingly.

3.1   The Risk Matrix Method

The risk matrix method is formulated as follows:

$$R = P \cdot L \tag{1}$$

where R is the risk value, and R's range is 1–25. P is the probability level of a dam break, and the value range is 1–5. L is the severity level of dam break consequences, and the value range is 1–5.

3.2   The Criteria for Risk Classification

The risk level classification of cascade reservoir groups based on the risk matrix method refers to the classification standard of the *Guidelines for Risk Prevention and Control of Cascade Reservoirs Safety* (NB/T10882-2021) [10]. The probability level of dam break is classified based on the severity of reservoir hazards, and the specific classification standards are shown in Table 1. The severity level of dam break consequences is classified based on the loss of life, and the classification criteria refer to the *Research on Risk Standards for the Reservoir Dams in China* [11]. Based on the four-level classification, an additional level has been added. A lower level of consequence severity has been added, as shown in Table 2.

According to the probability level and the severity of consequences of the dam break, the risk of the cascade reservoir group is divided into four levels: (1) Level IV, low risk, acceptable risk, with a risk value of 1–3; (2) Level III, general risk,

Table 1 The classification criterion for the probability level of dam break

Level	Qualitative description	The judgment basis		
		Safety appraisal conclusion	Engineering hazards	Management hazards
1	Highly unlikely to occur	"Class I dam"	General engineering hazards	Heavy management hazards
2	Unlikely to occur	"Class II dam"	1–2 heavy engineering hazards	1–2 serious management hazards
3	Basic unlikely to occur	"Class II dam"	3 or more heavy engineering hazards	3 or more severe management hazards
4	Likely to occur	"Class III dam"	1–2 serious engineering hazards	–
5	Very likely to occur	"Class III dam"	3 or more severe engineering hazards	–

Table 2 The classification criterion for the severity level of dam break consequences

Level	Qualitative description	Loss of life (person)
1	General	$L < 3$
2	More	$3 \leq L < 15$
3	Significant	$15 \leq L < 30$
4	More significant	$30 \leq L < 100$
5	Disastrous	$L \geq 100$

tolerable risk, with a risk value of 4–9; (3) Level II, high risk, unacceptable risk with a risk value of 10–16; (4) Level I, very high risk, with a risk value of 17–25.

4 The Engineering Case

Taking a cascade reservoir group composed of three medium-sized reservoirs, D, C, and H in a small and medium-sized watershed as a case study, a risk assessment of cascade reservoirs was carried out.

4.1 The Probability Analysis of Dam Break

The author has investigated the hidden dangers of D, C, and H reservoirs, and the results are shown in the second column of Table 3. Compared with Table 1, the probability level of dam break of each reservoir can be obtained, which is listed in the third column of Table 3.

Table 3 The investigation results of hidden dangers of cascade reservoirs

Reservoir	The description of hidden dangers	Level
D	The safety appraisal conclusion was "Class III dam". The reservoir is in the process of danger removal and reinforcement There were severe engineering hazards (1) The anti-seismic stability of the main dam slope does not meet the standards' requirements (2) The filling quality of the dam filling needs to be improved, the permeability coefficient of the dam body does not meet the standards, and the sand and pebble foundation of the dam foundation needs to be thoroughly cleared (3) The upstream dry block stone slope protection of the dam is seriously damaged, and the concrete of the wave wall on the dam crest is carbonized and cracked (4) The monitoring and management facilities need to be completed	4
C	The safety appraisal conclusion was "Class II dam" There were severe engineering hazards (1) The dam crest has been taken as a National Highway, but the dam crest as a highway has not been subject to the unique safety demonstration (2) Longitudinal and transverse cracks exist in the dam crest's pavement, drainage ditches on downstream slopes are partially damaged and blocked, and outlets of drainage prism and ditches are blocked (3) The concrete revetment on the left bank of the spillway discharge section has local settlement and collapse, and the drainage is blocked (4) The water stop of the bulkhead gate is damaged	3
H	The safety appraisal conclusion was "Class II dam" There were severe engineering hazards (1) The dam crest is asphalt concrete pavement, and as the trunk line of traffic, its safety has yet to be demonstrated in safety appraisal (2) The quality of dam filling could be better. Various project risks have occurred, and reinforcement has been carried out, but risks may still arise under high water levels (3) Some dam deformation monitoring facilities were damaged	3

4.2 The Analysis of Consequences of Dam Break

Assuming that the watershed experiences the check flood with a recurrence interval of 2000 years. The dam failure model is set up with three working conditions: (1) Condition 1: the D Reservoir experiences a dam failure; (2) Condition 2: the C Reservoir experiences dam failure; (3) Condition 3: the H Reservoir experiences a dam failure. As mentioned above, in Condition 1, the dam failure of the D reservoir caused the downstream C and H reservoirs to overflow one after another. In Condition 2, the dam failure of the C reservoir did not cause the downstream H reservoir to collapse. In Condition 3, the D and C reservoirs did not collapse, only the H reservoir did.

After the analysis and simulation of the dam-break flood, the calculation results are shown in Table 4. It can be seen that when encountering floods of the same

Table 4 The calculation results of dam break consequences

Condition	Inundated area (km^2)	Risk population (person)	Loss of life (person)	Level
1	146.18	324,424	206	5
2	58.92	91,738	102	5
3	39.64	52,458	75	4

magnitude, the consequences of cascade reservoirs failure caused by the dam failure of D Reservoir corresponding to Condition 1 are the most serious.

Dam-break loss of life is calculated by Dekay and McClelland method with the following formula:

$$L_{OL} = \frac{P_{AR}}{1 + 13.277\left(P_{AR}^{0.440}\right)e^{(0.759W_T - 3.790F + 2.223W_T F)}} \tag{2}$$

where P_{AR} represents the population at risk within the scope of flood inundation (people). W_T is the Alarm time (h). F is Flood severity, and the range is 0–1. F equals 1 for a high-degree severe flood and 0 for a low-degree severe flood.

4.3 The Risk Analysis of the Cascade Reservoir Group

The risk value and level can be obtained according to the possibility and life loss of dam break of D, C, and H cascade reservoirs. The calculation results are shown in Table 5. It can be seen that the risk of the D, C, and H cascade reservoirs is extremely high, and the risk is mainly determined by the D Reservoir. When the reinforcement project for reservoir D is completed, the probability of dam break will be reduced to 1. Although the severity of a dam break is 5, the risk value is only 5, which can be reduced to general risk. While reservoirs C and H are "Class II dam", they are defective and must be eliminated to reduce the risk to available or low risk.

Table 5 The risk calculation results of cascade reservoirs

Condition	The probability level of the dam break	The severity level of dam break consequences	Risk value	Risk level
1	4	5	20	Extremely high
2	3	5	15	High
3	3	4	12	High

5   Conclusions

(1) The risk matrix method is proposed to assess the risk of cascade reservoir groups, in which the possibility of dam break was qualitatively evaluated according to the severity of potential reservoir hazards. Additionally, the seriousness of dam break consequences was quantitatively evaluated according to the life loss and economic loss of dam break.

(2) The risk classification criteria of cascade reservoir groups and the possibility and severity of consequences classification criteria of dam break were proposed.

(3) The case study results of the D, C, and H cascade reservoirs show that the D reservoir mainly determines the risk of this cascade reservoir group. It is suggested to take engineering measures to eliminate hidden dangers in various reservoir projects and non-engineering actions (e.g., joint scheduling of cascade reservoirs and preparing emergency plans) to reduce the consequences of the dam break and the overall risk of cascade reservoirs.

(4) There are many cascade reservoirs in small and medium-sized watersheds in China. Once the cascade reservoirs collapse, the consequences are severe. The risk assessment method for the cascade reservoir group proposed in this paper is simple and feasible. For the cascade reservoir group assessed as extremely high and high risk, engineering and non-engineering measures are recommended to reduce the dam break risk and ensure the high quality and safe operation of the cascade reservoir group in China.

References

1. X.J. Meng, J.X. Chang, X.B. Wang, Y.M. Wang, Z.Z. Wang, Flood control operation coupled with risk assessment for cascade reservoirs. J. Hydrol. **572** (2019)
2. X.H. Peng, J.B. Sheng, L. Li, L.H. Liu, K.F. Zhou, Study on risk standard establishment of reservoir dams in China. J. Hydraul. Water Transp. Eng. **04**, 7–13 (2014)
3. Y.L. Li, S.L. Wang, L. Wang, X.H. Du, X.B. Zhou, Research progress on risk analysis of cascade reservoirs in river basins. Chin. Sci. Techn. Sci. **51**(11), 1362–1381 (2021)
4. X.H. Du, B. Li, Z.Y. Chen, Y.J. Wang, P. Sun, Study on safety criteria for design of extra-high dam and its cascade reservoir groups II: standard for safety factor of stability of high earth-rock dam slope. J. Hydraul. **46**(06), 640–649 (2015)
5. J.P. Zhou, X.B. Zhou, X.H. Du, F.Q. Wang, Study on dam risk control design for cascade reservoir groups. J. Hydropower **37**(01), 1–10 (2018)
6. Z.P. Xu, Analysis and consideration on dam break event in Michigan. Hydropower Water Resour. Express Rep. **41**(06), 8–13+51 (2020)
7. Fujian Society of Hydropower and Clean Energy Power Generation Engineering, *Lessons Learned and Countermeasures for Dam Breakage in Yong'an and Xinfa Reservoirs Inner Mongolia* (2021). http://www.hydropower.org.cn/showNewsDetail.asp?nsId=31171
8. P.Z. Lin, Y. Chen, Risk analysis of dam overflow for cascade reservoirs based on Bayesian network. Eng. Sci. Technol. **50**(03), 46–53 (2018)
9. Y. Zhang, T. Liu, X.W. Wu, Y. Yang, Simulation study of cascade reservoir collapse based on MIKE 11. Hydraul. Sci. Cold Zone Eng. **2**(01), 25–30 (2019)

10. State Energy Administration, *Guidelines for Safety Risk Prevention and Control of Cascade Reservoir Groups* (China Water Resources and Hydropower Press, Beijing, 2021)
11. X.H. Peng, Y.B. Cai, J.B. Sheng, L. Li, S.C. Zhang, K.F. Zhou, *Study on Risk Criteria of Reservoir Dams in China* (China Water Resources and Hydropower Press, Beijing, 2015)

Study on Influencing Factors of Hydraulic Efficiency of Clearwell

Shuo Zhang, Jiajiong Xu, Min Rui, and Jian Wang

Abstract The clearwell has the dual function of hydraulic regulation and disinfection contact. Improving hydraulic efficiency is an important measure to reduce the amount of disinfection by-products. The example of the clearwell was modeled as a whole. Then the influence of the main geometric dimensions on the hydraulic efficiency of the clearwell were studied by two-dimensional numerical simulation method, and the rationality of the design scheme was verified. The results showed that the length–width ratio had the most obvious influence on hydraulic efficiency, but the trend of t_{10}/T increasing was slowing down. The t_{10}/T value could be increased by reducing the number and width of bends and increasing the diameter of inlet pipe. The Pe under each working condition was greater than one, so the particle translation with the water flow was the main factor of mass transfer. The length–width ratio of the clearwell was 38 and the t_{10}/T was 0.54 which would meet the design requirement of not less than 0.5.

Keywords Clearwell · CFD · t_{10}/T · Length–width ratio · Inlet

S. Zhang (✉)
Research Institute, SMEDI, Shanghai 200092, P.R. China
e-mail: zhangshuo@smedi.com

J. Xu · J. Wang
Second Design and Research Institute, SMEDI, Shanghai 200092, P.R. China
e-mail: xujiajiong@smedi.com

J. Wang
e-mail: wangjian5@smedi.com

M. Rui
Shanghai Water Industry Design Engineering Co., Ltd., Shanghai 200092, P.R. China
e-mail: ruimin@smedi.com

© The Author(s) 2025
W. Wang et al. (eds.), *Hydraulic Structure and Hydrodynamics*, Lecture Notes in Civil Engineering 608, https://doi.org/10.1007/978-981-97-7251-3_3

1 Introduction

In all drinking water plants, the clearwell is an indispensable structure, which has the dual role of hydraulic regulation and disinfection contact. Reasonable design of clearwell is an important guarantee to ensure disinfection and reduce the amount of disinfection by-products [1]. There were obvious backflow phenomena on the back side of the diversion plate and the wall area of the clearwell, and these flow dead zones would seriously affect the disinfection efficiency [2]. Optimizing the hydraulic conditions of the clearwell could make the flow state in the clearwell as close as possible to the ideal push flow. The optimization of the inner structure could improve the hydraulic efficiency of the high cleanwell [3, 4]. The Surface Water Treatment Law of the United States adopts traditional CT value as a practical control parameter to predict the disinfection effect. Thus the t_{10}/T becomes an indicator to evaluate the hydraulic efficiency of a clearwell, reflecting the degree of short flow. The closer the value is to 1.0, the closer the flow state of a clearwell is to the push flow. The flow state would be better and the t_{10}/T would be higher.

Traditionally, tracer methods have been used to determine the relationship between the hydraulic efficiency of cleanwells and the internal structure of cleanwells [5]. In recent years, some scholars have conducted researches using NaCl as tracers in pilot tests. However, the determination of the clearwell time distribution function by tracer tests would be time-consuming and may not be possible due to field conditions. Computational fluid dynamics (CFD) simulation could easily, accurately and intuitively obtain the distribution of water flow state and velocity in a clearwell [6, 7]. The reveal velocity distribution and other information that could not be expressed by physical models, such as local velocity, short-circuit flow pattern, residence time of individual particles and running trajectory [8, 9]. Therefore, based on CFD, the influencing factors such as the geometric structure, length–width ratio and the number of partitions were studied, and the hydraulic parameters were analyzed to provide technical methods for the transformation, construction and design of the cleanwell.

2 CFD Parameter

2.1 Grid

The quadrilateral grid in map format was divided by Gambit into 35,000–51,000. Both the side grid spacing and the surface grid spacing were taken as 0.25 m, and the minimum volume 1.56E−02 was greater than zero.

2.2 Model Parameters

In fluent simulation, the water surface of a clearwell was regarded as a plane [10], so two-dimensional CFD modeling (no depth dimension) was conducted. The inlet was set to velocity inlet and the outlet was set to pressure outlet. The RNG k-ε turbulent flow model was used for the aqueous phase, and the Euler–Lagrange dispersive phase model was used for the granular phase. One thousand particles were loaded for tracking with the Discrete Phase Model (DPM).

2.3 Influencing Factor

The actual clearwell was 50 m long, 48 m wide, 4 m deep and the volume was 10,000 m^3. The corridor was 8 m wide, and the bend was 4 m wide. The inlet and outlet pipes were DN1000. Structure factors such as the setting of baffle, inlet and outlet conditions, and corridor angle had a great influence on t_{10}/T [11].

By adding the partition wall, the ratio of the total length and width of the corridor would change. In general, the greater the number of partitions, the greater the length–width ratio. For a clearwell of the same plane size, the partition wall numbers were 2, 3, 4, 5 and 11, and the length–width ratio was 9, 17, 26, 38 and 150 respectively.

The clearwell inlet could affect the energy conversion and velocity distribution uniformity of water. In the example, the inlet pipe adopted a single pipe DN1000 with a flow rate of 0.74 m/s. Other water inlet methods, such as double-pipe DN700, the addition of perforated wall, weir overflow or single pipe DN1500, were compared to analyze the influence of water inlet methods on hydraulic efficiency.

In the engineering example, the corridor was 8 m wide and the bend was 4 m wide. The influence of water flow characteristics caused by different width of bend on hydraulic efficiency was analyzed by contrast 8 m width of the bend.

By changing the layout of the clearwell from 48 and 50 m to 24 and 100 m, the number of turns could be reduced accordingly. The influence of the number of turns on hydraulic efficiency was studied.

According to the influencing factors, eleven working conditions were used for modeling and simulation, and the structural parameters of each working condition were shown in Table 1. The perforated wall was 5 m away from the water inlet. The sixty holes size was 250 and 250 mm. The hole spacing was 250 mm.

Table 1 The clearwell conditions under CFD

No.	Partitions (number)	Total plane length (m)	Total plane width (m)	Gallery width (m)	Length/ width	Inlet	Turning width (m)
1	2	50	48	16	9	DN1000	4
2	3	50	48	12	17	DN1000	4
3	4	50	48	9.6	26	DN1000	4
4	5	50	48	8	38	DN1000	4
5	5	50	48	8	38	DN1000 + perforated wall	4
6	5	50	48	8	38	2 × DN700	4
7	5	50	48	8	38	DN1500	4
8	5	50	48	8	38	Weir	4
9	5	50	48	8	38	DN1000	8
10	11	50	48	4	150	DN1000	4
11	2	100	24	8	38	DN1000	4

3 Results and Discussion

3.1 Length–Width Ratio

Different number of partitions would produce different length–width ratios. The hydraulic efficiency t_{10}/T could increase with the rise of length–width ratio, as shown in Fig. 1.

As the length–width ratio increases to more than 26, the growth of hydraulic efficiency would slow down significantly (Fig. 1). Measured by the t_{10}/T of 0.5,

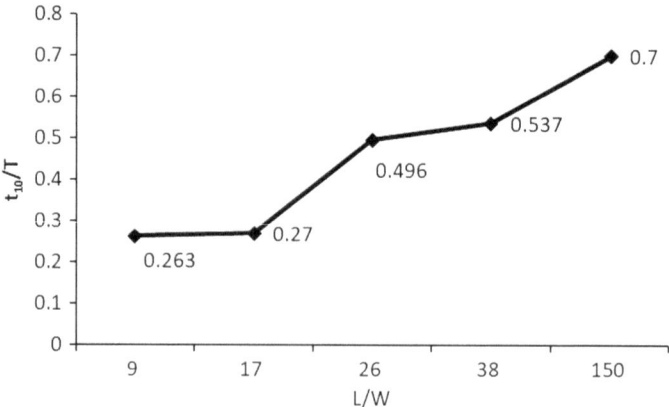

Fig. 1 Hydraulic efficiency at different length–width ratios

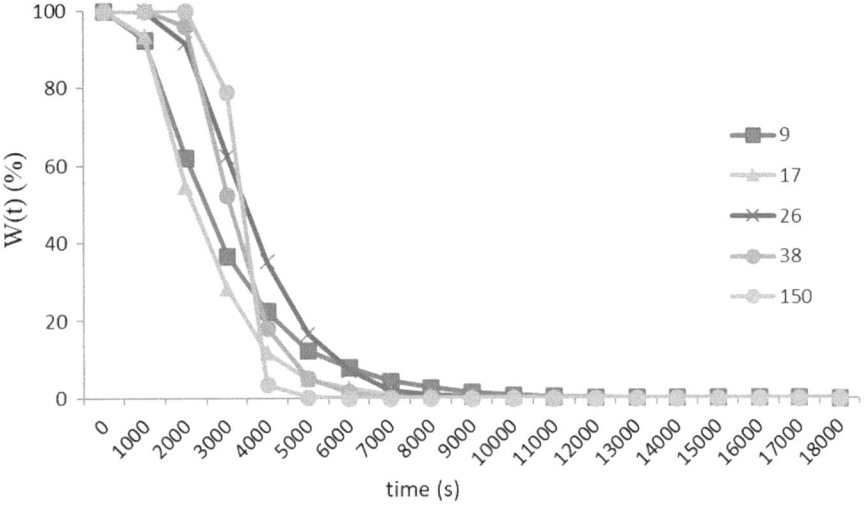

Fig. 2 Residence time distribution of different length–width ratios

according to the simulation trend line method, the length–width ratio should be 26 or more.

The residence time of particles was affected by the actual hydraulic conditions and often deviates from the theoretical residence time [12]. With the increase of length–width ratio, the W(t) curve became steeper and the particles flowed out more and more in a short time. When the length–width ratio was 150, particles were concentrated in the 2000–4000 s, and the time of concentrated outflow was significantly shortened, as shown in Fig. 2.

3.2 Inlet

The influence of different inlet on hydraulic efficiency was obvious because the initial velocity distribution was important. The difference of t_{10}/T between the maximum and minimum values was 0.2, as shown in Fig. 3.

Under the single DN1000, t_{10}/T could be more than under walls or spills. Under the DN1500, t_{10}/T was slightly better than under the DN1000 which was more economical. The hydraulic efficiency would decrease after the addition of the perforated wall, because the flow velocity through the perforated wall orifice was 0.14 m/s and the distance from the water intake was not far enough. The flow rate of two DN700 tubes was the same as that of single DN1000 tubes, but the t_{10}/T was low.

Among all working conditions, DN1500 had the longest t_{10}, as shown in Fig. 4. For the single DN1000, the residence time density was concentrated.

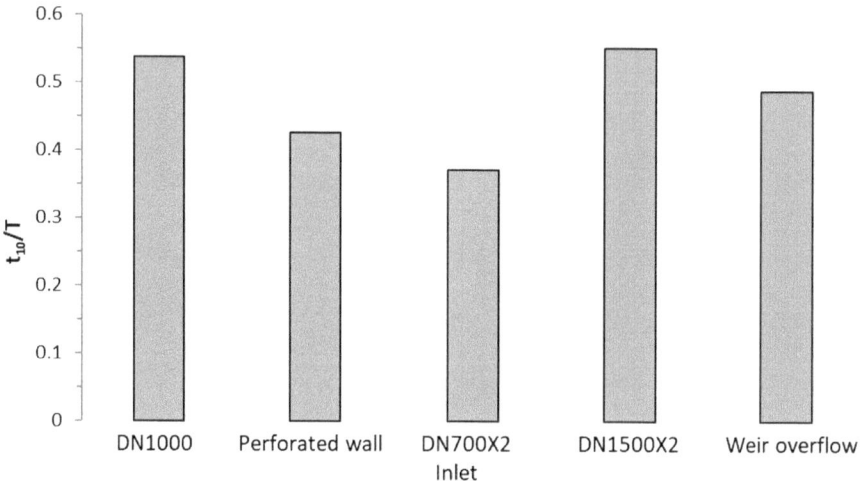

Fig. 3 Hydraulic efficiency under different inlet

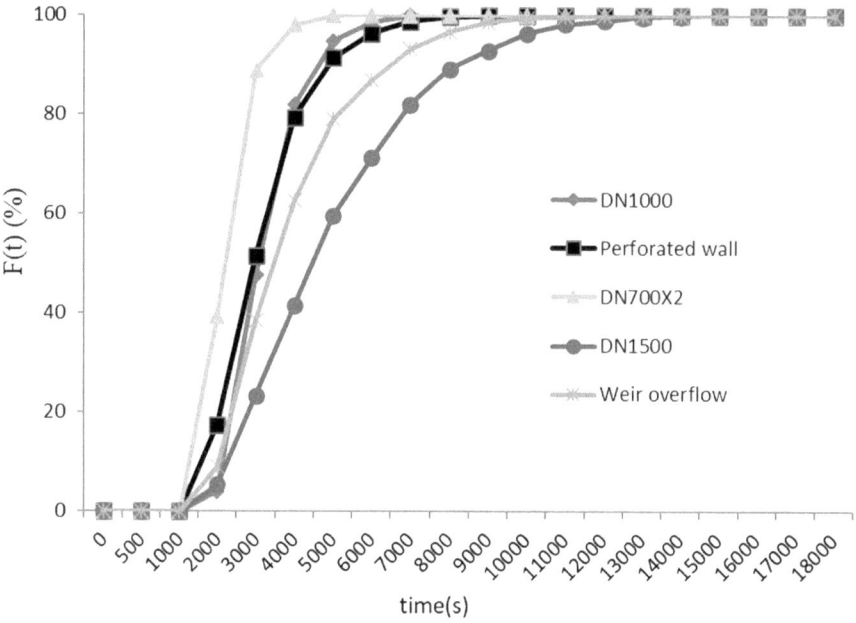

Fig. 4 Cumulative residence time distribution function under different import modes

3.3   Turning Width

When the width of the bend was increased to 8 m, the number of 3000–4000 s particles concentrated in the outflow was about 120, which slightly reduced than that in the 4 m bend.

When the curve was 4 m, the hydraulic efficiency was 0.537, and the t_{10}/T would decrease to 0.444 after the curve was expanded to 8 m.

3.4   Number of Turns

If the clearwell plane was changed from 50 and 48 m to 100 and 24 m, the number of curves was reduced from 5 to 2. The number of particles increased by about 40 when the outflow was concentrated at 3000–4000 s. As a result, hydraulic efficiency had increased. When the plane was 50 and 48 m, the hydraulic efficiency was 0.537. The t_{10}/T would slightly increase to 0.561 after it was changed to 100 and 24 m. The influence of the number of bends was significantly smaller than that of the length–width ratio, which indicated that the number of bends was not the most critical factor affecting the hydraulic efficiency.

3.5   Hydraulic Efficiency

The working conditions were sorted by t_{10}/T from small to large, as shown in Fig. 5.

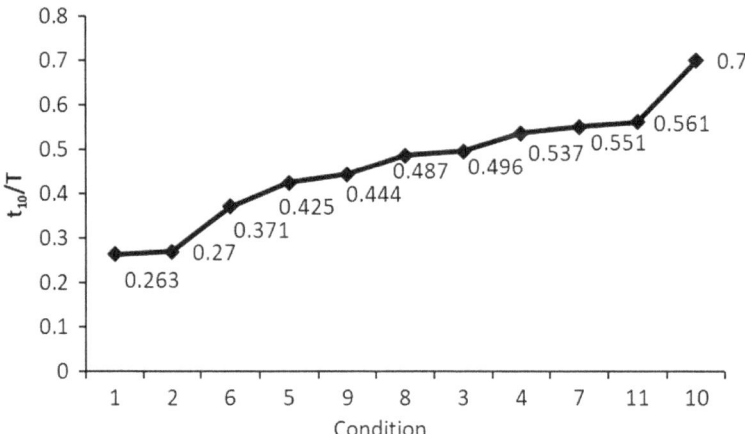

Fig. 5 Hydraulic efficiency under different working conditions

The length–width ratio had the most obvious influence on the hydraulic efficiency, and the length–width ratio of working condition 10 was as high as 150, and the t_{10}/T also could reach the good evaluation standard 0.7. When the number of partitions was less than 3, the hydraulic efficiency was obviously less than 0.30. Changing the total plane size of the clearwell, reducing the number of turns, and increasing the diameter of the inlet pipe also had a certain effect on improving the hydraulic efficiency. Thus the hydraulic efficiency of working conditions 11 and 7 was higher than that of the design working condition 4. For the design condition 4, the t_{10}/T was higher than 0.5, and similar to the condition 11 and 7. This effect was also acceptable in the water plant project.

3.6 Number of Turns

The mass transfer parameters mainly included Pe, dispersion number d and axial dispersion coefficient D, as shown in Table 2.

The Pe of each working condition was greater than one, so the particle translation with the water flow was the main factor of mass transfer. With the increase of length–width ratio, Pe value would increase and d would decrease, so the dispersion degree in the clearwell was weakened. The dispersion number of each working condition was much lower than 0.05, which indicated that even if the length–width ratio was less than 9, the axial dispersion degree was still low.

Table 2 Mass transfer performance under different working conditions

Condition	D	Pe	d
1	0.030	181.5422	0.0055
2	0.037	259.0322	0.0039
3	0.044	343.3306	0.0029
4	0.050	434.1844	0.0023
5	0.050	434.1844	0.0023
6	0.050	434.1844	0.0023
7	0.050	434.1844	0.0023
8	0.050	434.1844	0.0023
9	0.050	434.1844	0.0023
10	0.078	1110.7507	0.0009
11	0.050	434.1844	0.0023

Table 3 Hydraulic performance under different working conditions

Condition	MDI	1/MDI	t_1/T
1	4.94	20.24	0.17
2	3.69	27.10	0.17
3	2.68	37.31	0.22
4	2.00	50.00	0.38
5	3.00	33.33	0.30
6	1.98	50.51	0.26
7	3.56	28.09	0.37
8	3.16	31.65	0.39
9	2.28	43.86	0.25
10	1.28	78.13	0.59
11	1.84	54.35	0.42

3.7 Hydraulic Performance

The time of the first outflow particle is t_1. The Morrill dispersion Index (MDI), volumetric efficiency 1/MDI and short-circuit index t_1/T of each working condition were calculated and sorted by t_{10}/T from small to large, as shown in Table 3.

The length–width ratio had significant influence on MDI, volume efficiency and short circuit index. When the number of partitions increased, the volume efficiency would increase and the t_{10}/T would also increase. When the length–width ratio of working condition 10 was as high as 150, the t_{10}/T could reach 0.7. The volume efficiency was as high as 78%, and the short circuit index was about 0.6. This indicated that the working condition was close to the ideal thrust flow reactor, and the volume utilization was full.

Under the condition with fewer partition walls, the volume efficiency was lower than 0.5 and the t_1/T was small. The clearwell was close to the complete mixed reactor. For example, when the number of partitions was less than 3, the hydraulic efficiency was significantly less than 0.30. Under design condition 4, t_{10}/T was higher than 0.5, and the volume efficiency was 50%. And t_1/T was 0.38. The any MDI of operating conditions 4, 6, 10 and 11 was not greater than 2. According to the standards of the EPA, the clearwell was a flat thrust flow reactor with good efficiency.

4 Conclusions

Among the main influencing factors, the length–width ratio has the most obvious influence on the hydraulic efficiency, while the width and number of bends and the water inlet mode were not the most important factors affecting t_{10}/T.

The t_{10}/T increased with the increase of L/W, but the increase slowed down. The length–width ratio of the clearwell example was 38, and the t_{10}/T was 0.54, which could meet the design requirement of t_{10}/T not less than 0.5.

Reducing the number of bends and increasing the diameter of the inlet pipe could increase the t_{10}/T, but the increase amplitude and effect were not obvious.

The any Pe was greater than 1.0 and the dispersion number was lower than 0.05. The length–width ratio had significant influence on MDI, volume efficiency and short circuit index.

Acknowledgements This work was financially supported by the Shanghai Science and Technology Committee Program (No. 22dz1209103).

References

1. C.N. Hass, J. Joffe, Disinfection under dynamic conditions: modification of Hom's model for Decay. Environ. Sci. Technol. **28**(7), 1367–1369 (1994)
2. K. Shiono, E.C. Teixeira, Turbulent characteristics in a baffled contact tank. J. Hydraul. Res. **38**(6), 271–278 (2000)
3. R. Amini, R. Taghipour, H. Mirgolbabaei, Numerical assessment of hydrodynamic characteristics in chlorine contact tank. Int. J. Numer. Methods Fluids **67**(7), 885–898 (2011)
4. I.A. Hatmoun, P.F. Boulos, E.J. List, Using hydraulic model to optimize contact time. J. Am. Water Works Assoc. **90**(8), 77–87 (1998)
5. O. Charaigner, J. Cavard, Chlorine disinfection from CT requirement to clearwell design. Water Supply **17**(3/4), 191–197 (1999)
6. M.R. Templeton, R. Hofmann, R.C. Andrews, Case study comparisons of computational fluid dynamics (CFD) modeling versus tracer testing for determining clearwell residence times in drinking water treatment. J. Environ. Eng. Sci. **5**(6), 529–536 (2006)
7. E. Shin, J. Ryu, H. Park, Computational fluid dynamics modelling and analysis approach for estimating internal short-circuiting in clearwells. Water **13**(13), 1849 (2021)
8. A.I. Stamou, Improving the hydraulic efficiency of water process tanks using CFD models. Chem. Eng. Process. **47**(8), 1179–1189 (2008)
9. J. Zhang, A.E. Tejada-Mart Nez, Q. Zhang, Developments in computational fluid dynamics-based modeling for disinfection technologies over the last two decades: a review. Environ. Model. Softw. (8), 71–85 (2014)
10. E. Issakhanian, J.A. Saez, A. Helmns, C. Nickles, Full simulation of disinfection stage in a water recycling plant using low-cost, hybrid 3-dimensional computational fluid dynamics. Water Environ. Res. **91**(8), 177–184 (2019)
11. M.M. Bishop, J.M. Morgan, C.D.K. Jamison, Improving the disinfection detention time of a water plant clearwell. Drinking water and health: balancing risks. J. Am. Water Works Assoc. **85**(3), 68–75 (1993)
12. R.L. Porter, K.P. Stewart, N. Feagin, S. Perry, Baffling efficiency insights gained from tracer studies at 32 Washington treatment plants. AWWA Water Sci. (1), 1115 (2019)

Spatial Characteristics of Basic Storage Media of Reservoir-Induced Earthquakes in the Lower Reaches of Jinsha River—A Case Study in Baitan-Zhilixincun

Liu Yang, Zedong Du, Zhiren Feng, and Bo Jin

Abstract Reservoir-induced earthquakes have the characteristics of shallow focal depth, small magnitude, high frequency and density in the Baitan-Zhilixincun section of the lower reaches of the Jinsha River. The rock mass of the shallow strata is mostly carbonate rock, which is widely distributed. The shallow metamorphic rocks are mainly distributed at the depth of 10 km. In the concentrated area of reservoir-induced earthquakes, the number of secondary faults is relatively large, and they are mainly reverse faults. The stress and strain patterns of the shallow cracks are mainly unloading cracks in the rock mass on the left and right banks of the Baitan-Jiaopingdu section. The main storage characteristics of groundwater are bedrock fissure water and karst fissure water. The rapid impoundment of the reservoir makes the bedrock fissure.

Keywords The lower reaches of Jinsha River · Reservoir-induced earthquakes · Fault · Cracks · Lithology

L. Yang · Z. Feng (✉) · B. Jin
Institute of Engineering Mechanics, China Earthquake Administration, Harbin, China
e-mail: fzr0451@163.com

L. Yang
e-mail: 15805228856@163.com

B. Jin
e-mail: jinbo@iem.net.cn

Z. Du
China Three Gorges Construction Engineering Corporation, Sichuan, China
e-mail: du_zedong@ctg.com.cn

B. Jin
Heilongjiang Gongzhen Science and Technology Co. Ltd., Harbin, China

W. Wang et al. (eds.), *Hydraulic Structure and Hydrodynamics*, Lecture Notes in Civil Engineering 608, https://doi.org/10.1007/978-981-97-7251-3_4

1   Introduction

Special reservoir seismic monitoring should be carried out for the construction and operation of super large hydropower stations. Reservoir earthquakes refer to various types of earthquakes detected within the monitoring area. Reservoir-induced earthquakes are earthquakes detected within the monitoring area due to the impoundment of the reservoir. They are nurtured and occur under natural material conditions, including geological conditions above the depth of the source and hydrogeological conditions [1]. We refer to these two types of research objects as the basic medium of reservoir-induced earthquakes. Reservoir-induced earthquakes have the characteristics of small magnitude, high frequency, and shallow epicenter.

Reservoir-induced earthquakes are characterized by small magnitude, high frequency and shallow source. The small earthquake swarm in Xinfengjiang Reservoir from 1961 to 1999 [2], the Hujiaping earthquake with M_L 4.1 in the head area of the Three Gorges Reservoir in 2008 [3], and the Shanxi reservoir earthquake with M_L 4.6 in Wenzhou in 2006 [4], were all caused by the change of the characteristics of its basic storage medium due to the impoundment [5]. Scholars at home and abroad have less research on the mechanism of reservoir-induced earthquakes storage medium, with different emphasis. Guo analyzed the temporal and spatial characteristics of reservoir-induced earthquakes based on the focal mechanism solution of micro seismic swarms in the Three Gorges Reservoir area [6]. Dai explored the relationship between seismic activity and fault permeability structure in the Three Gorges Reservoir area based on water storage in different periods [7]. Bell established a uniform diffusion model to analyze the softening effect of permeability on rock strength [8]. This paper mainly takes the spatial area of the intensive zone of reservoir-induced earthquakes after impoundment as the research object, uses CAD and ORIGIN drawing to study the spatial characteristics of the reservoir medium. Then further study the internal mechanism between the basic storage medium of reservoir-induced earthquake and the activity of reservoir-induced earthquake.

2   Temporal and Spatial Distribution Characteristics of Reservoir-Induced Earthquakes in Reservoir Area

During the impounding period of the reservoir from January 2020 to March 2022, the reservoir-induced earthquakes were monitored. Seismosignal was used for seismic wave analysis, and 1395 reservoir-induced earthquakes were counted in Fig. 1 (All monitoring data in this paper are provided by China Three Gorges Construction Engineering Group), all of which were minor earthquakes with focal depths within 8 km. The reservoir-induced earthquakes in the reservoir area mostly occurred near the Tonggouchang-Jiaoxi secondary fault (A) and Tanglang-Yimen secondary fault (B) in Fig. 1. The CAD and ORIGIN software are used to visualize the data and the visual expression of the strata. For example, the three impounding and the fourth discharge of water in the reservoir area correspond to four cycles, as shown in Fig. 2. The intensity of reservoir-induced seismicity in Baitan-Zhilixincun section is 2–3.

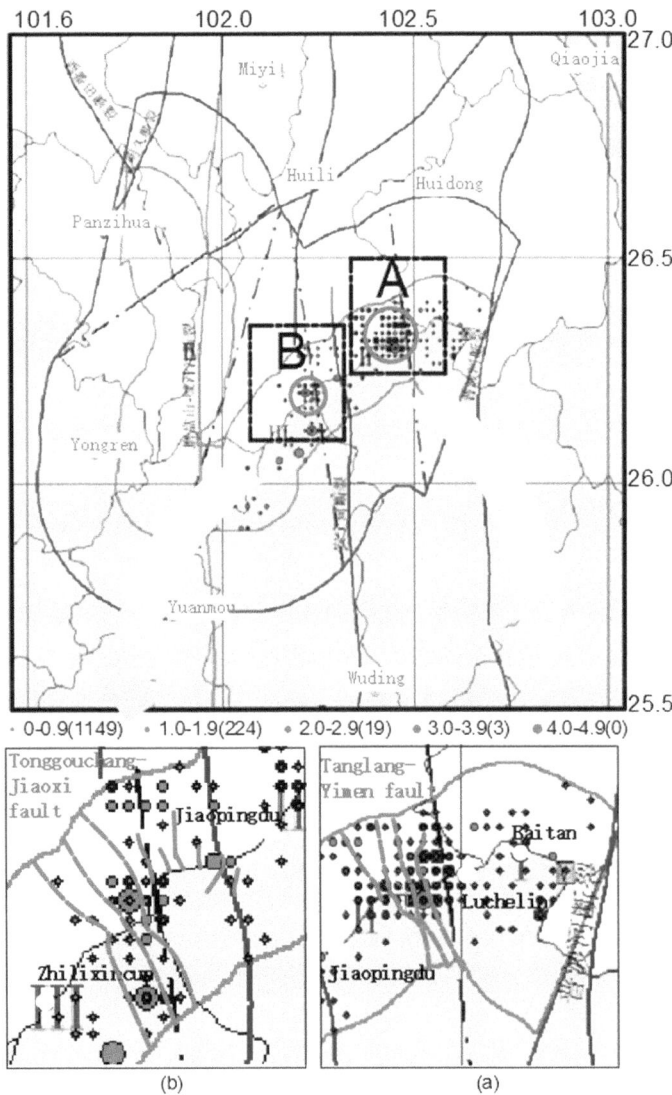

Fig. 1 Reservoir-induced earthquake within 10 km (red line) of the reservoir area during the impounding period from January 2020 to March 2022

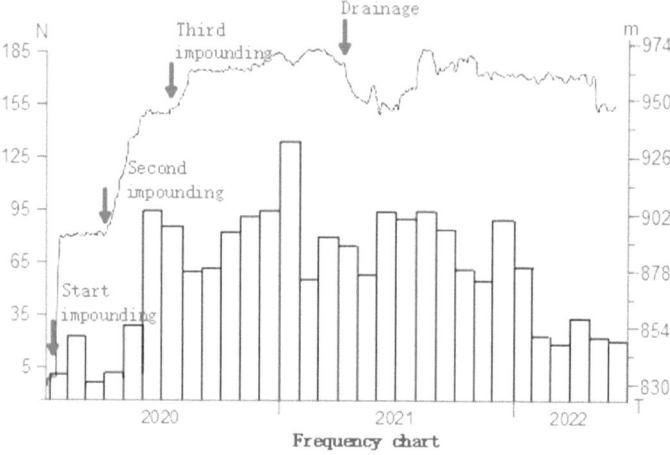

Fig. 2 Frequency map of reservoir-induced earthquakes and water level change map during impoundment of Baitan-Zhilixincun section within 10 km from 2020 to 2022

3 Spatial Characteristics of Basic Storage Medium of Reservoir-Induced Earthquake in Reservoir Area

3.1 Spatial Characteristics of Strata and Rock Mass

Geological conditions include stratum rock mass and rock mass structure. The strata buried more than 10 km in the reservoir area are composed of basement and cap rock. The basement is composed of P_{t1} (Mesoproterozoic Huili Group) deep metamorphic rocks, migmatites, magmatic complexes and P_{t1} shallow metamorphic rocks. The caprock consists of Z (Sinian system) glutenite, carbonate rock, K (Cretaceous), J (Jurassic), T_3 (Triassic system) mudstone, shale, sandstone and magmatic rock. Rock mass of surface strata in the reservoir area in Fig. 3.

The section from Baitan to Zhilixincun, with a surface elevation of $750 \sim 1500$ m, is composed of sedimentary rocks and epimetamorphic rocks. The coverage of sedimentary rocks accounts for 60%, mainly carbonate rocks, as shown in Fig. 3.

3.2 Spatial Characteristics of Rock Mass Structure

This paper only analyzes the faults and fissures in the neotectonic period.

Fault Zone, Fault

The spatial distribution and storage of neotectonic faults are important factors inducing reservoir earthquakes [9, 10]. Since the Cenozoic, the Qinghai Tibet Plateau

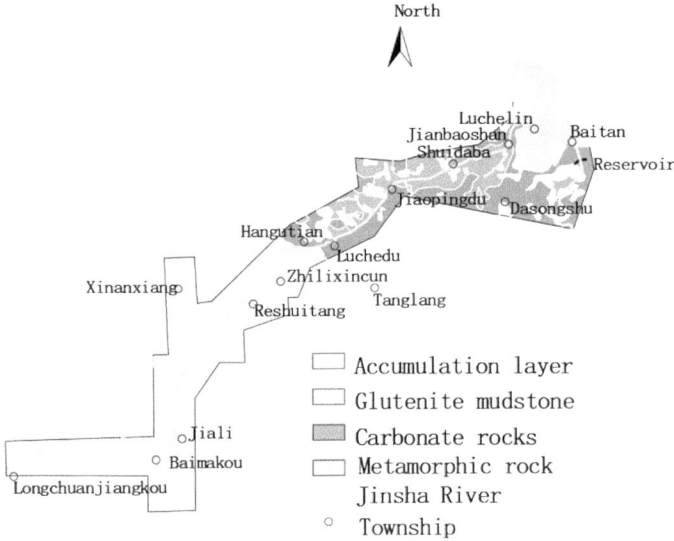

Fig. 3 Rock mass distribution of exposed strata in the reservoir

has been uplifted in a large area [11], and the reservoir area has been affected by the Kangdian rhombic fault block, forming a number of fault zones [12].

Through data analysis and field investigation, the main secondary faults in the reservoir area seismic monitoring area are shown in Table 1, and there are dozens of neotectonic faults and faults, as shown in Fig. 4. According to the frequency of reservoir-induced earthquakes and the spatial characteristics of rock mass structure, areas A and B in Fig. 1 are divided into two sections: Baitan-Jiaopingdu and Jiaopingdu-Zhilixincun, as shown in Table 2 and Fig. 1a, b.

Table 1 Characteristics and spatial distribution of main secondary faults in reservoir seismic monitoring area

Fracture	Length (km)	Direction	Tendency	Inclination angle (°)	Bandwidth	Main lithology of the fractured zone	Mechanical properties
Reshuitang fault	7	NW	SW	65 ~ 80	6 m	Carbonate rock	Thrust fault
Qingcaoping fault	32	EW	SE	40 ~ 50	32 km	Sandstone and carbonate rock	Thrust fault
Malutag fault	53	EW	SE	55 ~ 70	35 m	Carbonate rock and shallow	Thrust compression

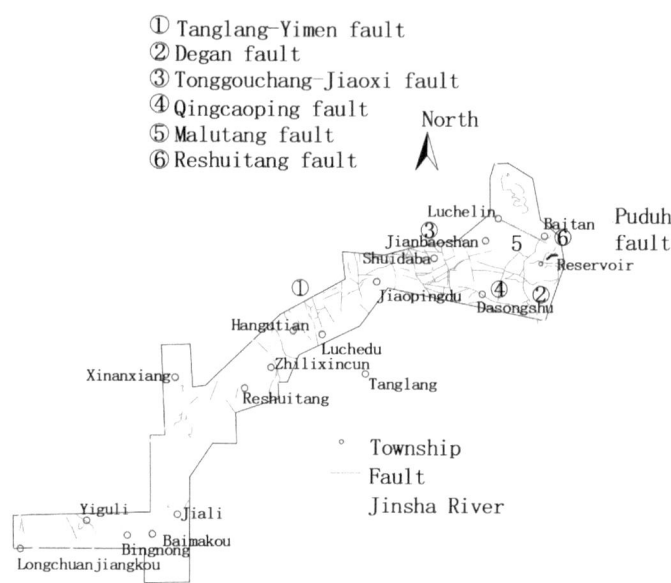

① Tanglang–Yimen fault
② Degan fault
③ Tonggouchang–Jiaoxi fault
④ Qingcaoping fault
⑤ Malutang fault
⑥ Reshuitang fault

Fig. 4 Plane distribution of neotectonic faults and faults in reservoir seismic monitoring area

Table 2 Statistics of fault density in various sections of the reservoir area in Baitan-Zhilixincun

Lot	Total length of fault (km)	Fault area (km^2)	Section area (km^2)	Proportion of spatial density (‰)	Corresponding to areas A and B
Baitan-Jiaopingdu	156.75	4.72	519	9.1	A area
Jiaopingdu-Zhilixincun	78.9	2.37	470	5.0	B area

According to the statistics, there are 60 faults in the reservoir area, including 56 faults with a length of more than 1 km. According to the field survey, the width of the fault zone is generally 30 m. There are 36 faults in the Baitan-Jiaopingdu section, including 6 normal faults. There are 24 faults, 1 normal fault and 17 reverse faults in Jiaopingdu-Zhilixincun section. The fault length, area and spatial density of the three sections are counted. The density proportion shown in Table 2 is the ratio of fault area to section area.

Fissure

On site investigation was conducted on the fissures in the rock mass at the left and right banks of Jinsha River in the Baitan-Jiaopingdu section. The storage and distribution of fissures are shown in Figs. 5, 6 and 7.

According to the survey, the crack length of the section is mostly less than 5 m, and rarely more than 10 m. The statistical average linear density is about 0.55 pcs/m. From Figs. 5, 6 and 7, the fracture trend on the left bank is Sn and that on the

Fig. 5 Rose diagram of fracture trend in Baitan-Jiaopingdu section

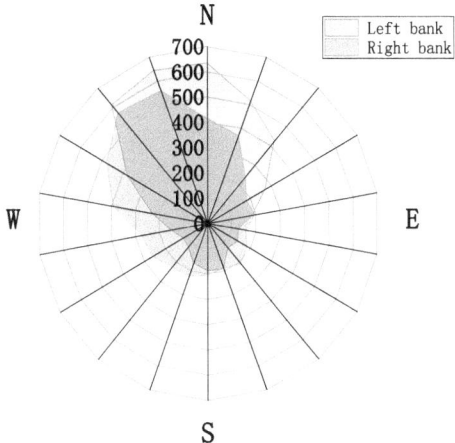

Fig. 6 Rose diagram of fracture tendency in Baitan-Jiaopingdu section

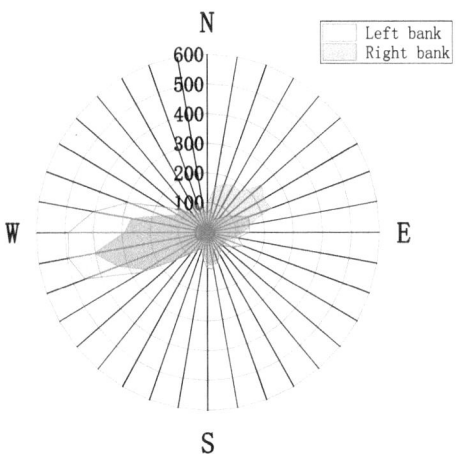

right bank is NW. Both sides of the Strait tend to be EW oriented. The dip angles of fractures on both banks are mostly 45 ~ 60° (medium dip angle), followed by 60 ~ 90° (steep dip angle), and 0 ~ 45° (slow dip angle) are not developed.

It can be seen from Table 3 that the stress–strain pattern of rock mass cracks on the left and right banks is mainly unloading cracks, with an elevation of more than 900 m. Its lower limit horizontal depth is generally 16 ~ 32 m, and the unloading crack below the elevation of 850 m is not obvious. The smaller the lower limit horizontal depth of the unloading fracture, the shallower the horizontal development depth of the fracture in the rock mass.

17 boreholes are arranged in the Baitan-Jiaopingdu river section [13], with an average depth of 7.40 m and the deepest depth of 14.40 m. Table 4 shows the analysis results of fracture characteristics of drill core. The crack width is mostly 2–8 mm.

Fig. 7 Inclination histogram
of Baitan-Jiaopingdu section

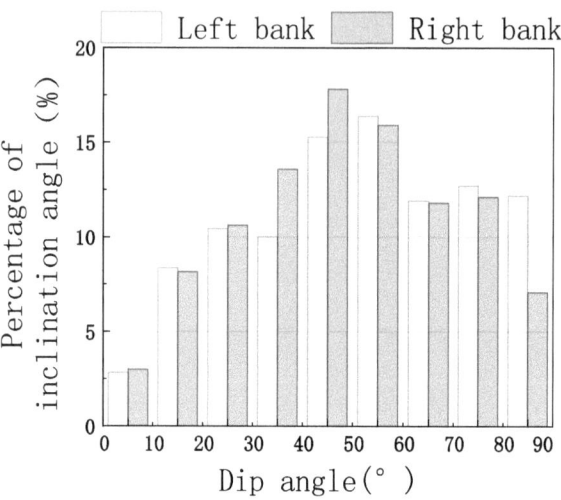

Table 3 Rock mass unloading fracture development in Baitan-Jiaopingdu section

Bank (m)	Distance from bank slope (m)	Lower limit horizontal depth (m)	Unloading fracture characteristics	Bank elevation (m)
Left	900 ~ 990	2–3	21–38	Slightly developed
	850 ~ 900	0–3	4	Agenesis
Right	900 ~ 960	1–2	12–21	Slightly developed, slightly loose rock mass
	850 ~ 900	1–2.5	7.5	Agenesis

The statistical linear density is mostly 1–2 strips/m. The rock mass in the shallow part of the riverbed is relatively broken and has dissolution.

3.3   Spatial Characteristics of Hydrogeological Conditions

The main storage characteristics of groundwater in the Baitan-Jiaopingdu section are bedrock fissure water and karst fissure water. The elevation above 640 m is dominated by karst fissure water, and the spatial distribution of karst development is shown in Fig. 8.

The borehole water pressure test was carried out at the karst location shown in Fig. 8, and the results are shown in Table 5 and Fig. 9. Before the impoundment of the reservoir, the permeable layers of the shallow rock mass are alternately distributed. The characteristics of permeability are as follows. Firstly, the sedimentary rocks in

Table 4 Development characteristics of gently inclined fractures in cores drilled at Jinsha River riverbed in Baitan-Jiaopingdu

Location	Borehole No.	Number of cracks (pcs)	Fracture state	Linear density (pcs/m)	Average linear density (pcs/m)
Left bank of riverbed	1	10	The width is 1 ~ 2 mm, and there is dissolution along the crack	2.0	1.34
	2	7	The width is 2 ~ 6 mm, and there is dissolution along the crack	1.3	
	3	13	It is generally 1 ~ 2 mm wide, with a maximum width of 4 mm. There is dissolution phenomenon along the crack in some parts, and the rock mass is relatively broken	1.1	
	4	7	It is generally 1 ~ 2 mm wide, with a maximum width of 5 mm. There is dissolution along the crack locally	1.6	
	5	8	It is 2 ~ 5 mm wide, with dissolution along the crack, and the local rock mass is slightly broken	0.7	
Middle of a river	6	–	Cracks are developed in some sections, with dissolution, and the rock mass is relatively broken	–	1.96
	7	7	The width is 1 ~ 3 mm, and there is dissolution along the crack	2.1	
	8	3	It is 2 ~ 4 mm wide and slightly eroded along the crack	1.3	
	9	20	It is generally 2 ~ 8 mm wide, with a maximum width of 30 mm. It has corrosion phenomenon along the crack and is broken locally	3.0	
	10	11	The width is generally 3 ~ 8 mm, and the maximum is 12 mm. There is corrosion along the crack	1.7	
	11	6	It is generally 1 mm wide and locally eroded along the crack	1.7	
Right bank of riverbed	12	–	It is 4 ~ 8 mm wide, with dissolution along the crack and broken rock mass	2.9	1.68
	13	6	2 ~ 8 mm wide	1.6	
	14	8	It is 2 ~ 4 mm wide, with dissolution along the crack and relatively broken rock mass	1.1	

(continued)

Table 4 (continued)

Location	Borehole No.	Number of cracks (pcs)	Fracture state	Linear density (pcs/m)	Average linear density (pcs/m)
	15	5	Local rock mass is broken, and the crack is generally 6 mm wide	1.3	
	16	2	It is 2 ~ 4 mm wide, with relatively developed steep fissures, obvious dissolution and partial fragmentation	0.3	
	17	4	It is generally 4 ~ 8 mm wide, with a maximum width of 12 mm. There is dissolution phenomenon along the crack, and the rock mass is broken	2.9	

Fig. 8 Distribution diagram of karst development in Baitan-Jiaopingdu section

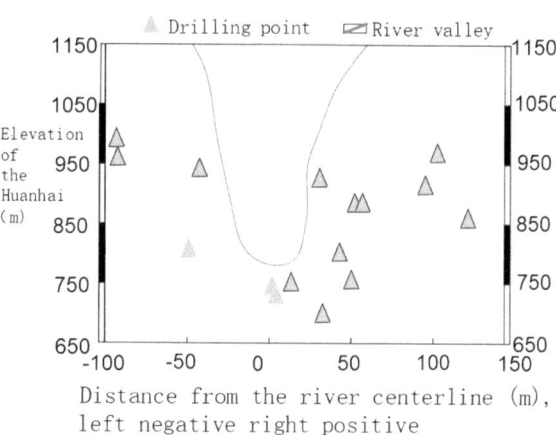

Distance from the river centerline (m), left negative right positive

the strata are higher than the epimetamorphic rocks. Secondly, the karst is smaller than the unloading crack between shallow rock masses. Thirdly, the infiltration rate and saturation rate under the valley are faster than those on both sides of the valley.

Table 5 Permeability analysis of rock mass in Baitan-Jiaopingdu section

Rock stratum	Poor permeability (%)		Bad permeability (%)
	Unloading	Karst	
Epimetamorphic rock	6.10	0.00	6.10
Carbonate rocks	3.60	1.68	72.50
Total	9.70	1.68	78.60
	11.38		

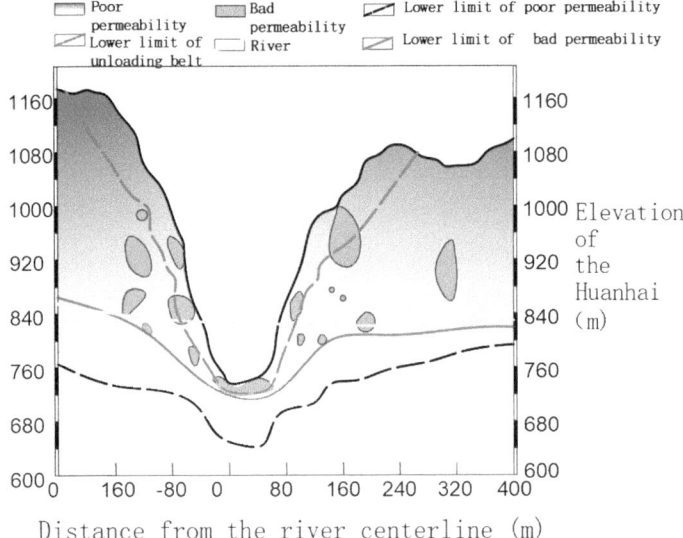

Fig. 9 Permeability distribution of strata in Baitan-Jiaopingdu section

The reservoir impoundment makes the directionality and nonuniformity of cracks in rock mass tend to be consistent, and increases the penetration degree of cracks. After the storage water forms an infinite increment, it forms a continuous amount of bedrock fissure water with increased degree. With the duration of water storage, the bedrock fissure water gradually evolves into pore water, thus forming a continuously increasing pore water pressure.

4 Conclusions

(1) The spatial distribution of earthquakes in Baitan-Zhilixincun reservoir in the lower reaches of Jinsha River is scattered. The distribution of reservoir-induced earthquakes is relatively centralized in Baitan-Jiaopingdu and Jiaopingdu-Zhilixincun. Reservoir-induced earthquakes in the reservoir area have the characteristics of shallow focal depth, small magnitude, high frequency and dense.

(2) In the two sections of Baitan-Jiaopingdu and Jiaopingdu-Zhilixincun, the shallow rock mass under the riverbed of Jinsha River is mainly carbonate rock, which is widely distributed. It is mainly epimetamorphic rock at the focal depth of 10 km.

(3) According to the statistics, there are relatively many secondary faults in the Baitan-Jiaopingdu and Jiaopingdu-Zhilixincun sections in the reservoir area, which are mainly reverse faults.

(4) According to the analysis, the fracture length of the section from Baitan to Jiaopingdu is mostly less than 5 m. The average linear density is about 0.55 m, and the dip angle of the fractures is mostly 45 ~ 60°. The stress–strain pattern of the shallow fractures in the rock mass of the left and right banks is mainly unloading fractures. The fractures are divided into caprock fractures and basement fractures.

(5) The main storage characteristics of groundwater in the Baitan-Jiaopingdu section are bedrock fissure water and karst fissure water. The rapid impoundment of the reservoir makes the bedrock fissure water gradually evolve into the bedrock porous media pore water.

Wudongde Dam is a national Megaproject and a national important tool. Applying research methods and achievements to earthquake prevention, disaster reduction, and emergency rescue work of giant hydropower stations will play an important technical support role in the construction and operation safety of these hydropower projects.

Acknowledgements The research described in this paper was funded by Institute of Engineering Mechanics, China Earthquake Administration, basic scientific research business projects of central public institutions (No. 2020B06; No. 2019C08). The Natural Science Foundation of Heilongjiang Province (No. LH2019E96).

References

1. D. Wei, K.W. Wang, W. Zhu, B.P. Jin, F. Zhang, Reservoir seismic migration law of Badong reservoir bank section in the Three Gorges Reservoir area. J. Eng. Geol. **26**(04), 915–929 (2018)
2. G.A. Guo, T.P. Liu, N.G. Qin, L.F. Chen, Analysis of the results of comprehensive mechanism solutions of small earthquakes in Xinfengjiang reservoir from 1961 to 1999. Acta Seismol. Sin. **26**(3), 261–268 (2004)
3. Y.T. Che, J.H. Chen, L.F. Zhang, C.L. Liu, W.H. Zhang, Study on induced earthquake of Hujiaping ms 4.1 reservoir in the head area of the Three Gorges reservoir. Earthquake **29**(4), 1–13 (2009)
4. F. Ye, J. Yuan, Geological structure characteristics of Shanxi reservoir in Wenzhou and its relationship with reservoir earthquakes, in *Geological Work Promotes Ecological Civilization Construction—Proceedings of the 2018 Academic Annual Meeting of Zhejiang Geological Society* (2018), pp. 353–355
5. H.K. Gupta, A review of recent studies of triggered earthquakes by artificial water reservoirs with special emphasis on earthquakes in Koyna, India. Earth-Sci. Rev. **58**(3–4) (2002)
6. W. Guo, C.P. Zhao, K.Z. Zuo, C. Zhao, Characteristics of seismicity before and after impoundment of Baihetan reservoir in the lower reaches of Jinsha River. Chin. J. Geophys. **65**(12), 4659–4671 (2022)
7. M. Dai, G.J. Wu, J. Liu, C.Y. Shen, S.A. Sun, X.L. Shen et al., Characteristics of apparent stress changes before and after the Badong earthquake in the Three Gorges Reservoir area. Seismol. Geol. **39**(04), 837–852 (2017)
8. M.L. Bell, A.J. Nur, Strength changes due to reservoir-induced pore pressure and stresses and application to lake Oroville. J. Geophys. Res. **83**(B9), 4469–4483 (1978)

9. Z. Long, H. Yao, S. Liu, X.J. Sun, 3D visualization analysis of Longtan Reservoir-induced earthquakes and active faults, in *Geo-Spatial Knowledge and Intelligence. GRMSE. Communications in Computer and Information Science*, ed. by H. Yuan, J. Geng, F. Bian, 699 pp.
10. W. Leith, D.W. Simpson, W. Alvarez, Structure and permeability: geologic controls on induced seismicity at Nurek reservoir, Tajikistan, USSR. Geology **9**(10), 440–444 (1981)
11. Y. Yao, J.B. Chen, S. Li, H.P. Song, J.L. Xie, Inversion of paleoseismic events along the Beiluntai fault in southern Tianshan, Xinjiang. J. Eng. Geol. **24**(06), 1278–1285 (2016)
12. S. Jang, Y.F. Wang, C. Tang, H.Y. Pan, K. Wang, Preliminary study on the mechanism of low-speed landslide activity in Jinpingzi II area in the lower reaches of Jinsha River. J. Eng. Geol. **25**(06), 1547–1556 (2017)
13. X.Q. Niu, B.X. Xing, Y.H. Weng, *Feasibility Study Report of Wudongde Hydropower Station on Jinsha River* (Changjiang Institute of Survey, Planning, Design and Research, Yunnan Province)

Spatio-temporal Distribution Patterns of Sediment Carrying Capacity in the Three Gorges Reservoir

Le Feng, Zhongwu Jin, Quanxi Xu, Ya Liu, and Yujiao Liu

Abstract Reservoir sedimentation profoundly impacts riverbed dynamics and engineering applications, posing a dual challenge by compromising riverbed evolution and critical water storage infrastructure. The escalating loss of reservoir capacity due to sediment accumulation imperils lives, properties, and global water sustainability, undermining the reservoir's safety role. Using the Three Gorges Reservoir (TGR) as a case study, we analyze extensive hydrological data from Cuntan, Qingxichang, Wanxian, and Miaohe stations (2002–2020). Our findings reveal: (1) Pre-impoundment, Qingxichang and Wanxian exhibit high sediment carrying capacities, which reduce noticeably during early impoundment. Subsequently, with the changes of pre-dam water levels, an overall decreasing trend of sediment-carrying capacity is observed, characterized by annual fluctuations, while Miaohe consistently maintains a lower level due to its proximity. (2) Wanxian experiences significant sediment carrying capacity reduction during constant water level periods of various operational stage, with up to 99.4% decrease post-impoundment. (3) Sediment carrying capacity varies annually primarily due to changes of reservoir pre-dam water levels, and secondarily, owing to incoming water and sediment fluctuations. Increased influx raises sediment carrying capacity, while reduced influx decreases it. Morphological cross-section configuration emerges as a dominant factor in sediment carrying capacity, outweighing dam proximity. Wide, shallow sections have lower capacity than narrow, deep ones. For narrow, deep sections, dam proximity's influence exceeds cross-sectional morphology.

Keywords Three Gorges Reservoir · Sediment carrying capacity · Reservoir siltation

L. Feng · Z. Jin (✉) · Q. Xu · Y. Liu · Y. Liu
Changjiang River Scientific Research Institute of Changjiang Water Resources Commission, Wuhan 430010, PR China
e-mail: zhongwujin@163.com

49

W. Wang et al. (eds.), *Hydraulic Structure and Hydrodynamics*, Lecture Notes in Civil Engineering 608, https://doi.org/10.1007/978-981-97-7251-3_5

1 Introduction

Reservoir sedimentation, exemplified by the Three Gorges Reservoir (TGR), situated at Sandouping, the upstream from Yichang on the Yangtze River, poses critical challenges to riverbed stability, infrastructure safety, and global water supply sustainability. As a vital water resource and hydropower project operating since 2003, TGR requires ongoing study of sediment accumulation to ensure facility safety and address global sedimentation challenges while supporting sustainable water resources.

Numerous studies have analysed sediment accumulation and sediment-discharge ratio in the TGR, improving our understanding of their spatial distribution and characteristics [1–3]. While some research has compared observation data with deposition predictions [1], there's limited exploration into the causes of sediment accumulation. A few studies have discussed morphological factors in sedimentation pre and pose-impoundment [4], mainly concerning fine sand accumulation (median grain size 0.01 mm) [5]. However, comprehensive research on sediment accumulation causes over the 20 years of TGR operation remains scarce.

Reservoir sedimentation results from complex interactions among river sediment transport, soil erosion, and reservoir dynamics. To better grasp sediment transport within water flow, it is essential to investigate variations in water flow's sediment carrying capacity during riverbed erosion and deposition processes, especially as water depth increases in the TGR.

This study, based on field measurements, analyzes the spatio-temporal distribution of sediment-carrying capacity in the TGR from pre-reservoir construction (2002) to post-construction period (2004–2020). It focuses on: (1) characterizing variations in sediment-carrying capacity, and (2) assessing the impacts of reservoir impoundment and flood discharge.

2 Region, Data and Methods

2.1 Region

TGR, a narrow gorge-type reservoir with a 3.93×10^{10} m^3 capacity and a normal storage level at 175 m, operates in distinct phases. These include the cofferdam power generation phase (2003–2005), where the water level in front of the dam was maintained at 135 m during the flood season and 139 m during the dry season; the initial operation phase (2006–2007), characterized by a flood season water level of 144 m and a dry season level of 156 m in front of the dam; the experimental phase at 175 m (2008–2020), and the subsequent normal storage phase at 175 m (after 2021), with a flood season water level of 145 m and a dry season level of 175 m in front of the dam.

Following the 175 m experimental impoundment of the TGR, the backwater extends upstream to the vicinity of Jiangjin, creating a fluctuating backwater area

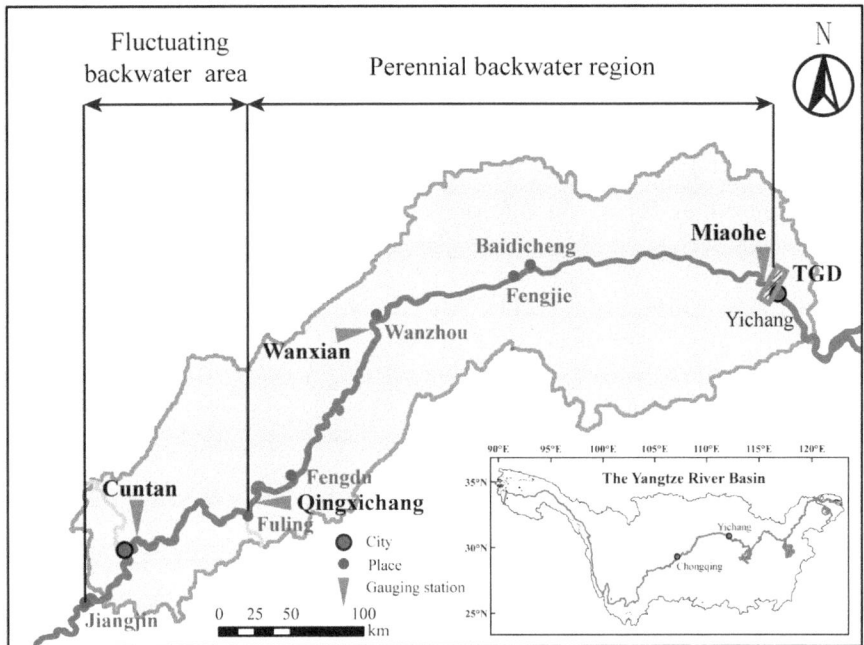

Fig. 1 Illustration of the study area in the Three Gorges Reservoir Region

from Jiangjin to Fuling, which constitutes roughly 26.3% of the total length. The perennial backwater region extends from Fuling to the dam.

In this study, the reach from Cuntan to Miaohe, with a total length of approximately 620 km, was selected as the research area, which essentially covers the main river channel of the TGR (Fig. 1). Based on the river channel's topography and geomorphic features, the research area is divided into three sections: Cuntan to Fuling, Fuling to Fengjie, and Fengjie to Miaohe. The Cuntan-Fuling section, characterized by terrain fluctuations between gorges and wide valleys. River width varies from 250 to 1500 m. The Fuling-Fengjie section, features a broad river valley, with widths ranging from 1500 to 2000 m and undulating hills on both sides. The Fengjie-Miaohe section, includes a canyon stretch from Baidicheng to Miaohe, with a typical width of 200–300 m and the narrowest point being just over 100 m.

2.2 Data and Methods

The study focuses on understanding the causes of sediment accumulation in the TGR reservoir area by analyzing the spatio-temporal variations in sediment-carrying capacity since its operation. Hydrological, topographic, and bed sediment data were sourced from Changjiang Water Resources Commission (CWRC) and China Three

Gorges Corporation (CTG). The study used daily average flow, water level, and sediment concentration data from four hydrological stations (Cuntan, Qingxichang, Wanxian, and Miaohe) for 2002–2020. Topographic data from monitoring sections along the Cuntan to Miaohe reach (2002–2020) were also employed. Bed sediment data from the four stations for 2004–2009 were used, as shown in Table 1.

The study of water flow sediment-carrying capacity dates back to Gilbert's flume experiments [6]. In China, research began in 1947 [7], and by 1954, experiments at the Nanjing Hydraulic Research Institute established empirical equations [8]. Over time, numerous scholars proposed formulas, including the Einstein formula [9], Zhang Ruijin [10], Bagnold [11], Dou Guoren [12], Zhang Hongwu [13], Han Qiwei formulas [14]. For this study, the Zhang Ruijin formula was selected due to its simplicity, ease of application, and suitability for low sediment-concentration rivers (less than 100 kg/m^3). It has been extensively applied in the Yangtze River and has been well-validated for its applicability in the TGR [5, 15]. The basic structure of the formula is as follows:

$$S* = k\left(\frac{U^3}{gR\omega}\right)^m \tag{1}$$

where U = cross-section averaged flow velocity, m/s; g = gravitational acceleration, m/s^2; R = hydraulic radius, m; ω = average settling velocity of suspended load, m/s, calculated by $\omega = \sum_{i=1}^{n}(p_i\omega_i)$, where p_i = proportion of the ith grain size, ω_i = settling velocity of the ith grain size, m/s; and k and m = model coefficients, which is usually 0.02 and 0.92 in the TGR, respectively.

Based on the research achievements of the China Institute of Water Resources and Hydropower Research, Yangtze River Scientific Research Institute, and relevant studies both domestically and internationally regarding the TGR, it has been found that setting m to 0.92 and k to 0.02 in the formula effectively describes water–sediment interaction and riverbed scour-deposition variations [5, 15]. Therefore, in this study, we adopt these values for m and k.

Table 1 Data range of Cuntan, Qingxichang, Wanxian, and Miaohe

Station	Range of topographic, discharge, water level, and sediment concentration data	Range of particle size distribution data
Cuntan	2010 ~ 2012, 2017 ~ 2018, 2020	2020
Qingxichang	2002, 2004 ~ 2012, 2015, 2017 ~ 2018, 2020	2004 ~ 2020
Wanxian	2002, 2004 ~ 2012, 2015, 2017 ~ 2018, 2020	2004 ~ 2020
Miaohe	2003 ~ 2012, 2015, 2017 ~ 2018, 2020	2004 ~ 2018

3 Spatio-temporal Distribution Characteristics of Sediment Carrying Capacity

3.1 Temporal Distribution Characteristics of Sediment Carrying Capacity

To analyze the variation characteristics of sediment-carrying capacity during different water storage stages in the TGR, Fig. 2 presents the annual average sediment-carrying capacity changes for Cuntan, Qingxichang, and Wanxian stations from 2002 to 2020. (Miaohe station was established in April 2003.)

Before reservoir impoundment, Qingxichang and Wanxian exhibited high sediment-carrying capacity, which decreased significantly during the initial stage of impoundment. Subsequently, as the pre-dam water level gradually rose, sediment-carrying capacity decreased overall with annual fluctuations, with Miaohe consistently having the lowest sediment-carrying capacity due to its proximity to the dam.

Overall, sediment-carrying capacity in the reservoir area ranged from 0.001 to 0.131 kg/m^3, influenced by impoundment. Before impoundment, Wanxian had the highest sediment-carrying capacity at 0.71 kg/m^3. After impoundment, Qingxichang peaked in 2004 at 0.131 kg/m^3, Wanxian and Miaohe in 2005 at 0.073 kg/m^3 and 0.011 kg/m^3, respectively. The reduction in sediment-carrying capacity for these three stations before and after impoundment was 8%, 90%, and 25%, respectively.

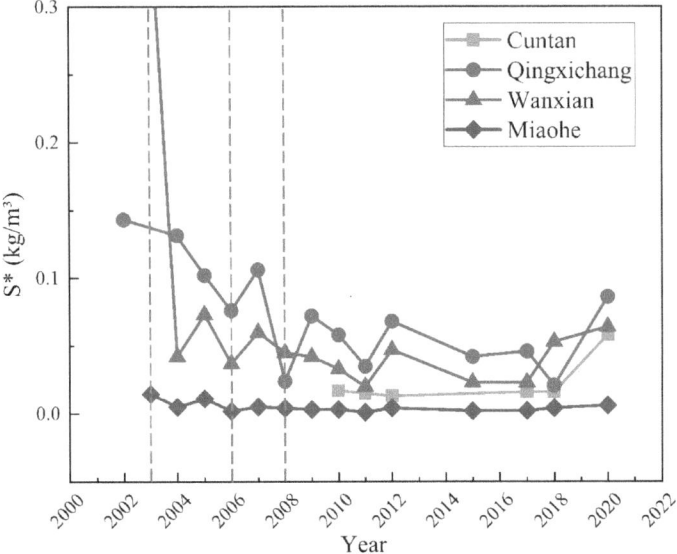

Fig. 2 Annual average sediment-carrying capacity changes (Cuntan ~ Miaohe)

3.2 Spatial Distribution Characteristics of Sediment Carrying Capacity

To manage sedimentation in the reservoir area, the TGR follows a four-part annual schedule: the abatement period (April–May), the flood season (June–September), the storage period (October–November), and the period of constant water levels (December–March). To evaluate how water level regulation affects sediment-carrying capacity during these periods, we conducted an analysis. It involved plotting the relationship between the annual average sediment carrying capacity at four stations and the four aforementioned stages (2002–2020). This analysis is shown in Fig. 3.

In Fig. 3, we observe a downstream decrease in sediment carrying capacity in the TGR during its four stages, with Cuntan being the exception. Prior to impoundment, Wanxian held the highest sediment carrying capacity, while post-impoundment, this distinction shifted to Qingxichang. Furthermore, post-impoundment, sediment carrying capacity decreases significantly between Qingxichang and Wanxian, with a 70% deduction during both the storage and abatement periods, and a 36% decrease during the flood season. Moving downstream from Wanxian to Miaohe, there is an 89% reduction during the flood season, while during other periods, the reduction surpasses 94%.

4 Key Influencing Factors of Sediment Carrying Capacity

4.1 Key Influencing Factors of Sediment Carrying Capacity Temporal Distribution

To further analyze sediment carrying capacity temporal variations in the TGR, we studied annual changes in flow discharge, sediment concentration, flow velocity, and water depth during the flood season at Cuntan, Qingxichang, Wanxian, and Miaohe stations from 2002 to 2020. Figure 4 illustrates the results. The first three factors trend align with sediment carrying capacity. Water depth, influenced by cross-section adjustments, exhibits intricate year-to-year variations that require further discussion within specific time intervals.

After the impoundment, the peak sediment carrying capacity of Qingxichang was observed in 2004, while that of Wanxian and Miaohe occurred in 2005. This variation can be attributed to the distant location of Qingxichang from the dam, resulting in minimal water storage influence, thus manifesting a certain degree of hysteresis in sediment carrying capacity changes.

Furthermore, sediment carrying capacity fluctuated with changing water and sediment conditions. For example, in the dry year 2006, peak-season average discharge at Qingxichang, Wanxian, and Miaohe was 13,159 m^3/s. In 2007 and 2008, increased

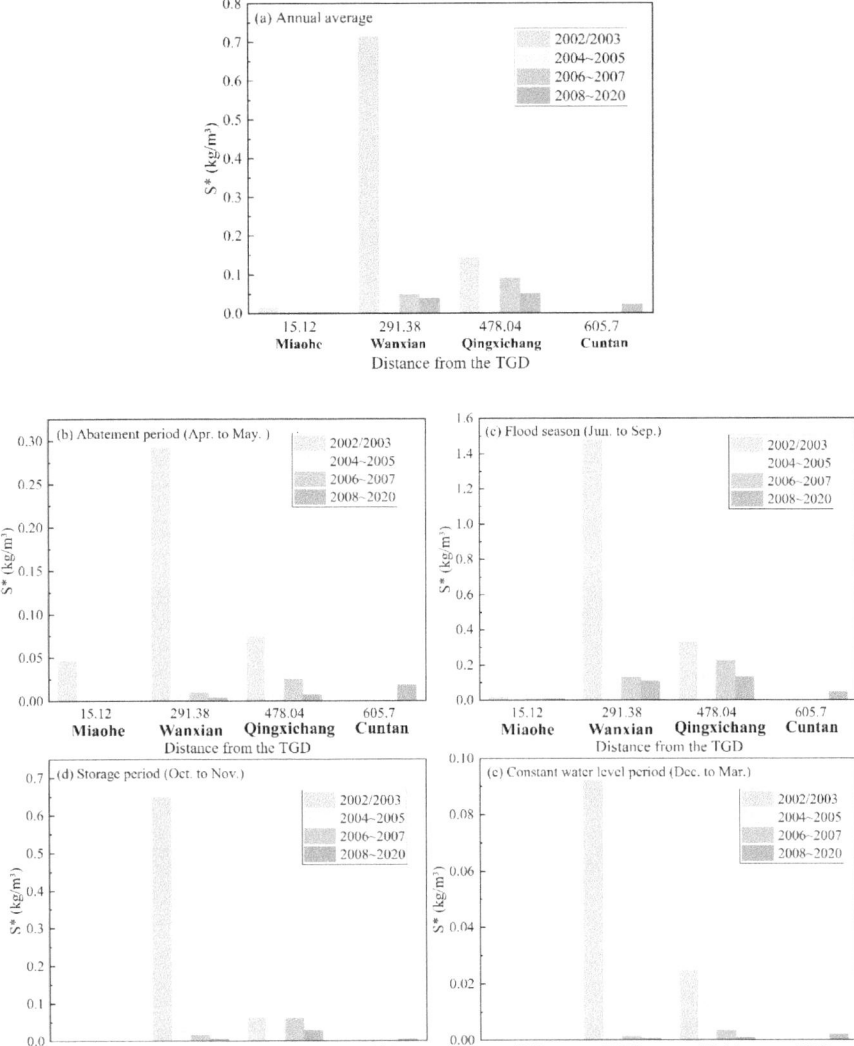

Fig. 3 Spatial distribution of sediment carrying capacity (Cuntan ~ Miaohe)

water and sediment raised the peak-season average discharge to 24,582 m³/s. Concurrently, peak-season average flow velocity rose from 1.16 to 1.50 m/s, resulting in an amplified sediment carrying capacity, vice versa.

Figure 5 illustrates the changes in sediment-carrying capacity at four hydrological stations under different water level operation phases. Prior to impoundment, Qingxichang and Wanxian stations peaked in August, Miaohe in May. Under the 135–139 m operational scheme, Qingxichang's peak was in September, Wanxian and Miaohe in

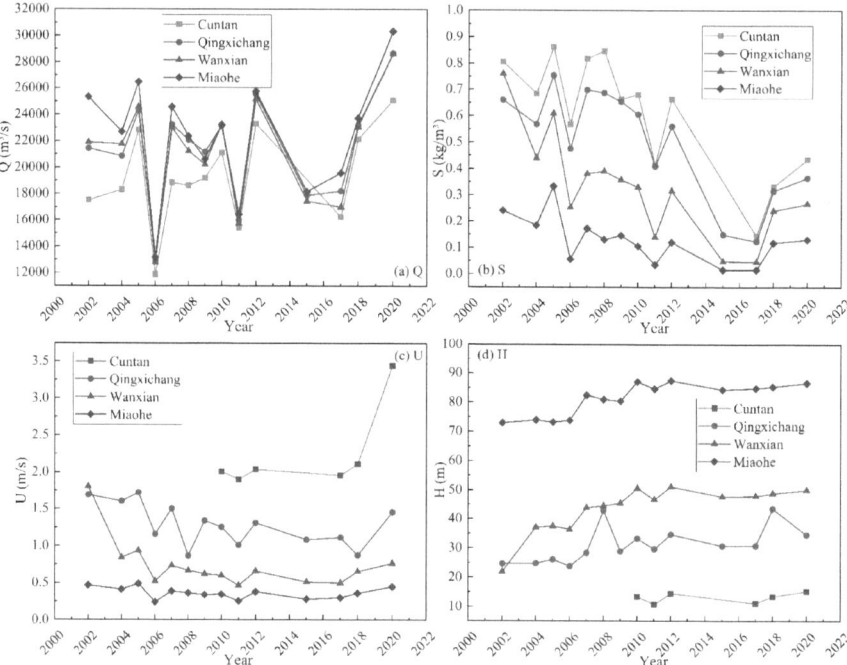

Fig. 4 Yearly changes in key influencing factors of sediment carrying capacity

July. Similarly, under the 145–156 and 175 m phases, all three stations peaked in July. In addition, after the 175 m program, Cuntan station began experiencing the influence of water returns, causing its peak to shift from August to September in 2020.

4.2 Key Influencing Factors of Sediment Carrying Capacity Spatial Distribution

Post-impoundment, sediment carrying capacity in the TGR is expected to decrease downstream due to pre-dam water level changes. Notably, from 2010 to 2020, Qingxichang and Wanxian consistently surpassed Cuntan in sediment carrying capacity, both during the flood season (June–September) and the storage period (October–November). Because the section morphology will directly affect the water depth, velocity and other factors, the section shape map of four stations from 2002 to 2020 is drawn in Fig. 6. At the 175 m normal storage water level, mean width-to-depth ratios ($\sqrt{B}/H$) are 1.53, 0.53, 0.46, and 0.29 for Cuntan, Qingxichang, Wanxian, and Miaohe, respectively. This suggests section morphology has a greater impact on sediment carrying capacity than distance from the dam. Narrower and deeper sections

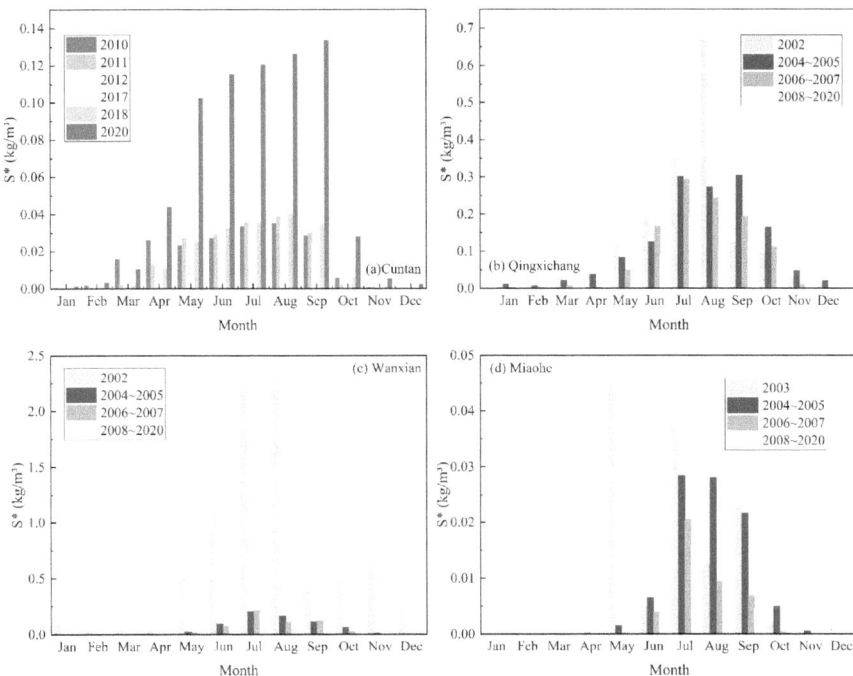

Fig. 5 Sediment carrying capacity distribution at different water levels

have higher sediment carrying capacities than wider, shallower ones. Conversely, for narrow, deep sections, the distance from the dam's influence on sediment carrying capacity exceeds that of section morphology.

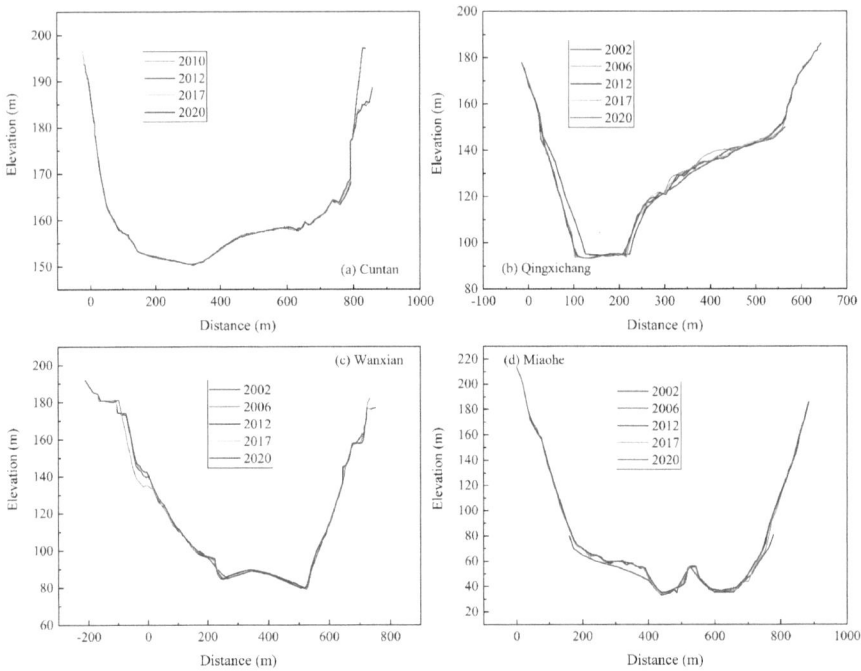

Fig. 6 Sections of Cuntan, Qingxichang, Wanxian, Miaohe stations

5   Conclusion

This study examines sediment carrying capacity variations in the TGR using water and sediment data from four hydrological stations (Cuntan, Qingxichang, Wanxian, and Miaohe) spanning 2002–2020. The research evaluates spatial and temporal changes before and after TGR implementation, quantifying sediment carrying capacity reduction and identifying main influencing factors. Key findings are as follows:

(1) Sediment carrying capacity was higher in Qingxichang and Wanxian pre-impoundment, with a substantial reduction upon initial impoundment. Over time, it gradually declined with the gradual rise of the pre-dam water level during different operational stages, experiencing yearly fluctuations. Miaohe exhibited consistently lower sediment carrying capacity due to its proximity to the dam. Water storage influenced overall reservoir sediment capacity (0.001 ∼ 0.131 kg/m^3). Wanxian exhibited the highest sediment carrying capacity during flood season pre-impoundment (1.48 kg/m^3), while Miaohe demonstrated the lowest sediment carrying capacity during the constant water level period (5.39 × 10^{-5} kg/m^3). Post-impoundment, Qingxichang cofferdam power generation phase displayed the greatest sediment carrying capacity (0.25 kg/m^3), whereas

Miaohe's 175 m experimental water storage exhibited the lowest sediment carrying capacity during the constant water level period (1.57×10^{-5} kg/m^3).

(2) The sediment carrying capacity during the constant water level period at Wanxian exhibited the greatest decrease before and after the impoundment. As the pre-dam water level gradually rose, the reductions in sediment carrying capacity during the cofferdam power generation phase, initial operation phase, and the 175 m experimental storage phase at Wanxian Station were 98.9%, 98.4%, and 99.4%, respectively, compared to baseline levels.

(3) The factors affecting annual variations in sediment carrying capacity within the reservoir area are complex. Primarily, the pre-dam water level plays a pivotal role; higher water levels reduce sediment carrying capacity. Moreover, water storage's impact lessens with distance from the dam. Monthly peaks in sediment carrying capacity vary, with more remote areas experiencing greater delays. Yearly changes in incoming water and sediment also influence sediment carrying capacity: increased inflow boosts it, while decreased inflow lowers it. Among factors affecting sediment carrying capacity, cross-section morphology bears greater influence than proximity to the dam. Wide and shallow cross-sections demonstrate significantly lower capacity than their narrow and deep counterparts. For the latter, the impact of distance from the dam outweighs that of cross-section morphology.

Acknowledgements This work was financed by the Major Science and Technology Project of the Ministry of Water Resources (SKS-2022075), the Key Project of the National Natural Science Foundation of China (NSFC) (U2040218), the Central-Level Public Welfare Basic Scientific Research Operating Expenses Project (CKSF2023397/HL, CKSF2023359/HL).

References

1. S. Ren, B. Zhang, W.-J. Wang et al., Sedimentation and its response to management strategies of the Three Gorges Reservoir, Yangtze River, China. CATENA **199**, 105096 (2021)
2. S. Liu, D. Li, D. Liu et al., Characteristics of sedimentation and sediment trapping efficiency in the Three Gorges Reservoir, China. CATENA **208**, 105715 (2022)
3. X. Li, J. Ren, Q. Xu et al., Impact of cascade reservoirs on the delayed response behaviour of sedimentation in the Three Gorges Reservoir. J. Geogr. Sci. **33**(3), 576–598 (2023)
4. Y. Yang, J. Zheng, L. Zhu et al., Influence of the Three Gorges Dam on the transport and sorting of coarse and fine sediments downstream of the dam. J. Hydrol. **615**, 128654 (2022)
5. W. Li, S. Yang, Y. Xiao et al., Rate and distribution of sedimentation in the Three Gorges Reservoir, Upper Yangtze River. J. Hydraul. Eng. **144**(8), 05018006 (2018)
6. G.K. Gilbert, E.C. Murphy, The transportation of debris by running water. Professional paper (1914)
7. Research N I O H, *Project Plan for Flume Experiments on the Limit Sediment Concentration in Silt-Laden Flow* (National Institute of Hydraulic Research, Nanjing, 1947)
8. Research N I O H, *Report on Experiment Equipments for Suspended Sediment Laden Flow* (Nanjing Institute of Hydraulic Research, Nanjing, 1954)
9. H.A. Einstein, The bed-load function for sediment transportation in open channel flows (1950)

10. R. Zhang, *River Sediment Dynamics* (China Water & Power Press, Beijing, 1961)
11. R.A. Bagnold, An approach to the sediment transport problem from general physics. Professional paper (1966)
12. G. Dou, All-sand modeling similarity law and design examples. Hydro-Sci. Eng. **03**, 3–22 (1977)
13. H. Zhang, Calculation formula of Yellow River sediment carrying capacity. Yellow River (11), 7–9+61 (1992)
14. Q. Han, *Siltation of Reservoirs* (Science Press, Beijing, 2003)
15. J. Zhou, Discussion on the reliability of sediment computations employed in the feasibility studies of Three Gorges Project. J. Hydroelectr. Eng. (01), 33–39+32 (2005)

Application of Multi-objective Intelligent Management and Control Platform for Cascade Reservoirs—Case Study on Nam Ou River in Laos

Yang Li, Jianxin Zhou, Rong Dai, Tianqing Li, and Baiyin Yang

Abstract In the construction and operation of cascaded reservoir groups, higher requirements have been put forward for multi-objective optimization scheduling of cascaded reservoir groups, aiming to promote benefits and eliminate hazards through joint scheduling, leverage comprehensive benefits, ensure basin safety, and meet the needs of digital transformation of hydropower and water conservancy. Based on the research of integrated methods for hydrological forecasting, multi-objective scheduling, and decision-making of basin cascade reservoirs, this study constructs a multi-objective intelligent control system that serves hydrological forecasting, cascade scheduling, flood control, and risk assessment of cascade reservoirs, and applies it to the development of cascade hydropower in the southern European river basin of Laos.

Keywords Cascade reservoir group · Multi objective optimization scheduling · Intelligent control · Nam Ou River

1 Introduction

The cascade reservoirs in a watershed are the main engineering measures to promote the efficient development and utilization of hydropower resources, ensure flood control safety in the watershed, and enhance the comprehensive benefits of irrigation

Y. Li · R. Dai · B. Yang
Planning Department, China Renewable Energy Engineering Institute, Beijing, China
e-mail: yangby@creei.cn

J. Zhou
Technical Management Department/Information Center, Powerchina Resources Limited, Beijing, China

T. Li (✉)
Water Resource Management Research Center, Kunming Engineering Corporation Limited, Kunming, China
e-mail: 1074761688@qq.com

© The Author(s) 2025
W. Wang et al. (eds.), *Hydraulic Structure and Hydrodynamics*, Lecture Notes in Civil Engineering 608, https://doi.org/10.1007/978-981-97-7251-3_6

water supply, ecological environment, and shipping safety in the watershed. In order to enhance the comprehensive utilization demand of water resources for cascade reservoirs in the watershed, many experts and scholars have applied hydrological forecasting technology and multi-objective scheduling technology of cascade reservoirs to the operation and management of cascade reservoirs, with fruitful research results. In terms of hydrological forecasting research, the land atmosphere coupled hydrological forecasting technology can effectively improve the accuracy of runoff and flood forecasting and extend the foresight period by coupling numerical weather models with land surface hydrological models. This technology is one of the main development trends in hydrological forecasting [1]. Li et al. [2] predicted winter rainfall in southern China by analyzing the land atmosphere coupling relationship over Western Europe. Li et al. [3] proposed a deep learning hydrological model PCA-LSTM that integrates the spatiotemporal characteristics of meteorological elements, and applied it in the source area of the Yellow River, improving the accuracy of runoff simulation. In terms of multi-objective scheduling research for cascaded reservoir groups, the main research hotspot is the study of multi-objective scheduling rules and scheduling model solving algorithms. Xu et al. [4] considered flood control, power generation, and water supply, and studied the multi-objective joint scheduling method of the downstream cascade of the Jinsha River and the Three Gorges Gezhouba Dam; Khorshidi et al. [5] established a Leader Pollowe multi-objective optimization model, which solved the problem of difficult coordination between the water storage loss target during the dry season of the reservoir and the agricultural water demand target; Lan et al. [6] proposed an improved multi-objective firefly algorithm and applied it to the multi-objective scheduling of power generation and ecology in the Three Gorges Reservoir of the Yangtze River; Zhang et al. [7] proposed a new improved multi-objective moth flame optimization algorithm based on R-dominance (R-IMOMFO), which obtained a set of solutions with good convergence and strong distribution in the multi-objective optimization scheduling model of cascade reservoir groups; Ma et al. [8] used an 8-reservoir cascade system in a certain watershed as a calculation example to test the application effect of PDPoS (Parallel DP Algorithm based on Spark) under different conditions by adjusting model parameters. In terms of algorithm solving and scheduling decision-making, research is often conducted by improving the TOPSIS decision model. In terms of multi-objective scheduling, relevant research also includes considering the uncertainty of hydrological forecasting and conducting risk assessment on multi-objective scheduling schemes. Zhu et al. [9] simultaneously considered the uncertainty of the quantitative and weight values of various dimensional indicators, SMAA by quantifying implementation uncertainty Stochastic Multicriteria Acceptance Analysis introduces the TOPSIS framework to form the SMAA-TOPSIS decision model, which is solved based on Monte Carlo simulation algorithm. Cascade dispatch centers have been established and put into use in the Dadu River Basin and the Three Gorges section of the Yangtze River in China, but the application of intelligent control systems that integrate forecasting, regulation, and risk assessment is relatively limited. This study focuses on the joint dispatch of cascade reservoir groups for the benefit and pest control of construction and operation. Based on the research of integrated methods for hydrological fore-casting, multi-objective scheduling of cascade reservoirs in the basin, and risk assess-ment, a multi-objective intelligent control system has been constructed to serve the hydrological forecasting, cascade scheduling, flood control, and risk assessment of cascade reservoirs, in order to leverage comprehensive benefits, ensure basin safety, and meet the needs of digital transformation of hydropower and water conservancy.

2 Materials and Methods

Based on the higher requirements for multi-objective optimization operation of basin cascade reservoirs, this study is based on the study of land atmosphere coupled hydro-logical forecasting, using multi-objective optimization operation as a means, and with prediction, operation, and flood control safety as the core, conducting research on the integrated method of basin cascade reservoir group prediction, operation, and flood control. Based on the development of relevant systems, a multi-objective intel-ligent control system for cascade reservoir group has been integrated and formed, and applied to the Nam Ou River basin in Laos.

2.1 Hydrological Forecasting Based on Land–Atmosphere Coupling

The research is based on hydrological forecasting, using radar, satellite, ground stations, and regional climate prediction products as the basic data, and based on precipitation prediction data products of different forecast periods (real-time, short-term, medium-term, and long-term), using hydrological numerical simulation tech-nology to establish multiple lumped and distributed hydrological models. It can provide runoff and flood processes with different foresight periods (next 6 h, 3 days, 7 days, and 45 days) for multi-objective scheduling of cascade reservoirs in the basin.

(1) There are multiple sources of meteorological forecast results: Firstly, the basin meteorological forecast issued by the meteorological department can be the average rainfall of the basin or the point rainfall forecast that coincides with the location of the rainfall station; Secondly, the meteorological department releases meteorological forecast data; The third is foreign numerical weather forecasting products, such as CFS.

(2) Hydrological forecasting model. It mainly includes four distributed hydro-logical models: VIC, EasyDHM (Easy Distributed Hydrological Model) [10], HEC-HMS [11], and CREST (Distributed Production Concentration Coupling Model), and lumped models, such as Xin'anjiang model and so on.

(3) Hydrological forecasting of land atmosphere coupling. The study selects a numerical weather forecast model built in relevant research areas, and the hydrological model used for coupling includes four improved distributed hydrological models in this study. The coupling method is unidirectional coupling, which is a hydrological model driven by precipitation forecast results and optimized parameters, in order to extend the flood prediction period. Among them, the data format requirements and conversion methods for land air coupling mainly rely on the precipitation forecast output results and the hydrological model using the same time step, and the time system is Universal Time (UTC). Based on the same coordinate system, the center coordinate of the runoff generation and concentration calculation unit is used to search for the precipitation forecast output grid containing this coordinate point, and the precipitation information of the grid is assigned to the runoff generation and concentration unit of the hydrological model, Implemented correspondence between two resolution grids.

2.2 Research on the Integrated Method of Multi Objective Operation and Decision Making for Cascaded Reservoir Groups in Watershed

The actual operation tasks of a cascade reservoir group in a basin include power generation, flood control, ecology, water supply, navigation, etc. The flood control task is related to the flood control safety of reservoir engineering and downstream flood control objects, and usually has strict scheduling regulations. Therefore, in multi-objective scheduling research, flood control tasks are added as constraints to the model and not considered as separate goals. Extract, classify, and functionalize the targets from the perspective of model establishment, and construct joint operation targets for cascade reservoir groups.

From the basic principle of multi-objective scheduling, different combinations of competing objectives in a single scheduling task can establish a multi-objective scheduling model, such as multi-objective scheduling of power generation (considering the maximum and minimum output of power generation); The combination of different scheduling task objectives can also form multi-objective scheduling models, such as power generation ecological multi-objective, power generation shipping multi-objective, power generation water supply shipping multi-objective scheduling, etc. This time, based on the established objective system and considering different combinations of objectives, a multi-objective scheduling model for cascaded reservoir groups is constructed.

In terms of multi-objective scheduling algorithms, in order to improve the computational efficiency of model and algorithm solving, this study combines the multi-objective bee colony algorithm solving mechanism and introduces the Fork/Join parallel computing framework into the entire algorithm optimization process. The multithreaded parallel computing mode is used to decompose the multi-objective scheduling calculation tasks, greatly improving the efficiency of solving multi-objective scheduling models for cascade hydropower stations [8]. In response to the multi-objective and multi scheme decision-making problem of cascade hydropower stations in the Nanou River Basin, this study adopts TOPSIS decision-making technology. By identifying the optimal and worst schemes among the limited schemes, the distance between each evaluation object and the optimal and worst schemes is calculated, and the relative similarity between each evaluation object and the optimal scheme is obtained, which serves as the basis for evaluating the advantages and disadvantages. The details are as follows:

(1) A multiple attribute decision making method based on entropy weighted vague set and generalized set pair analysis:

$$
P_p(\tilde{\mu}) = \left(\frac{2a^l}{b^r + c^r} - \frac{c^r + \gamma b^r}{a^l + (1 - \gamma)b^l}, \frac{2a^m}{b^m + c^m} - \frac{c^m + \gamma b^m}{a^m + (1 - \gamma)b^m}, \\
\frac{2a^r}{b^l + c^l} - \frac{c^l + \gamma b^l}{a^r + (1 - \gamma)b^r} \right) \tag{1}
$$

(2) TOPSIS method:

$$
Z^+ = \left(Z_1^+, Z_2^+, \ldots, Z_m^+ \right) = \left\{ \max_i Z_{ij} \,|\, j = 1, 2 \ldots, m \right\} \\
Z^- = \left(Z_1^-, Z_2^-, \ldots, Z_m^- \right) = \left\{ \min_i Z_{ij} \,|\, j = 1, 2 \ldots, m \right\} \tag{2}
$$

2.3 Multi-objective Intelligent Control System for Cascade Reservoir Groups

The research content mainly includes: platform framework construction, data system construction, data collection and management research, and multi-objective scheduling intelligent control system for cascade reservoir groups.

Platform framework construction, adopting the development concept of MVC, designing the system architecture to improve development efficiency; The functions of each module are implemented based on microservices, and the system interfaces between each module and other systems are designed in a RESTful manner. The construction of a data system adopts a multi-dimensional and scalable temporal data repository table design, mainly including rainfall and runoff data, power plant scheduling and operation process, and power generation data. The main types are actual data and published scheme data. Research on data collection and management, organizing and storing attribute data, hydrological characteristics, dynamic indicators, characteristic curves, and other data of the power station, integrating water and rain data and monitoring data, achieving data synchronization between

Safety Zone II and Safety Zone III. Data can be queried, edited, added, deleted, and other operations through web form. The research on the intelligent control system for multi-objective scheduling of cascade reservoirs is based on integrated technology methods and systems. Through data exchange and sharing, data permission management, and process management of geographic information, hydrology, meteorology, and cascade scheduling, it achieves intelligent forecasting and scheduling, and flexible and convenient interaction between data.

2.4 Research Area

The Nam Ou River is the largest tributary on the left bank of the Mekong River in Laos, originating from the border mountain range between Jiangcheng County, Yunnan Province, China and Fengshali Province, Laos. The entire river basin covers an area of 25,634 km^2 and the length of the river is 475 km. The Nam Ou River has a well-developed water system in Laos, with 11 major tributaries.

According to the plan, the Nam Ou River is divided into Nam Ou River Class I to Nam Ou River Class VII hydropower stations from bottom to top, with Nam Ou River Class VII water based on hydrology, as shown in Fig. 1.

The power station is the leading reservoir for hydropower development planning, which has a multi-year regulation capacity; The regulating performance of the sixth and fifth cascade hydropower stations on the Nam Ou River has reached seasonal regulation, while the regulating performance of the other cascade hydropower stations is daily regulation. All units of the first phase of the Nam Ou River cascade hydropower station project were completed and fully put into operation by the end of May 2016; The main works of each cascade hydropower station in the second phase of the project began construction in the first half of 2016. In September 2021, all the hydropower stations in the Nam Ou River Basin were completed and fully put into operation, and the basin's joint debugging and transportation began. As of now, the total installed capacity of the entire basin of the project has reached 1.272 million kW, with an average annual power generation of about 5 billion kWh.

According to the analysis, the scheduling and operation of cascade hydropower stations on the South European River involve demands in power generation, flood control, ecology, water supply, shipping, and other aspects. By analyzing the primary and secondary relationships between various scheduling objectives and their transformation methods, the objectives are refined, classified, and functionalized, and a multi-objective joint scheduling objective system for cascade hydropower stations is proposed. On this basis, a multi-objective joint scheduling model for cascade hydropower stations is established that considers the maximum comprehensive benefits of basin power generation, flood control, ecology, water supply, and shipping, to achieve decision-making and optimization of joint optimization scheduling schemes for cascade hydropower stations, thereby providing a theoretical basis for the evaluation of joint scheduling operation schemes for cascade hydropower stations in the basin.

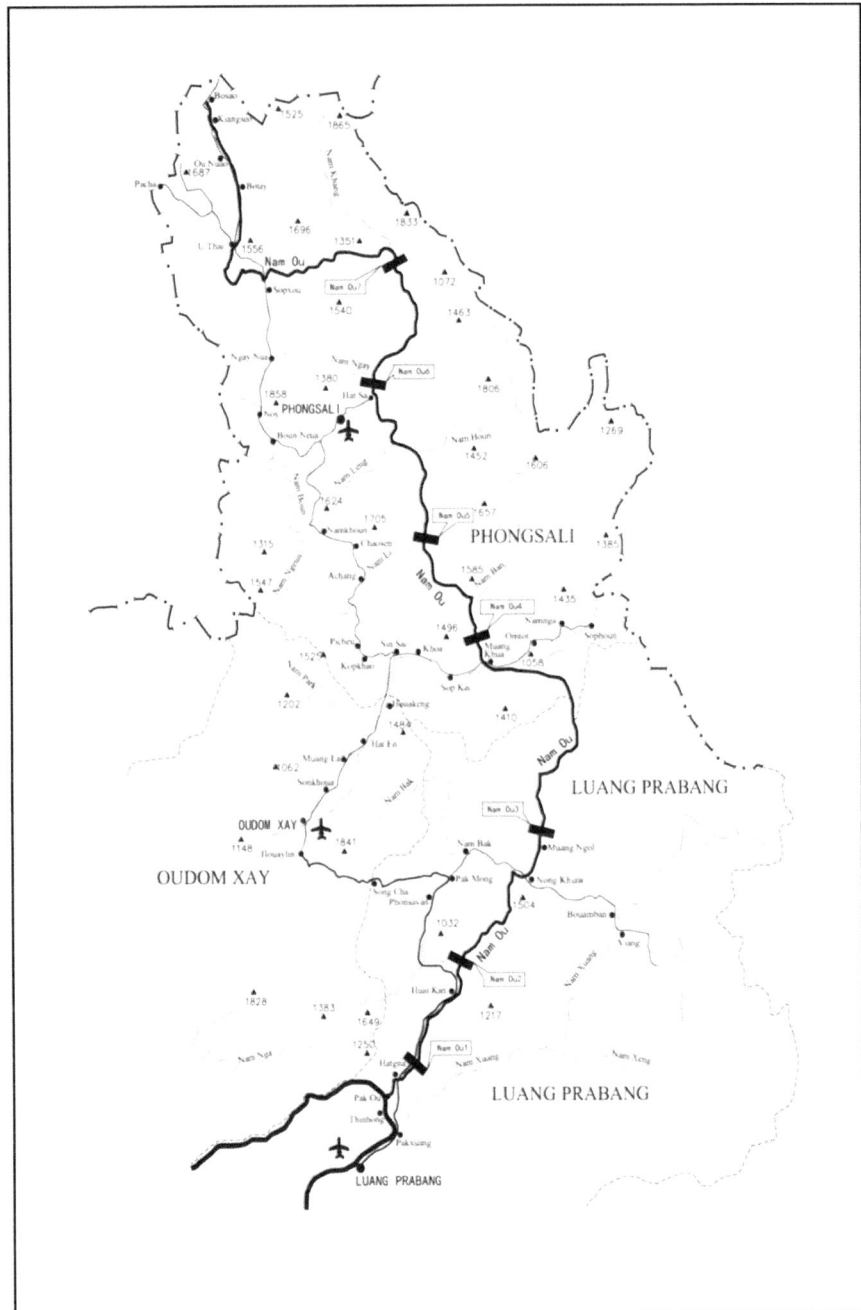

Fig. 1 Cascade development in the Nam Ou River basin of Laos

3 Results and Discussion

3.1 Results

On the basis of meteorological and hydrological forecasting, this study investigates individual scheduling schemes for each reservoir, designs conventional and experiential scheduling modules, and achieves multiple control modes such as water level, flow rate, output, inflow and outflow balance, scheduling diagram, and flood control scheduling parameters, achieving the goal of flexible application of power generation, flood control, and ecological scheduling in each cascade from top to bottom. On this basis, a multi-objective joint optimization scheduling and risk decision support system for the cascade reservoirs of the Southern European River has been constructed, as shown in Fig. 2.

The multi-objective intelligent scheduling platform developed through research has been applied to the cascade scheduling of the Nam Ou River Basin, and was officially put into trial operation by the end of 2021. The water consumption rate of power generation in the southern European river basin in 2022 decreased by 6.87% compared to 2021, and the power generation increased by 185 million kWh. It effectively dealt with a 5-year flood and avoided cascade power stations and personal and property disasters for people along the coast during the flood season. Based on the research results and experience of optimizing reservoir operation, the optimized operation model can increase the economic benefits of hydropower in the basin by more than 3%. Among them, the operation and dispatching of cascade VII power station with the strongest regulation capacity in the basin are as shown in Fig. 3.

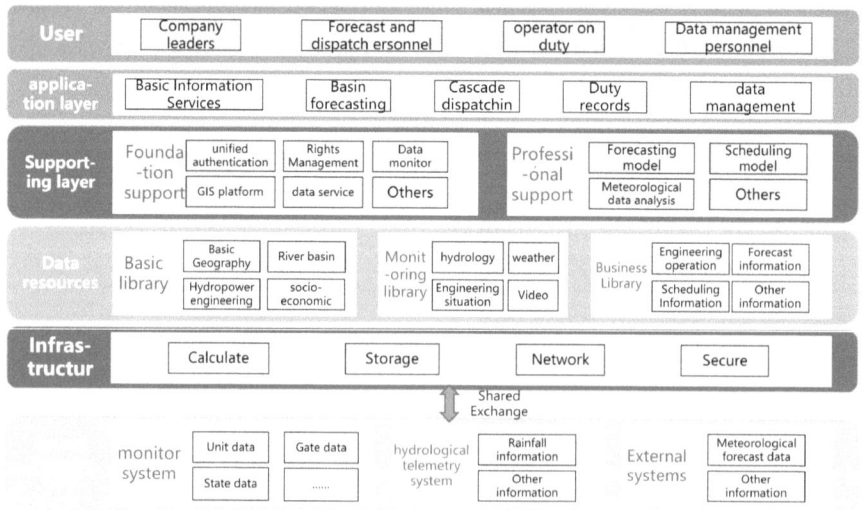

Fig. 2 Intelligent platform architecture diagram

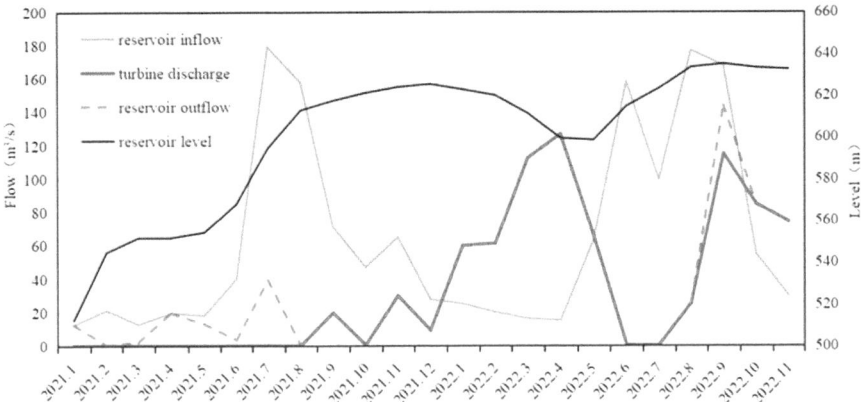

Fig. 3 Operation and dispatch of the seven hydropower station

The platform provides statistical analysis and export functions for power station operation data, meteorological forecast data, hydrological forecast data, and scheduling plan data, greatly reducing the workload of power station staff, ensuring the accuracy of data, and improving work efficiency. Through the construction of this platform, various types of information can be uniformly collected and managed, improving the efficiency of information organization, reducing manpower investment, and making it intuitive and convenient to use this information data.

3.2 Discussion

(1) We should further deepen our research results, improve the accuracy of water and rain forecasting, the calculation accuracy, speed, and time series length of reservoir scheduling, and serve the cascade power market scheduling in the basin; On the basis of the above research, improve the calculation module for dam break evolution, and further enhance the early warning ability of natural disasters in the basin cascade; Study the adaptability of hydropower station operation under climate change scenarios.

(2) The presentation format of the results and enhance human–computer interaction capabilities. We should be further optimized. With the support of data visualization and 3D image display technology, the expressive power of research results is further enhanced.

4   Conclusions

The research has formed a platform for meteorological and hydrological forecasting and multi-objective joint scheduling of cascade hydropower stations in the Nanou River Basin. The platform comprehensively considers the needs of cascade power generation, flood control, ecology, and other aspects of the basin. Through meteorological and hydrological forecasting and multi-objective joint scheduling functions, the economic and social benefits of cascade hydropower stations have been improved, and the water distribution of main and tributaries, upstream and downstream has been coordinated, which can significantly improve flood control and power generation benefits. Based on the research results and experience of optimizing reservoir operation, the optimized operation model can increase the economic benefits of hydropower in the basin by more than 3%.

The platform system couples meteorological forecast information, which can extend the hydrological forecast prediction period. Combined with the warning mechanism, it can effectively reduce the risks during the operation period of hydropower projects, improve the operational efficiency of hydropower projects. Based on the research results and experience of optimizing reservoir operation, under a multi-objective risk controllable operation plan, the long-term economic benefits are significant.

It is recommended to continue optimizing and adding cutting-edge models, enriching the model library, and further improving the functional level of the platform. It is recommended to conduct timely research on the scheduling strategy of cascade hydropower stations on the South European River in the context of climate change, actively respond to the challenges brought by climate change, and ensure the long-term benefits of the project.

Acknowledgements Study on meteorological and hydrological forecasting and multi-objective joint scheduling intelligent system research for cascade hydropower stations on the Nam Ou River in Laos (DJ-PTZX-2018-05).

References

1. T. Li, R.L. Xia, B. Li et al., Research on flood forecasting and its prospect. China Rural Water Hydropower **12**, 107–114 (2022). https://doi.org/10.12396/znsd.220245
2. J.Y. Li, F. Li, H.J. Wang, Subseasonal prediction of winter precipitation in southern China using the early November snowpack over the Urals. Atmos. Ocean. Sci. Lett. **13**(6) (2020)
3. B. Li, F.Q. Tian, Y.K. Li et al., Development of a spatiotemporal deep-learning-based hydrological model. Adv. Water Sci. **33**(6), 904–913 (2022)
4. B. Xu, P.A. Zhong, Y.T. Chen et al., The multi-objective and joint operation of Xiluodu cascade and Three Gorges cascade reservoirs system. Sci. Sin. Technol. **47**(8), 823–831 (2017)
5. M.S. Khorshidi, M.R. Nikoo, M. Sadegh et al., A multi-objective riskbased game theoretic approach to reservoir operation policy in potential future drought condition. Water Resour. Manage. **33**(6), 1999–2014 (2019)

6. H.G. Lan, Y.H. Li, Q.Q. Li et al., Research on reservoir multi-objective operation decision based on improved multi-objective firefly algorithm. Yangtze River **53**(9), 195–201 (2022)
7. Z.D. Zhang, H. Qin, L.Q. Yao et al., Improved multi-objective moth-flame optimization algorithm based on R-domination for cascade reservoirs operation. J. Hydrol. **581**, 124431 (2020)
8. Y.F. Ma, P.A. Zhong, B. Xu et al., Multidimensional parallel dynamic programming algorithm based on Spark for large-scale hydropower systems. Water Resour. Manage. **34**(11), 3427–3444 (2020)
9. F. Zhu, P.A. Zhong, Y. Sun et al., Real-time optimal flood control decision making and risk propagation under multiple uncertainties. Water Resour. Res. **53**(12), 10635-10654 (2017)
10. X.H. Lei, W.H. Liao, Y.Z. Jiang et al., Distributed hydrological model EasyDHM I: theory. J. Hydraul. Eng. **7**, 786–794 (2010)
11. Z.X. Xu, L. Cheng, Progress on studies and applications of the distributed hydrological models. J. Hydraul. Eng. **41**(9), 1009–1017 (2010)

Analysis of the Influence of Dewatering of Subway Foundation Pit on the Displacement of Surrounding Strata

Jianjun Wang, Yongzhou Jian, Tangwei Xue, and Kunyong Zhang

Abstract In order to study the influence of subway foundation pit on the displacement of surrounding strata under the condition of dewatering. Based on the finite element modeling of a subway foundation pit project under construction in Nanjing, this paper discusses the influence of excavation only on the ground settlement around the subway foundation pit and the influence of dewatering foundation pit on the ground settlement around the subway foundation pit. Through the comparison of the two cases, the influence law of the horizontal and vertical displacement of the surrounding strata caused by the precipitation is analyzed. It is concluded that the vertical displacement of the surrounding strata caused by the foundation pit is larger in the case of the dewatering, and only the horizontal displacement of the excavation foundation pit is considered to be smaller.

Keywords Foundation pit dewatering · Foundation pit excavation · Stratum displacement · Numerical simulation

J. Wang
CCCC Tunnel Engineering Company Limited, Beijing, China

Y. Jian
CCCC-SHB Fourth Engineering Co., Ltd., Luoyang, China

T. Xue (✉) · K. Zhang
Institute of Geotechnical Engineering, Hohai University, Nanjing, China
e-mail: 1219522710@qq.com

K. Zhang
e-mail: ky_zhang@hhu.edu.cn

© The Author(s) 2025
W. Wang et al. (eds.), *Hydraulic Structure and Hydrodynamics*, Lecture Notes in Civil Engineering 608, https://doi.org/10.1007/978-981-97-7251-3_7

1 Introduction

With the development of China's cities, the first- and second-tier urban rail transit network is gradually improved, the underground routes criss-cross, the deeper the foundation pit of the station, there are more and more factors to consider. Groundwater is one of the most basic problems in deep foundation pit engineering, and it is also the key factor that causes the foundation pit to lose its stability.

At present, scholars at home and abroad have made fruitful research results on the settlement caused by foundation pit dewatering. In the 1920s, Meinzer discovered and defined land subsidence when studying groundwater movement [1]. Then Gambolati predicted the influence of precipitation on the surrounding land surface settlement based on linear boundary element method [2]. Xie et al. [3] studied the change of stress state of layered soil caused by dewatering excavation of foundation pit, and explored the calculation method of soil layer stress change and surrounding surface settlement. By assuming that foundation pit construction only induces one-dimensional steady flow of groundwater, he derived the theoretical calculation formula of effective soil stress and surface settlement around the pit. Zhang [4] investigated and analyzed relevant data at home and abroad about foundation settlement of surrounding buildings caused by foundation pit precipitation, adopted numerical calculation method to deeply discuss the influence of foundation pit precipitation in wide station yard of high-speed railway on settlement of composite foundation near subway. Gao et al. [5] calculated the settlement amount of different soil layers after foundation pit dewatering based on the general situation of Luoyang metro station project. Zhou et al. [6, 7] simplified the settlement formula based on the mechanism of surface settlement caused by precipitation, considering the influence of groundwater level changes on soil body weight. Liu et al. [8] converted the soil pore water pressure into external load and acted on the ground in the form of additional stress in the numerical model, analyzed the size of the surface settlement at different positions away from the foundation pit, and classified the settlement influence grade during the foundation pit dewatering process.

According to the relevant literature reading, it is found that there are many researches on the formation settlement caused by foundation pit dewatering. However, the settlement of deep foundation pit caused by precipitation needs further study, and the rule of horizontal displacement of surrounding strata caused by precipitation is less studied. Taking a subway foundation pit project in Nanjing as an example, this paper studies the influence of precipitation on the surrounding strata during the excavation of deep foundation pit through numerical simulation analysis, so as to provide reference for the foundation pit dewatering project in the flood-flood-area of the Yangtze River.

2 Theoretical Analysis of Ground Settlement Caused by Foundation Pit Dewatering

According to the principle of effective stress, the change of stress of water in soil will lead to the change of effective stress between soil particles and then the subsidence of the ground. However, the physical properties of soil are very complicated, and the seepage in soil plays an important role in the ground settlement.

In the dewatering process of foundation pit, the seepage and stress field of soil around the dewatering well change greatly, which in turn breaks the equilibrium state of the initial seepage field, because pumping will lead to the seepage movement of groundwater. Among them, the main reason of soil consolidation and compaction due to water loss is the additional stress caused by seepage force, and its direction of action is basically the same as the additional stress in this process. The final formation deformation caused by precipitation includes both vertical subsidence and lateral deformation of deep soil.

When a reasonable dewatering scheme is used for foundation pit dewatering, the loss of fine particles caused by pumping is small, and the surface settlement around the foundation pit is mainly the consolidation settlement caused by pumping. The soil stress changes caused by foundation pit dewatering can be calculated according to the Terzaghi effective stress principle and consolidation theory. The total soil stress before foundation pit dewatering is shown in Formula (1):

$$\sigma = \sigma' + \mu \tag{1}$$

Formula: σ—Total soil stress, kPa; σ'—Soil effective stress, kPa; u—Pore water pressure, kPa.

It can be seen from the formula that the change of soil effective stress has a direct influence on the change of soil volume and the change of shear strength. According to the Terzaghi effective stress principle, the total soil stress in the dewatering project is always constant. After the precipitation, the pore water between the particles is extracted, and the pore water pressure is reduced, so that the effective stress on the soil continues to increase. Under the action of external load, the soil particle skeleton will undergo compaction deformation due to the action of load, and then cause vertical deformation of soil layer. It can be seen that, from the numerical point of view, the increase of soil effective stress is equal to the negative increase of pore water pressure, as shown in Formula (2).

$$\Delta\sigma' = \sigma - \mu + \Delta\mu \tag{2}$$

Therefore, the additional effective stress after foundation pit dewatering is:

$$\Delta p = \Delta\sigma' = \gamma_w(h_0 - h_1) \tag{3}$$

Formula: Δp—Additional effective stress of soil after precipitation, kPa; h_0—The height of the water table before precipitation, m; h_1—The height of the water table after precipitation, m.

After precipitation, due to the change of stress state, the resulting formation settlement is often calculated by layer-summation method.

3   Finite Element Modeling

3.1   Project Overview

The foundation pit project of a subway in Nanjing is in the overall direction of southwest to northeast. The station length is 398 m, the station is an underground two-storey island station, the standard section is a single-column and double-span box-type frame structure, the platform width is 11 m, the standard section width is 20.1 m, the total length is 398 m, the standard section floor buried depth is 16.35 m, the roof covered with soil is about 3 m, and the small mile end well floor buried depth is 20.4 m. Great mileage end well bottom buried 18.0 m. The landform types of the stations all belong to the Yangtze River floodplain with relatively flat terrain. The current ground elevation of the stations ranges from 5.35 to 6.91 m.

3.2   Model Building

In this project, the shape of foundation pit is shaped like a special trumpet, and the most attention is paid to the deformation of the position of the trumpet.

Half of the excavation of the model is taken on the principle of symmetry. The length of the whole model is 120 m and the thickness is 50 m. SIGMA/W module in GeoStudio software was used to numerically simulate the foundation pit model, and the whole model was divided into 1.1 m quadrilateral grids. Two dimensional modeling is carried out for the layered excavation of soil and the setting of supporting structure (underground continuous wall) involved in the construction of the deep foundation pit.

In actual projects, the distribution of soil layers is relatively complex, with uneven thickness, inconsistent buried depth, and unclear soil layer boundaries. Even if it is the same soil layer, its properties may not be the same. Therefore, it is necessary to simplify the soil layers in the model, and the parameters of each soil layer are shown in Table 1.

There are two models: one is the pure excavation model, which only considers the influence of the foundation pit excavation on the stratum displacement, and no precipitation; the precipitation model, before the foundation pit excavation, the water

Table 1 Soil layer parameters

Geotechnical designation	Argument				
	γ (kN/m³)	Φ (°)	K 0	v	E_{s1-2} (MPa)
Miscellaneous fill	19	15	0.5		5.55
Plain fill	19	10	0.6	0.38	4.84
Muck	19	3	0.75	0.35	3.26
Silty clay	18.73	7.5	0.45	0.31	3.55
Silt mixed with silt	18.75	29.7	0.38	0.28	12.17
Silty fine sand	18.4	29.5	0.4	0.26	11.64

level at the initial water level at the left boundary from the surface to the bottom of the model. The foundation pit model is shown in Figs. 1 and 2.

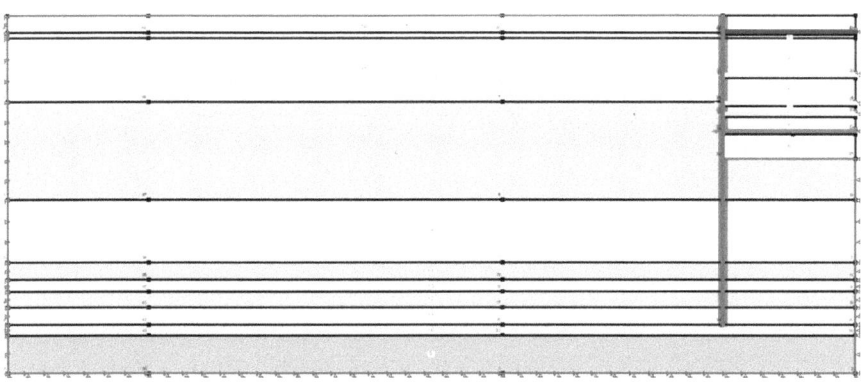

Fig. 1 Model of excavation only

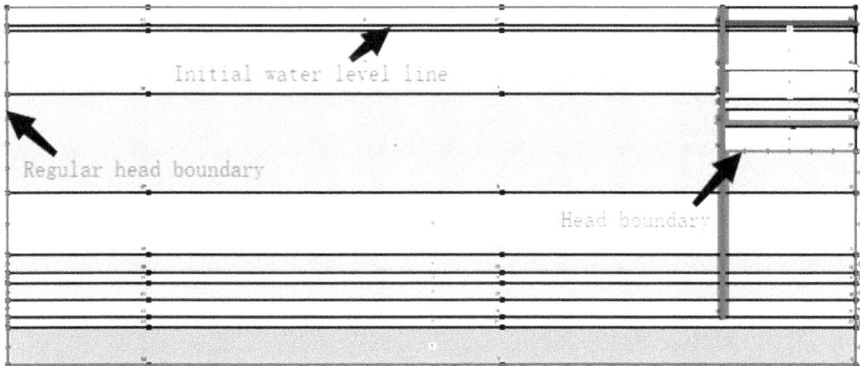

Fig. 2 Dewatering foundation pit model

4   Analysis of Calculation Results

4.1   *Only Excavation Model Analysis Is Considered*

According to the flared section of the foundation pit, the foundation pit is modeled only under the condition of excavation, and the displacement cloud map as shown in Figs. 3 and 4 is obtained.

The cloud map of the X and Y directions of the root foundation pit draws the formation settlement curve of the foundation pit in each excavation stage, as shown in Fig. 5.

After foundation pit excavation, the surface displacement and settlement of the surface displacement and settlement of the surface of the pit continue to expand with the continuous excavation, and the settlement value is also increasing. After

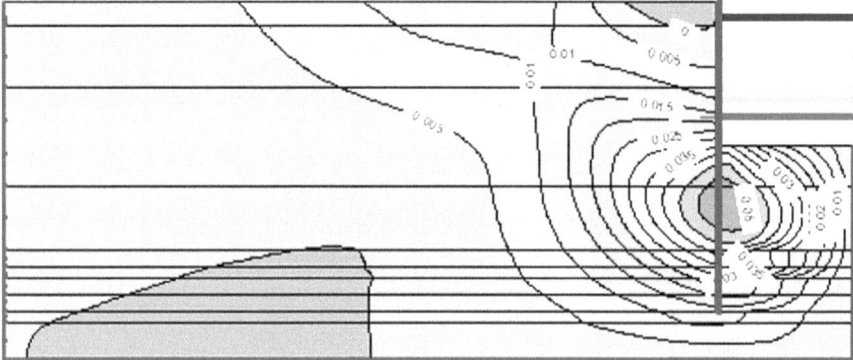

Fig. 3 Only considers the displacement in Y direction of the excavated foundation pit

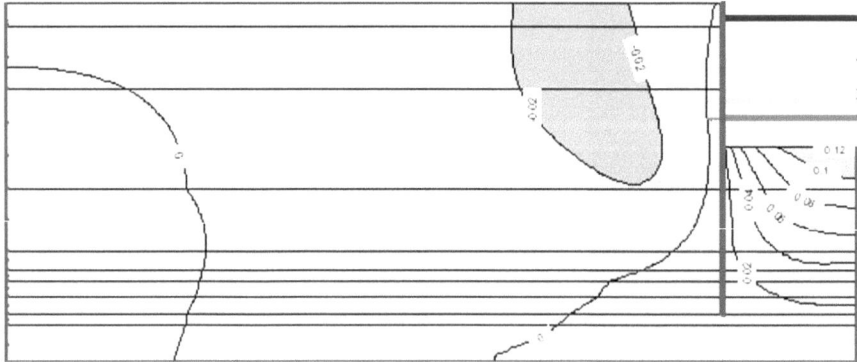

Fig. 4 Only considers the displacement in X direction of the excavated foundation pit

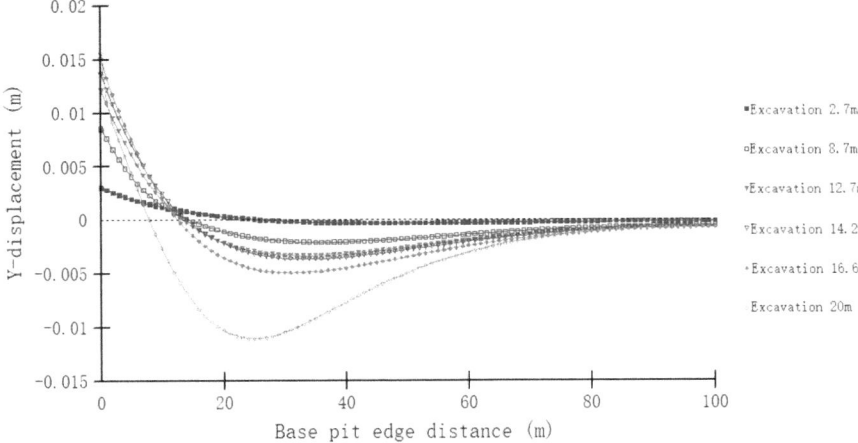

Fig. 5 Only considers the surface settlement curve around excavation

the last excavation, the settlement of the surrounding strata is slightly greater, with a maximum of 11 mm, at a distance of twice the depth of the foundation pit.

The horizontal displacement of strata is analyzed below, as shown in Fig. 6.

As can be seen from the figure, the horizontal deformation curve of the foundation pit connecting wall presents a "bulging" shape with small ends and large middle, indicating that the support and structural stiffness are sufficient to limit its displacement. It can be seen from the simulation results that the horizontal displacement of the ground wall reaches a maximum of 33 mm at a depth of nearly 23 m from the surface.

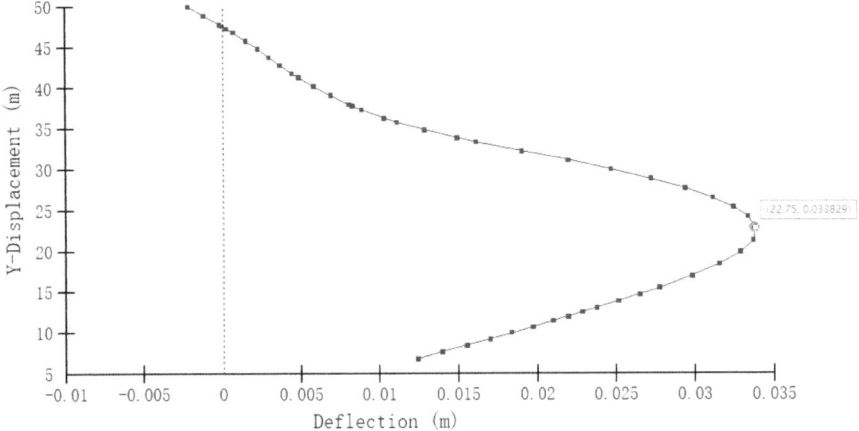

Fig. 6 Only considers the deformation of the excavated ground wall

4.2 Model Analysis of Dewatering Foundation Pit

According to the dewatering scheme of foundation pit, the dewatering of foundation pit is simulated, and Figs. 7 and 8 is obtained.

Compared with excavation, the displacement of foundation pit in Y and X direction becomes larger under the condition of precipitation. The settlement curve of surrounding strata is drawn according to the displacement cloud map of dewatering foundation pit. As shown in Fig. 9.

According to the surface settlement curve around the dewatering foundation pit, it can be seen that the settlement amount of the surrounding stratum increases with the decrease of the water level under the condition of dewatering. When the precipitation is 21 m, the settlement of the surrounding stratum reaches the maximum value of 29 mm at 23 m away from the foundation pit.

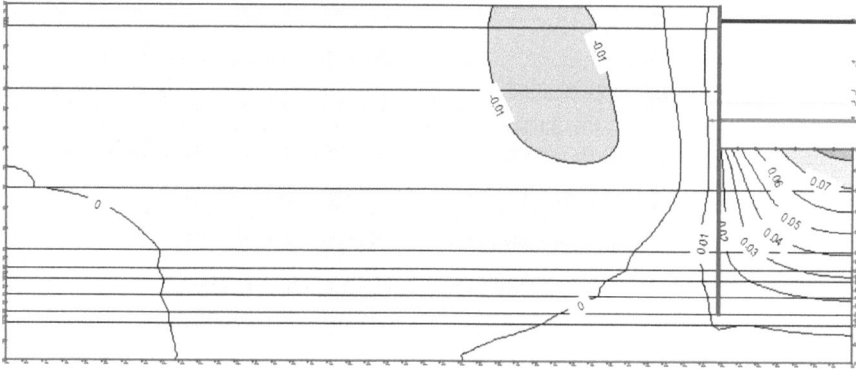

Fig. 7 Displacement in Y direction of dewatering foundation pit

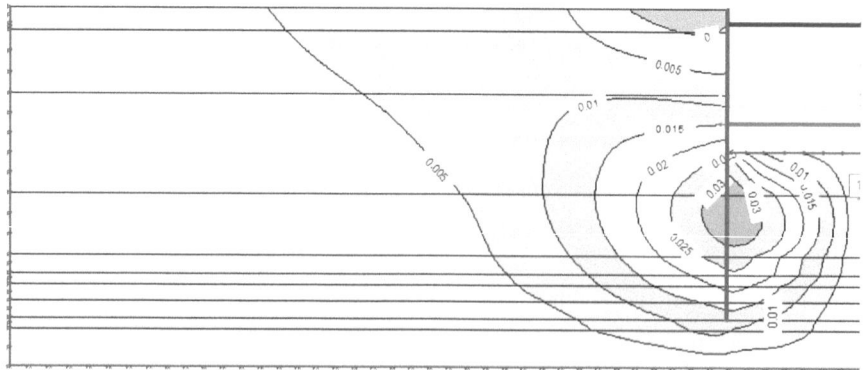

Fig. 8 Displacement of dewatering foundation pit in X direction

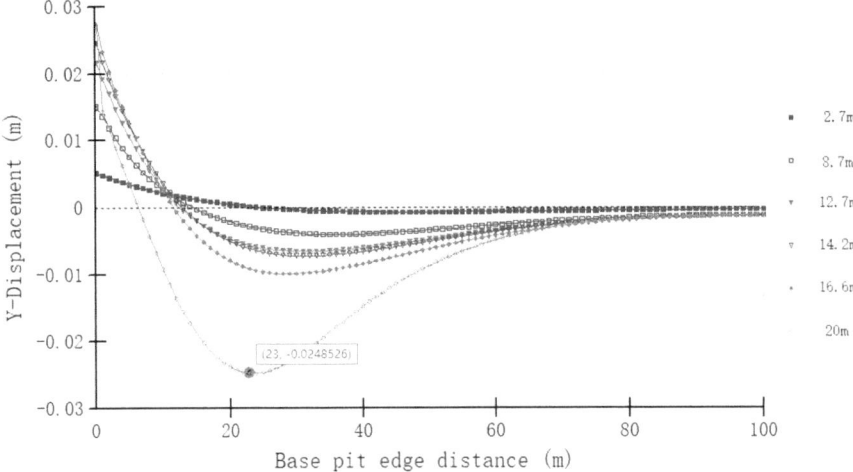

Fig. 9 Surface settlement curve around dewatering foundation pit

As can be seen from Fig. 10, the horizontal deformation curve of dewatering foundation pit ground wall is also relatively typical. It can be seen from the simulation results that the horizontal displacement of the ground wall reaches a maximum of 65 mm at a depth of nearly 21 m from the surface. This result is larger than that only considering excavation, which is mainly due to formation deformation caused by horizontal seepage force.

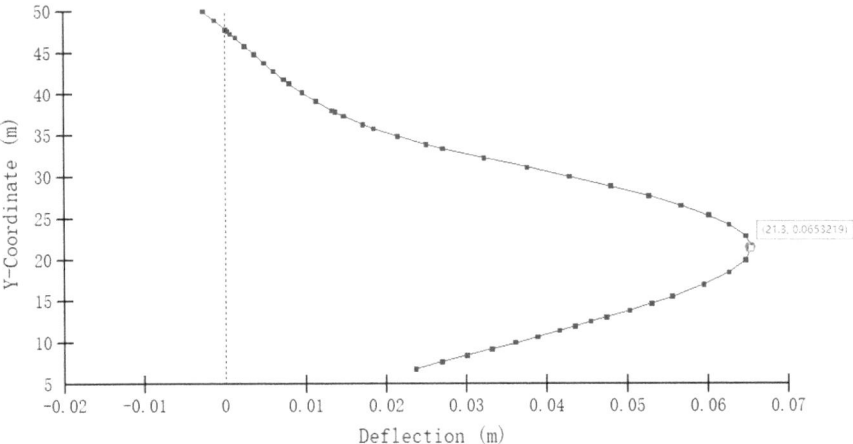

Fig. 10 Dewatering foundation pit ground wall deformation

5 Conclusions

Based on the analysis of the calculation results of the surrounding surface displacement and the foundation wall deformation of the foundation pit in two cases, the conclusions are as follows:

(1) When the foundation pit is completed, the ground displacement presents a typical settling trough. The stratum displacement is caused by dewatering and excavation of foundation pit, and the stratum displacement of foundation pit is larger when dewatering is considered.

(2) Precipitation has a great influence on the deformation of the ground wall. There is horizontal lateral movement at the bottom of the ground wall, which is mainly caused by the formation deformation caused by the horizontal seepage force. Therefore, it is necessary to set the inner support in time during the excavation of the dewatering foundation pit, which can effectively reduce the lateral deformation of the soil envelope caused by excavation.

(3) For the construction of deep large foundation pit in the water-rich sand layer area like the floodplain of the Yangtze River, it is necessary to make a detailed precipitation plan for the foundation pit, and consider the possible precipitation accidents to prevent the collapse, excessive deformation of the envelope structure, or unnecessary impact on the surrounding buildings. Precipitation should also consider the possibility of accidents such as gushing and flowing sand.

References

1. M. Zhao, *Study on the Influence of Excavation Dewatering on the Settlement of Surrounding Buildings* (Shandong University, 2010)
2. G. Gambolati, F. Sartoretto, A. Rinaldo et al., A boundary element solution to land subsidence above 3-D gas/oil reservoirs. Int. J. Numer. Anal. Methods Geomech. **11**, 489–502 (1987)
3. K. Xie, C. Liu, H. Ying et al., Analysis of surface settlement caused by excavation dewatering in stratified soil. J. Zhejiang Univ. (Eng. Technol. Ed.) (03), 9–12+21 (2002)
4. G. Zhang, *Effect of Foundation Pit Depreciation on Settlement of High-Speed Railway Composite Foundation* (Beijing Jiaotong University, 2018)
5. X. Gao, D. Li, J. Lei, Analysis of soil layer subsidence caused by precipitation in water-rich deep sand and gravel stratum, in *MCC General Institute of Building Research Co., Ltd. Proceedings of the 2020 Industrial Building Academic Exchange Conference (Second Volume)* (MCC Building Research Institute Co., Ltd., Industrial Building Magazine, 2020), p. 5
6. X. Chen, A. Zhou, B. Wang, Y. Hu, Calculation method of uniform settlement of pile foundation induced by foundation pit dewatering. Sci. Technol. Eng. **19**(23), 199–205 (2019) (in Chinese)
7. A.-Z. Zhou, Y. Hu, X. Sui, Study on calculation method and potential threat of settlement of site outside pit induced by negative capillary water pressure. Sci. Technol. Eng. **20**(02), 779–784 (2020)
8. K. Liu, Calculation and analysis of foundation settlement caused by water level drop in deep foundation pit. Jilin Water Resour. **01**, 16–19 (2021)

Seismic Safety of a High Geomembrane Faced Soft Rockfill Dam on Overburden Subjected to Strong Earthquake

Libo Wang, Weijun Cen, Dongliang Wang, and Jie Tang

Abstract 3D dynamic finite element calculation is carried out for a 161 m high geomembrane faced soft rockfill dam on overburden in a strong earthquake area. The dynamic response of the dam is obtained. The horizontal dynamic displacement is relatively more significant. The maximum vertical permanent deformation and deflection of geomembrane occur at the dam crest near the maximum dam cross section. The tensile strain of geomembrane increases significantly at the dig-fill junction and the reverse arc section. The maximum tensile strain of geomembrane is 1.52%, and the safety factor of geomembrane tensile strain is greater than the allowable value. Even subjected to the strong earthquake, the geomembrane slab can effectively coordinate the deformation with the dam body and deep overburden and the dam can be in a safe operation state.

Keywords Geomembrane faced soft rockfill dam · Overburden · Dynamic response · Strong earthquake

L. Wang · W. Cen (✉)
Hunan Provincial Key Laboratory of Key Technology on Hydropower Development, Changsha, China
e-mail: hhucwj@163.com

College of Water Resources and Hydropower Engineering, Hohai University, Nanjing, China

L. Wang
e-mail: wlb666@hhu.edu.cn

D. Wang · J. Tang
PowerChina Huadong Engineering Corporation Limited, Hangzhou, China
e-mail: wang_dl2@hdec.com

J. Tang
e-mail: tang_j3@hdec.com

1 Introduction

The number and height of dams have been greatly improved with the booming of dam industry in China [1]. Currently, there are fewer suitable sites for building high dams and large reservoirs. The remaining projects to be developed often face complex and harsh dam-building conditions such as deep overburden, high seismic intensity, soft dam materials and lack of impervious clay, thus concrete dams or conventional earth-rock dams are not the best choice.

Compared with clay and asphalt concrete, geomembrane has the advantages of low permeability, strong adaptability to deformation, convenient construction and low cost [2]. If a rockfill dam adopts flexible geomembrane as the impervious barrier, it may be well adapted to the large earthquake reciprocating deformation and avoid the impervious barrier damage when subjected to strong earthquake [3]. In addition, the geomembrane is especially suitable for impervious barrier of high rockfill dams filled with soft rock on the overburden [4], which can effectively coordinate the deformation with the dam body and overburden. Thus, the geomembrane faced rockfill dam is especially suitable for the above complex and harsh dam-building conditions. At present, geomembranes have been used in seepage control of more than 160 rockfill dams internationally, among which more than 40 dams have been built in China. However, most of the geomembrane faced rockfill dams are faced with low water heads, and are not built on the overburden in strong earthquake areas. Therefore, it is of great significance to investigate the seismic safety of high geomembrane faced soft rockfill dam built on the overburden subjected to strong earthquake.

A pumped storage power station in southwest China is located in a strong earthquake area where the basic seismic intensity is VII and the design intensity is VIII. The upper reservoir dam of the power station is a geomembrane faced soft rockfill dam, with a maximum dam height of 161 m, a dam crest elevation of 3786 m, on a 30 ~ 68 m thick overburden. The filling material of the dam are mainly divided into cushion layer, transition layer, rockfill I and rockfill II from upstream to downstream. The rockfill materials are mainly composed of slate whose saturated compressive strength and softening coefficient are low. The geomembrane is also used as impervious body at the bottom of the reservoir, which is connected with the geomembrane of dam surface in a reverse arc section. In this study, 3D dynamic finite element method is carried out to demonstrate the feasibility of the construction of the high geomembrane faced soft rockfill dam on deep overburden subjected to strong earthquake from the perspective of stress, deformation and seismic response. It provides reference for the design and construction of similar projects.

2 Calculation Model and Calculation Conditions

2.1 Finite Element Model

According to the topographic and geological conditions of the dam site area and dam design, the calculation domain is reasonably selected. Figure 1 shows the 3D finite element model of the high geomembrane faced soft rockfill dam. The finite element mesh is mainly composed of eight-node hexahedral elements with 19,333 nodes and 21,697 elements, including 631 geomembrane elements.

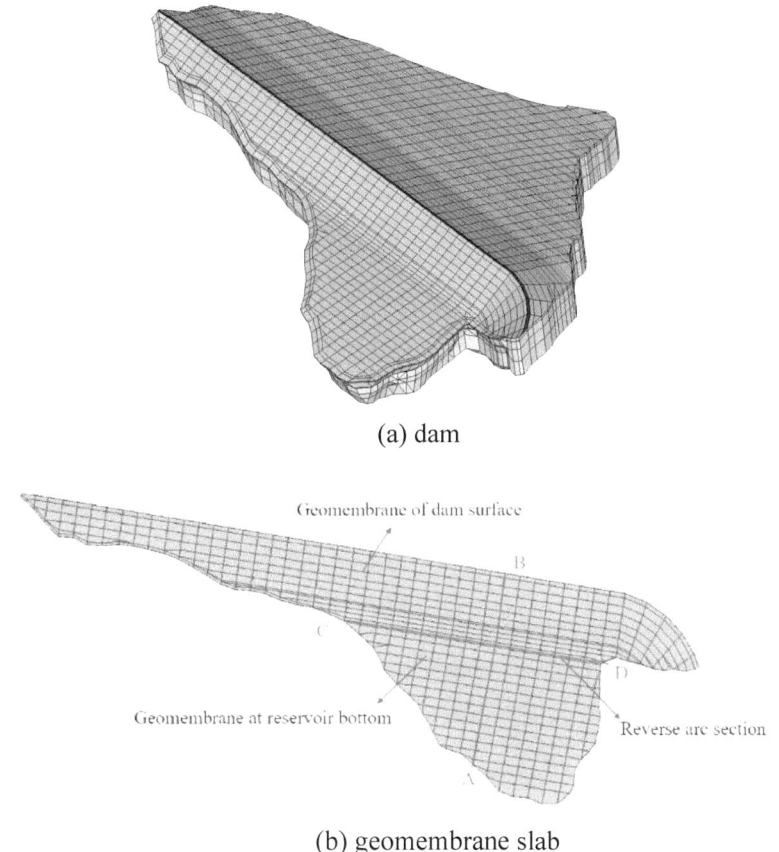

(a) dam

(b) geomembrane slab

Fig. 1 3D finite element mesh of the geomembrane faced soft rockfill dam

Table 1 Dynamic calculation parameters of dam materials

Materials	K_2	λ	n	c_1	c_2	c_3	c_4	c_5
Cushion layer	1950	0.21	0.38	0.75	0.86	0	6.99	0.85
Transition layer	2100	0.22	0.38	0.77	0.88	0	6.95	0.87
Rockfill I	2035	0.21	0.33	0.73	0.82	0	6.90	0.90
Rockfill II	1875	0.23	0.35	0.84	0.83	0	7.40	0.91

Table 2 Mechanical properties of the geomembrane

Density (g/cm^3)	Yield strength (MPa)	Yield strain (%)	Breaking strength (MPa)	Breaking strain (%)
0.94	11.52	13.00	11.46	610.32

2.2 Calculation Parameters of Dam Materials

The equivalent nonlinear viscoelastic model is used to simulate the dynamic stress–strain relationship of the dam materials, and the Shen Zhujiang model is used for earthquake permanent deformation [5]. Table 1 shows the results of large-scale indoor triaxial tests for dynamic calculation parameters of dam materials.

2.3 Properties of the Geomembrane

Smooth HDPE geomembranes with a thickness of 1.5 mm was employed in the soft rockfill dam. Table 2 shows the mechanical properties of the geomembrane.

2.4 Dynamic Interface Model of Geomembrane-Sandy Gravel

The cyclic secant shear stiffness K is defined as the ratio of the shear stress to the corresponding shear-displacement amplitude u, i.e.:

$$K = \frac{\tau}{u} = \frac{1}{a + bu} \tag{1}$$

where, a and b are the model parameters obtained by experiments.

The following model is widely used in soil dynamics:

$$\lambda = \lambda_{ult}\left(1 - \frac{K}{K_{max}}\right) = \lambda_{ult}\left(1 - \frac{a}{a + bu}\right) \tag{2}$$

Table 3 The design standard response spectrum parameters

A_{max}	β_m	T_0	T_1	T_g	C	α_{max}
396	2.6	0.04	0.10	0.50	0.9	1.030

where the ultimate damping ratio λ_{ult} is related to the asymptotical value under an theoretically infinitely large shear-displacement amplitude u.

To reasonably take into account the damping behavior of a geomembrane-sandy gravel interface under small cyclic shear displacements, the following modified model [6] for the pressure dependent damping ratio is proposed based on (2):

$$\lambda = \frac{\lambda_0}{1 + ku} + \left(\lambda_{ult} - \frac{\lambda_0}{1 + ku}\right)\left(1 - \frac{a}{a + bu}\right)^{(\alpha_1 + \alpha_2 \sigma n)} \tag{3}$$

where, λ_0 is the damping ratio for $u = 0$; λ_{ult} is the ultimate value of the damping ratio for $u \to \infty$; n is the vertical stress; k, α_1 and α_2 are constitutive parameters.

2.5 Simulation of Dam Filling by Stages

This calculation simulates the dam filling and water storage process in detail. The 1st–47th steps simulate the dam filling from the bottom to the top. The 48th step simulates the placement of geomembrane. The 49th–58th steps simulate the reservoir storing water to the normal water level of 3780 m.

2.6 Input Seismic Parameters

The seismic intensity at the dam site is VII, and the peak ground acceleration is 0.396 g. Table 3 shows the parameters of the design standard response spectrum. Figure 2 shows the time histories of acceleration for seismic input.

3 Results and Analysis

3.1 Dynamic Displacement of Dam

Figure 3 shows the 3D dynamic displacement distribution of the dam. The dynamic displacement increases with the increasing elevation and decreases from the valley center to two banks. The maximum dynamic displacement occurs near the top of the dam with the maximum cross section, which are 39.37, 40.86 and 27.36 cm at valley

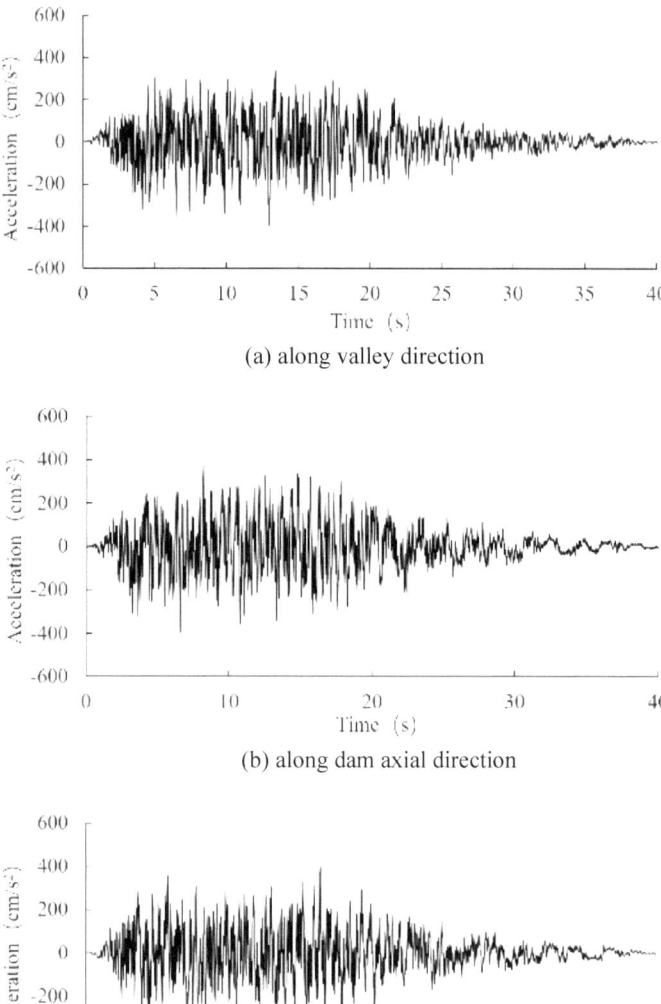

(a) along valley direction

(b) along dam axial direction

(c) along vertical direction

Fig. 2 Time histories of acceleration

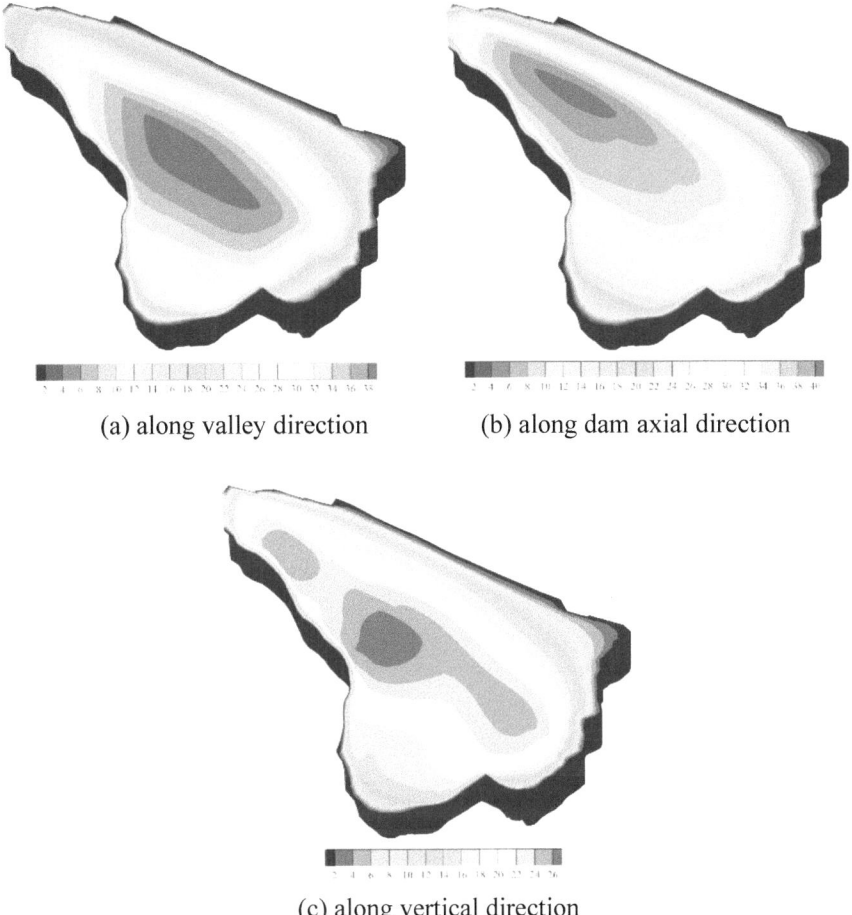

(a) along valley direction   (b) along dam axial direction

(c) along vertical direction

Fig. 3 3D dynamic displacement distribution of the dam (unit: cm)

direction, dam axial direction and vertical direction, respectively. The horizontal dynamic displacement of the dam is relatively more significant.

3.2 Dynamic Response of the Geomembrane

The geomembrane is deformed with the dam body and the overburden subjected to the earthquake, and the subordinate deformation and strain are generated. Figure 4 shows the vertical permanent deformation and deflection distribution of geomembrane at dam surface and reservoir bottom. The vertical permanent deformation of geomembrane increases with the increasing elevation which is in concordance with

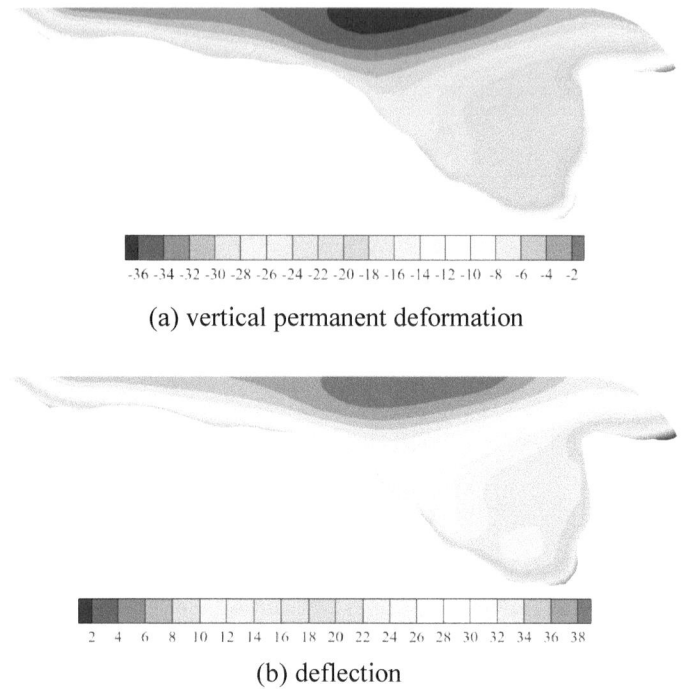

(a) vertical permanent deformation

(b) deflection

Fig. 4 Vertical permanent deformation and deflection distribution of geomembrane (unit: cm)

the deformation of the dam body. The maximum vertical permanent deformation and deflection of geomembrane are 36.83 and 39.79 cm, respectively. They are located at the dam crest near the maximum dam cross section.

Figure 5 shows the principal tensile strain distribution of the geomembrane. The tensile strain is relatively larger at the boundary of dig-fill junction in reservoir and the reverse arc section. The maximum tensile strain of geomembrane is 1.52%, which occurs on the dam surface at the right bank. According to (4), the value of tensile strain safety factor k is 8.6, which is greater than the allowable value of 5.0. Thus, the geomembrane is safe subjected to the strong earthquake.

$$k = \frac{\varepsilon_y}{\varepsilon_{max}} \qquad (4)$$

where, ε_y is the yield strain of geomembrane, its value is 13.0 in Table 2; ε_{max} is the maximum tensile strain of geomembrane.

Figure 6 shows the principal strain distribution of the geomembrane along particular paths. Figure 1b shows the particular paths, in which the path AB is perpendicular to the dam axis, and the path CD is located at the reverse arc section and parallel to the dam axis.

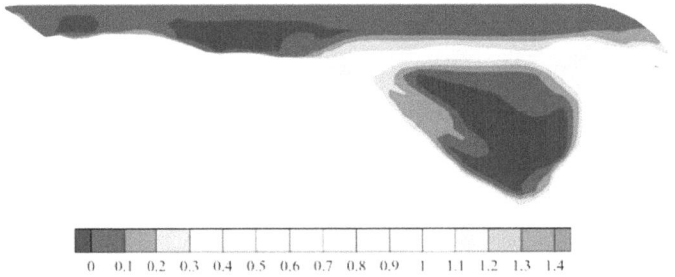

Fig. 5 Principal tensile strain distribution of the geomembrane (unit: %)

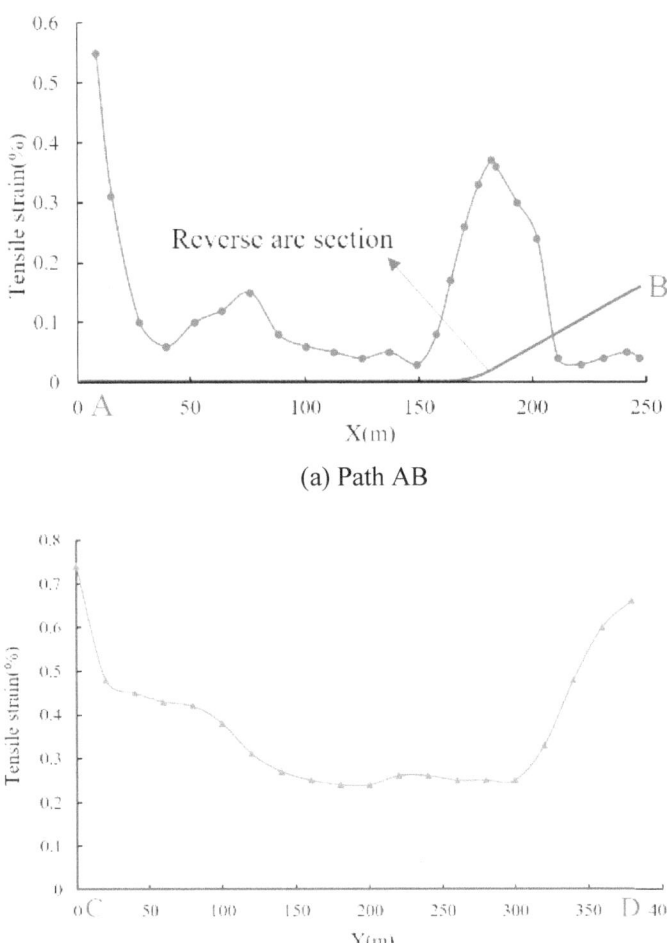

(a) Path AB

(b) Path CD

Fig. 6 Principal strain distribution of geomembrane along particular paths

The tensile strain of geomembrane increases significantly at the dig-fill junction and the reverse arc section, while the strain in other parts is small and the change range is not significant. The geomembrane at the reverse arc section is subjected to water pressure to produce tensile deformation to both sides, which results in relatively larger tensile strains. The vertical depth of the filling material at the dig-fill junction is different, which is easy to produce uneven deformation, resulting in a significant increase in the tensile strain of the geomembrane. Although, the tensile strain safety factor of the geomembrane is greater than the allowable value, it is necessary to diminish the tensile strain of geomembrane as much as possible in view of the above causes. Therefore, some reasonable engineering measures can be taken such as increasing the compaction quality of rockfill materials at the cut-and-fill junction during construction and improving the quality of backfill materials.

4   Conclusions

3D dynamic finite element method is carried out for a high geomembrane faced soft rockfill dam on overburden subjected to strong earthquake. The following conclusions are drawn from the analysis of dynamic response of dam body and geomembrane:

(1) The dynamic response of the dam increases with the increasing elevation, and the whipping effect near the dam crest is significant. The maximum dynamic displacement along valley direction, dam axis direction and vertical direction are 39.37, 40.86 and 27.36 cm, respectively.

(2) The maximum vertical permanent deformation and deflection of geomembrane occur at the dam crest near the maximum dam cross section. The maximum tensile strain of geomembrane is 1.52%, which is located on the dam surface at the right bank. The tensile strain safety factor of geomembrane is greater than the allowable value. Thus, the geomembrane is safe subjected to the strong earthquake.

(3) The tensile strain of geomembrane increases significantly at the dig-fill junction and the reverse arc section. Some engineering measures can be taken to diminish the tensile strain, such as paying attention to the compaction quality of rockfill materials at the cut-and-fill junction during construction and improving the quality of backfill materials.

(4) Subjected to the strong earthquake, the geomembrane can effectively coordinate the deformation with the dam body and deep overburden. Therefore, it is feasible to build high geomembrane faced soft rockfill dams on the overburden in strong earthquake area.

Acknowledgements   The authors wish to thank the Open Research Fund of Hunan Provincial Key Laboratory of Hydropower Development Key Technology (Grant No. PKLHD202004).

References

1. H. Ma, F. Chi, Major technologies for safe construction of high earth-rockfill dams. J. Eng. **2**, 498–509 (2016)
2. W.W. Müller, *HDPE Geomembranes in Geotechnics* (Springer Berlin Heidelberg, Berlin, Heidelberg, 2007)
3. D. Ding, J. Ren, Analysis of stress deformation characteristics of composite geomembrane rock-fill dam with core wall. Appl. Mech. Mater. **470**, 962–965 (2013)
4. H. Wu, L. Feng, Z. Teng, Y. Shu, Model test on mechanical and deformation property of a geomembrane surface barrier for a high rockfill dam. Appl. Sci. Basel **11**(23), 11505 (2021)
5. W. Cen, E. Bauer, L. Wen, H. Wang, Y. Sun, Experimental investigations and constitutive modeling of cyclic interface shearing between HDPE geomembrane and sandy gravel. Geotext. Geomembr. **47**, 269–279 (2019)
6. G. Yang, K. Liu, Y. Liu, Research on the maximum anti-seismic capability of high earth rock-fill dam under strong earthquake. Disaster Adv. **6**, 9–15 (2013)

Analysis of Seepage Monitoring Data of Anhua Reservoir Dam in Zhuji

Haiyun Wei, Haizhen Huang, Chaojie Zhang, and Li Lu

Abstract Through the analysis of seepage monitoring data of Anhua reservoir dam, the anti-seepage effect of dam reinforcement project is evaluated. The analysis of monitoring data shows that the osmometer water level behind the inclined clay wall of the dam is relatively stable and has little correlation with the reservoir water level. Compared with the upstream reservoir water level, the water level of osmometer in each section falls between 7.31 and 9.30 m at high water level, which indicates that the anti-seepage effect of clay inclined core wall of dam body is better. The water head of the concrete cut-off wall in the connection section between the right dam head and the wing wall of the spillway gate is reduced by 6.65 m, and the potential difference between the front and rear of the wall is 74.7%, which indicates that the anti-seepage effect of the connection between the dam and the wing wall of the spillway gate is good. There is no obvious correlation between the water level of the pressure gauge pipe around the dam on the left bank and the water level of the reservoir, which is mainly affected by rainfall infiltration and mountain groundwater, indicating that there is no seepage around the dam.

Keywords Clay inclined wall · Seepage in dam body · Seepage around dam · Osmotic potential · Regression analysis

1 Introduction

Anhua Reservoir is located in Zhuji City, about 2 km west of Anhua town in the upper reaches of Puyang River main stream. The reservoir has a rainwater collection area of 635.2 km^2, a main stream length of 51.6 km, and a total storage capacity

H. Wei · H. Huang (✉) · C. Zhang
Key Laboratory of Hydraulic Disaster Prevention and Mitigation of Zhejiang Province, Zhejiang Institute of Hydraulics and Estuary (Zhejiang Institute of Marine Planning and Design), Hangzhou, China
e-mail: hhaizhen@163.com

L. Lu
Zhejiang Guangchuan Engineering Consulting Co., Ltd., Hangzhou, China

© The Author(s) 2025
W. Wang et al. (eds.), *Hydraulic Structure and Hydrodynamics*, Lecture Notes in Civil Engineering 608, https://doi.org/10.1007/978-981-97-7251-3_9

of 58.38 million m^3. Anhua Reservoir is a medium-sized water conservancy project which mainly focuses on flood control and detention and takes into account farmland irrigation. Anhua Reservoir is an important control project in the flood control system of Puyang River basin. The dam is a clay inclined shell dam with a crest height of 34.40 m and the top of the wave wall of 35.40 m. The maximum dam height is 16.3 m, the crest length is 256 m and the width is 10.0–15.0 m. The top height of the clay inclined wall is about 33.10 m, which is firmly connected with the wave wall. The water trough at the bottom of the wall is excavated to the bedrock and treated with curtain grouting. The penetration ratio of the wall and the bottom of the perforated concrete cutoff wall (the overlapping part between the cutoff shaft) at the original right dam head and the spillway wing wall is larger. In 2015, curtain grouting was used for vertical anti-seepage treatment of dam foundation. A C15W6 anti-seepage wall was set 1 m behind the original concrete casing in the connection section between the right dam head and the wing wall of the spillway gate. The curtain grouting at the right dam head was connected with the curtain grouting at the bottom of the left wing wall of the spillway gate to form a closed anti-seepage system.

In order to ensure the normal and safe operation of the project, Anhua Reservoir has set up observation items such as reservoir water level, dam body seepage and seepage around the dam, and the seepage observation has realized automatic monitoring [1, 2]. This paper focuses on the analysis of seepage observation data of the dam to evaluate the effect of dam seepage prevention and reinforcement [3].

At present, a large number of research results have been obtained in the analysis of seepage monitoring data of dams. Ji and Salmasi et al. carried out seepage inversion analysis and numerical simulation analysis of inclined wall rockfill dam, and suggested the addition of engineering measures such as cureproof wall and Geotechnical Membrane, and the addition of dam seepage observation facilities [4, 5]. Xu and Xu et al. carried out research on analysis methods of seepage monitoring data of earth-rock DAMS, and introduced monitoring data regression analysis, seepage inversion analysis methods and monitoring point layout optimization methods [6, 7]. These research results have certain reference value.

The anti-seepage structure of Anhua reservoir dam is different from the common clay inclined core wall. Curtain grouting is used to prevent seepage of the foundation and abutment of the clay inclined core wall, and the anti-seepage wall is used to strengthen the back side of the original concrete casing well connecting the right dam head and the spillway gate wing wall. Therefore, it is necessary to comprehensively evaluate the anti-seepage and reinforcement effect of dam.

2 Layout of Dam Seepage Monitoring Instrument

2.1 Seepage Monitoring of Dam Body

Four seepage monitoring sections including $0 + 060.00$ m, $0 + 120.00$ m, $0 + 180.00$ m and $0 + 240.00$ m are arranged in the dam, and 3, 3, 3 and 2 piezometric pipes are set respectively, totaling 11 dam piezometric pipes. The osmometers are suspended in the piezometric pipes to realize the automatic monitoring of seepage flow in the dam body.

2.2 Seepage Monitoring Around the Dam

A seepage monitoring section around the dam was arranged in the mountain on the left bank of the dam, and one piezometric pipe was set at $0 + 051.00$ m, $0 + 056.00$ m and $0 + 061.00$ m under the dam, respectively, for a total of three piezometric pipes around the dam. The osmometers are suspended in the piezometric pipes.

3 Analysis of Seepage Monitoring Data

3.1 Osmometer Water Level Calculation Principle

In this project, GK4500S type osmometer produced by American Kikang Company is used, which is converted by the frequency of the measured vibration string to obtain the seepage pressure. The water level of the osmometer can be obtained by measuring the seepage pressure head in the dam and the buried elevation of the osmometer. The water level of the measuring point is equal to the sum of the pressure head of the osmometer and its buried elevation. The formula for calculating the water pressure above the measured probe could be expressed as Eq. (1):

$$P = G(R_0 - R_1) + K(T_1 - T_0) \tag{1}$$

where P is measured water pressure; G and K are calibration coefficients, which can be obtained from the calibration table of each branch osmometer; R0 and T0 are the readings before the osmometer is buried; R1 and T1 are the readings after the osmometer is buried.

3.2   Analysis of Seepage Observation Data of Dam Body

(1)   The dam seepage pressure process line

Table 1 gives the statistics of the characteristic value of the water level of the dam osmometer. The process line of dam osmometer water level is shown in Fig. 1.

The observation results showed that the pressure measuring pipe of the dam body is located in the sand gravel behind the inclined wall of the dam clay, and the water level of the osmometer is relatively stable. Sand gravel is a highly permeable area, and a small amount of water permeated through the inclined wall of clay can quickly dissipate. Compared with the upstream reservoir water level, the osmometer water level has a large drop. From 2017 to 2021, the reservoir water level ranges from 20.31 to 25.67 m, the osmometer water level drops from 7.31 to 9.30 m at high water level, and from 1.78 to 2.80 m at low water level, indicating that the clay inclined wall of the dam body has a good anti-seepage effect.

(2)   Regression analysis of dam osmometer water level

Figure 2 shows the regression fitting process line of osmometer water level, reservoir water level and rainfall at typical measuring points (CG21 and CG51).

Table 1   The characteristic value of the water level of the dam osmometer

Osmometer number	Buried location of osmometer		Osmometer water level (m)			Water head reduction (m)	
	Axis section pile number (m)	Flow direction pile number (m)	Max	Min	VAR	Max	Min
CG21	0 + 060.00	0 + 004.00	19.42	16.87	2.55	8.32	2.80
CG22		0 + 011.00	20.18	16.78	3.40	7.68	2.08
CG23-1		0 + 021.00	19.98	17.20	2.78	7.85	2.27
CG23-2			19.80	17.21	2.59	7.91	2.46
CG31	0 + 120.00	0 + 004.00	19.95	17.30	2.65	7.80	2.37
CG32		0 + 011.00	20.39	17.79	2.60	7.31	1.78
CG33-1		0 + 021.00	20.25	17.33	2.92	7.86	2.08
CG33-2			19.78	17.30	2.48	7.94	2.47
CG41	0 + 180.00	0 + 004.00	20.22	17.14	3.08	7.96	1.95
CG42		0 + 011.00	19.97	17.36	2.61	7.75	1.96
CG43-1		0 + 021.00	19.95	17.38	2.57	7.72	1.95
CG43-2			19.98	17.89	2.09	7.34	2.03
CG51	0 + 240.00	0 + 004.00	19.82	17.57	2.24	7.37	1.82
CG52		0 + 011.00	17.79	15.69	2.10	9.3	3.57

Water head reduction is the drop of osmometer water level relative to upstream reservoir water level

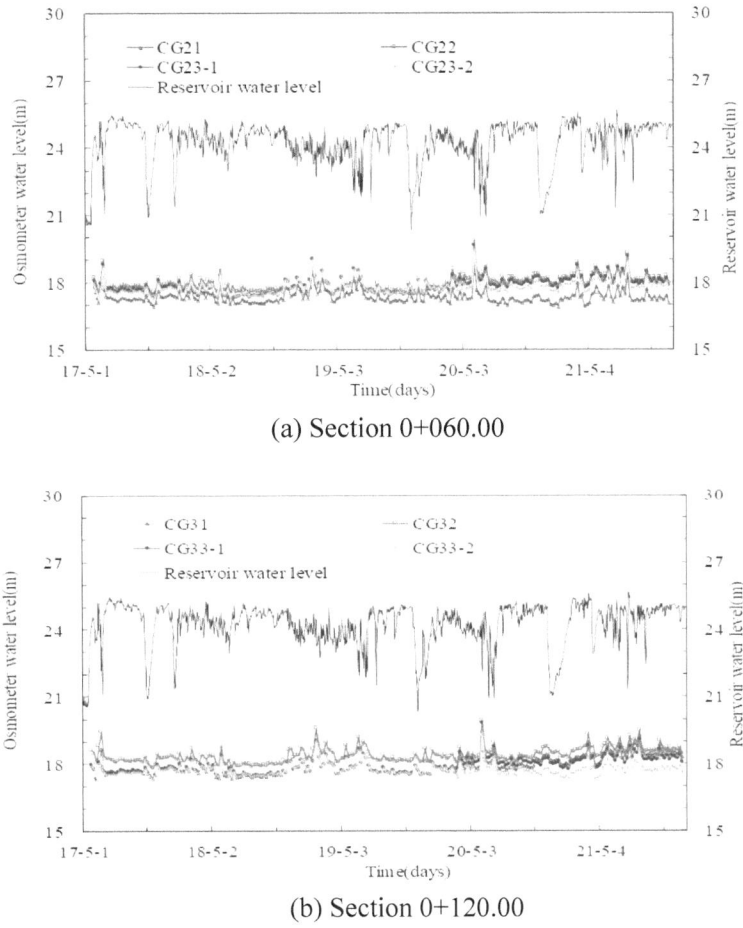

(a) Section 0+060.00

(b) Section 0+120.00

Fig. 1 The osmometer water level process lines

The observation results showed that the high water level of the osmometers behind the inclined wall of the dam is concentrated in the period of heavy rainfall, but the correlation between the osmometer water level and the reservoir water level and rainfall is not obvious [8]. The osmometers are located in the sand gravel and foundation behind the inclined clay wall of the dam body, and the foot of the downstream dam is close to the discharge canal (Puyang River). The frequent opening of the Anhua discharge gate causes the fluctuation of the osmometer water level.

(3) Seepage potential and wetting line of dam body

Potential refers to the percentage of an osmometer head in the total seepage head in the seepage field [9, 10]. Seepage potential could be expressed as Eq. (2):

$$F_i = (h_i - H_2)/(H_1 - H_2) \times 100\% \qquad (2)$$

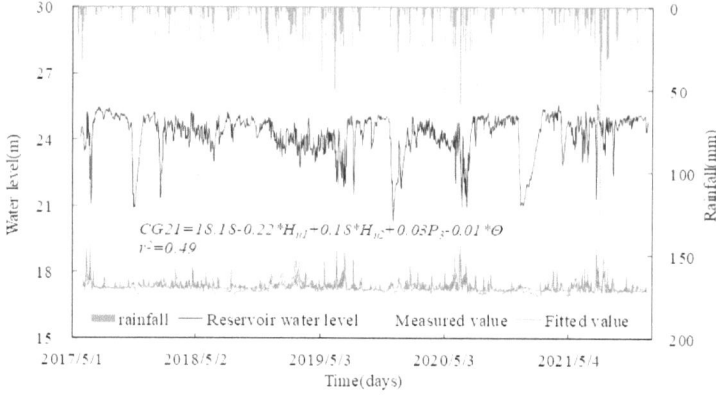

(a) Water level at CG21 of Section 0+060.00

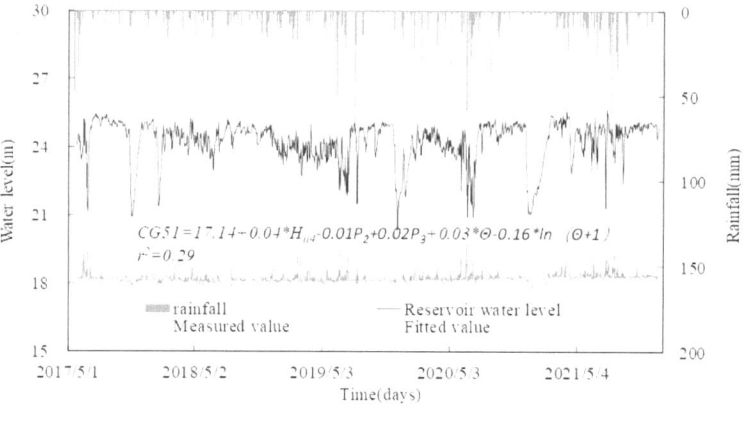

(b) Water level at CG51 of Section 0+240.00

Fig. 2 Regression fitting process line of osmometer water level, reservoir water level and rainfall at typical measuring points

where F_i is the potential of No. i measuring point; h_i is the water level of No. i measuring point; H_1 and H_2 are upper and lower water levels.

Taking the relatively stable high reservoir water level in 2021 as an example [11], the seepage potential characteristic value of the dam is calculated. The characteristic values of dam osmotic potential when the reservoir water level is 25.01 m on November 13, 2021 are shown in Table 2. The wetting line of the dam body at high water level in 2021 is shown in Fig. 3.

Under the high water level, the water level difference between the front and the back of the four cross-section clay inclined walls (dam 0 + 060.00, dam 0 + 120.00, dam 0 + 180.00 and dam 0 + 240.00) was 6.45–7.67 m, and the potential difference between the front and the back of the walls was 74.7–96.1%. It can be seen that

Table 2 The characteristic value of seepage pressure potential of dam body

Osmometer number	Axis section pile number (m)	Osmometer water level (m)	Seepage potential (%)	Water level difference after clay inclined wall (m)	Potential difference after clay inclined wall (%)
CG21	0 + 060.00	17.34	/	7.67	93.1
CG22		18.22	6.9		
CG23-1		18.12	5.6		
CG23-2		17.93	2.9		
CG31	0 + 120.00	18.56	11.6	6.45	88.4
CG32		18.73	13.9		
CG33-1		18.35	8.7		
CG33-2		17.86	2.0		
CG41	0 + 180.00	18.00	3.9	7.01	96.1
CG42		17.98	3.6		
CG43-1		18.00	3.9		
CG43-2		18.13	5.7		
CG51	0 + 240.00	18.36	8.8	6.65	74.7
CG52		/	/		

On November 13, 2021, the reservoir water level is 25.01 m, and the downstream outbound water level is 17.57 m

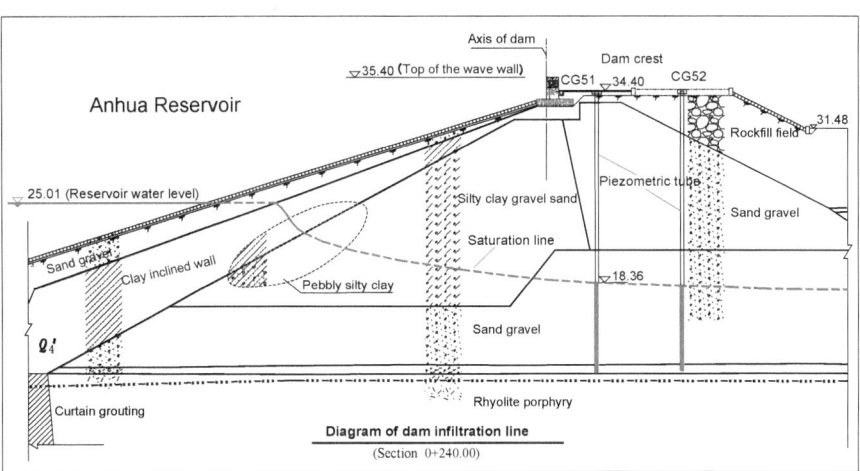

Fig. 3   Diagram of dam infiltration line

the osmometer water level of the clay inclined wall of the four sections is reduced greatly. The infiltration line of the dam body drops rapidly in the clay inclined core wall and silty clay gravel sand of the dam body. The anti-seepage effect of the clay inclined walls is better.

3.3 Seepage Evaluation of the Joint Section of Right Dam Head and Spillway Gate Wing Wall

The connecting section between the right dam head and the spillway wing wall is 23 m long, and 18 concrete columns with a diameter of 1.3 m were used in the previous construction. The thickness of the lap between the concrete columns is very small, resulting in a large permeability drop. In order to solve the above problems, in 2015, a C15W6 cutoff wall was set 1 m after the original concrete casing, with a thickness of 0.6 m and a depth of 0.5 m below the bedrock.

According to the monitoring data analysis results of osmometer in this part (Section 0 + 240.00): Taking the reservoir water level of 25.01 m on November 13, 2021 as an example, the water head of the concrete seepage prevention wall in the connection section between the dam and the spillway wing wall is reduced by 6.65 m, and the potential difference between the front and rear walls is 74.7%. The osmometer water level behind the cutoff wall is reduced greatly, and the seepage prevention effect is better.

3.4 Analysis of Seepage Around Dam on Left Bank

The process line of water level in the pressure gauge pipe around the dam on the left bank is shown in Fig. 4. The process line of water level in the pressure gauge pipe and rainfall is shown in Fig. 5.

At high water level, the water level of pressure measuring pipes (CG11, CG12, CG13-1, CG13-2) was 7.78 ~ 8.67 m higher than the reservoir water level, which was related to rainfall infiltration. At low water level, the water level of the pressure measuring pipe is 4.42 ~ 5.10 m lower than the water level of the reservoir, which is mainly affected by the groundwater in the mountain. In general, there is no obvious correlation between the water level of the pressure gauge pipe around the dam on the left bank and the reservoir water level, which is mainly affected by rainfall infiltration and mountain groundwater, and there is no seepage around the dam.

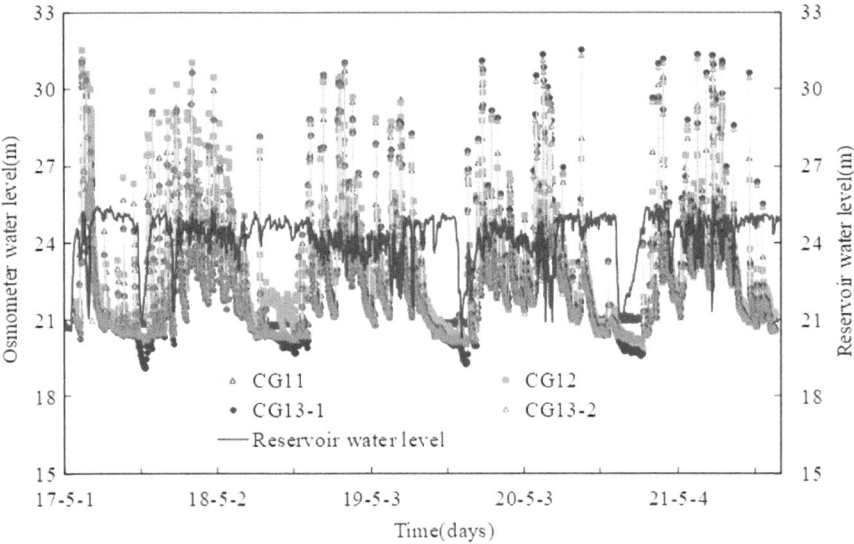

Fig. 4 Water level process line of pressure gauge pipe around dam on left bank

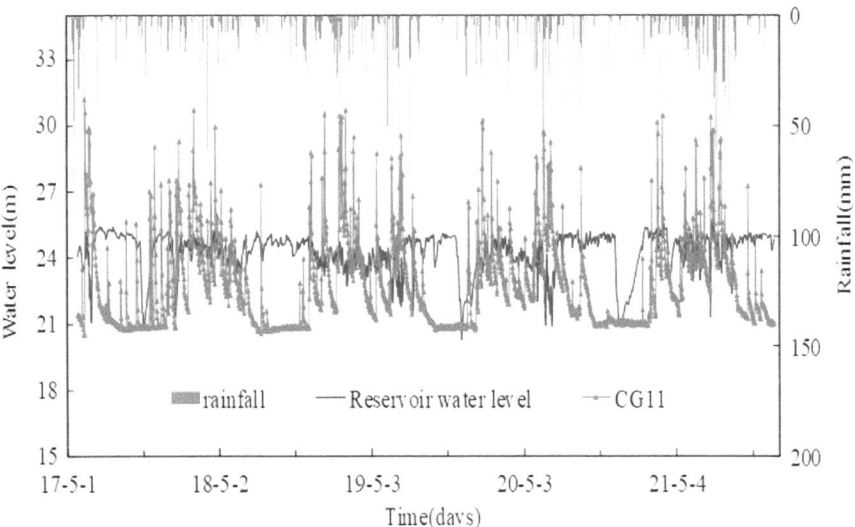

Fig. 5 The rainfall and water level process line of pressure gauge pipe around dam on left bank

4 Conclusions

The analysis of seepage monitoring data of Anhua Reservoir reinforcement project in Zhuji City indicates that the dam is running normally and the anti-seepage measures of reinforcement project are effective.

(1) Under the high water level, the water level difference between the front and the back of the four cross-section clay inclined walls was 6.45–7.67 m, and the potential difference after the clay inclined wall was 74.7–96.1%. It can be seen that the osmometer water level of the clay inclined wall of the four sections is reduced greatly. The infiltration line of the dam body drops rapidly in the clay inclined wall and silty clay gravel sand of the dam body. The anti-seepage effect of the clay inclined walls is better.

(2) According to the monitoring data analysis results, the water head of the concrete seepage prevention wall in the connection section between the dam and the spillway wing wall is reduced greatly, and the potential difference between the front and rear cutoff wall is greatly, and the seepage prevention effect is better.

(3) There is no obvious correlation between the water level of the pressure gauge pipe around the dam on the left bank and the reservoir water level, which is mainly affected by rainfall infiltration and mountain groundwater, and there is no seepage around the dam.

Acknowledgements This work was supported by Science and Technology Plan Project of Department of Water Resources of Zhejiang Province (RB2026 and RB2021) and the "unveiling list" major science and technology project (ZIHE23G001).

References

1. *Technical Specification for Earth-Rockfill Dam Safety Monitoring (SL551-2012)* (China Water & Power Press, Beijing)
2. J. Dunnicliff, G.E. Green, *Geotechnical Instrumentation for Monitoring Field Performance* (America John Wiley & Sons, New Jersey, 1993)
3. S. Ding, Z. Cai, Summary of seepage research for embankment dam. Yangtze River **39**(2), 33–36, 108 (2008)
4. Z. Ji, X. Duan, Inverse analysis on the seepage of rock-fill dam with inclined clay-core wall. Dam Saf. **5**, 53–58 (2011)
5. F. Salmasi, R. Norouzi, J.P. Abraham et al., Effect of inclined clay core on embankment dam seepage and stability through LEM and FEM. Geotech. Geol. Eng. **38**(S2), 1–16 (2020)
6. W. Xu, *Study on Seepage Monitoring Data Analysis of Earth-Rockfill Dam* (Zhejiang University, 2005)
7. X. Xu, X. Xu, C. Zhan et al., Study on the analysis method of embankment dam seepage monitoring data. Zhejiang Hydrotechn. **207**, 21–23 (2016)
8. Z. Wu, *Safety Monitoring Theory & Its Application of Hydraulic Structures* (Higher Education Press, Beijing, 2003)
9. G. Liang, M. Zheng, B. Sun et al., Analysis model and method of seepage observation data for earth rock-fill dams. J. Hydraul. Eng. **180**, 83–87 (2012)

10. C. Mao, X. Duan, *Numerical Computation in Seepage Flow and Programs Application* (Hohai University Press, Nanjing, 1999)
11. L. Lam, D.G. Fredlund, Saturated-unsaturated transient finite element seepage model for geotechnical. Adv. Water Resour. **7**(3), 132–136 (1984)

Application of IPv6 in the Intelligent Sensing Technology of Flood and Waterlogging Prevention

Weichang Chen, Sangang Wei, Bing Zhang, and Binpei Zhang

Abstract Due to the special landform, complex river system and rapid development of urban agglomerations in the Guangdong-Hong Kong-Macao Greater Bay Area, the task of flood and waterlogging prevention in the Greater Bay Area is heavy and its responsibility is important. In order to effectively improve the effectiveness of flood and waterlogging prevention, this paper analyzes current problems facing flood and waterlogging prevention in the Greater Bay Area, and proposes building an intelligent sensing IoT network of Flood and Waterlogging Prevention based on IPv6 and IoT technology, making the flood warning's response time be reduced from 60 ms to less than 30 ms, to provide more efficient, more precise and more secure network support and data protection for flood and waterlogging prevention.

Keywords Guangdong-Hong Kong-Macao Greater Bay Area · Front-end equipment of flood and waterlogging perception · IoT perception · 6LoWPAN gateway · IPv6 single stack network

1 Introduction

The Pearl River Basin, to which the Greater Bay Area belongs, is interlaced and complex, with abundant precipitation, uneven time and space, abundant water in flood season, and flood pooling to the Pearl River Delta region; The Greater Bay Area's long coastline, intricately distributed hills, basins, terraces and plains form a relatively closed topography of "surrounded by mountains on three sides, facing the sea on one side, confluence of three rivers, and having eight estuary". The northwest and east are high on three sides, while the south is low, therefore, mountain torrents can enter the city directly. At the same time, the waterlogging in the city center's low-lying area is difficult to discharge because of the tidal backwater in southern part. As a result, the Greater Bay Area faces the intertwined threat of flooding and

W. Chen (✉) · S. Wei · B. Zhang · B. Zhang
The Pearl River Water Resources Research Institute of the Pearl River Water Resources Commission of the Ministry of Water Resources, Guangzhou 572100, China
e-mail: 835498867@qq.com

W. Wang et al. (eds.), *Hydraulic Structure and Hydrodynamics*, Lecture Notes in Civil Engineering 608, https://doi.org/10.1007/978-981-97-7251-3_10

waterlogging from the upstream Pearl River Basin's flood, heavy rains, storm surges in the South China Sea and so on.

As the urbanization rate of the Greater Bay Area reached 87.7% in 2019, the heat-island effect and rain-island effect became significant, and the short-duration heavy rainfall occurred frequently. Urbanization also changes the underlying surface and the hydrophysical properties of the surface, resulting in significant changes in the pattern of runoff and sink flow, which makes the causes of flood and disaster-causing mechanism more complicated. Then, with the high intensity of urban development, the governance space is severely limited, which increases the difficulty of flood and waterlogging prevention during the rainstorm in the city.

Therefore, under the background of high tension of the urban land, how to greatly improve the urban waterlogging control standards? How to identify the action degree of various disaster-causing factors to support the macro decision-making of flood and waterlogging control during the rainstorm in the city? How to effectively warn and improve the efficiency of public travel in the context of high-density passenger flow? These problems are the three key technological problems that the prevention and control of flood and waterlogging during the rainstorm in the city in the Guangdong-Hong Kong-Macao Greater Bay Area mainly face. To solve these problems, water conservancy predecessors did a series of studies, the governance theory of flood and waterlogging during the rainstorm in the high-density city based on the holistic view of basin system proposed by Wenlong Chen in "Causes and prevention strategies on urban flood disasters in the Guangdong-Hong Kong-Macao Greater Bay Area" reveals the main causes of flood-waterlogging disasters during the rainstorm in the city of the Greater Bay Area, establishes the holistic view of flood-waterlogging disasters during the rainstorm in the city, illuminates the disadvantages of traditional urban flood-waterlogging control model under the background of high-intensity development, put forward a new idea of urban flood control planning and design of "unified goal, unified Rainfall pattern, Multidimensional co-governance, and system that is up to the standard." Reference [1] provides theoretical support for the planning of urban flood and waterlogging control. Wang, Huang et al.'s study on the temporal and spatial evolution of extreme rainfall in Zhusanjiao Area under the background of high urbanization, reveals the mechanism of intensified flooding in the Greater Bay Area under the background of high urbanization [2, 3]. Hu, Song et al. researched and developed a Hydrological and hydrodynamic two-way coupling flood mathematical model under complex urban underlying surface conditions [4]. The Pearl River Hydraulic Research Institute developed a series of miniaturized urban flood monitoring equipment, multi-model forecasting and precise targeted early warning technology, which promoted urban waterlogging control to achieve significant results.

2 Existing Problems of Flood and Waterlogging Prevention in the Greater Bay Area

The current flood and tide disaster prevention work is often plagued by the following problems:

(1) Difficulties in maintenance of storm surge monitoring:

The on-site environment of storm surge monitoring in the Pearl River Estuary is very harsh. Equipment is affected by typhoon rainstorm, marine salt fog corrosion, ship collision and so on, the failure rate is generally high. And due to the limitation of sea conditions and weather, the timeliness of on-site operation and maintenance cannot be guaranteed. Meanwhile, the cost of operation and maintenance is high.

(2) The delay of urban waterlogging monitoring data is large, and the network packet loss is serious:

The process of urban waterlogging often rises and falls sharply, which requires high real-time data. However, in the event of rainstorm, the monitoring equipment often fails to report the waterlogging data in time due to the congestion of communication network.

(3) The equipment status of the water conservancy project site is not visible, and there is no direct communication between the sites:

There are many monitoring elements in water conservancy projects, and the monitoring stations are widely distributed. However, the public communication network coverage in the area where water conservancy facilities such as reservoirs are located is poor. The monitoring stations cannot be directly connected to the network, and point-to-point connections cannot be directly established between the stations, and collaborative linkage cannot be carried out.

3 Application of IPv6 Technology in Flood and Waterlogging Prevention Business

3.1 Application of IPv6 Front-End Technology

At present, most of the IoT sensing front-end devices use IPv4 network communication. Due to the limitation of the number of IPv4 addresses, IPv4 cannot configure a unique IP address for each mobile terminal. The device can only use the local IP address of the operator's mobile network, and cannot access the device remotely and directly. In this situation, it is impossible to actively obtain the current status data of the device, and it is difficult to predict the possible problems of the device and realize the online operation and maintenance of the device, which greatly increases the frequency and difficulty of operation and maintenance. After adopting IPv6, all devices can be seamlessly connected to the IPv6 network, and all nodes have the world's only IP address. Each front-end device can directly and efficiently exchange data, make upper and lower linkage, and improve the multi-factor collaborative linkage and thorough perception of river and reservoir floods.

3.2   Integrated Application of Flood IPv6 Gateway System

With the rapid development of IoT technology and the urgent requirements of intelligent water conservancy for the network transmission rate and transmission volume of water conservancy information, more and more water conservancy monitoring, detection and early warning equipment have the need to access the network. However, most of the current water conservancy terminal information acquisition devices [5] use a serial port as the communication interface, which does not have a network interface and cannot directly connect to the network. And most of the devices that can connect to the networks [6] still only support the IPv4 protocol. In the process of practical application, there are the following problems:

(1) In extreme rainy weather, there are problems such as congestion of special lines leading to the suspension of public communication network services [7], which delays or even loses the information transmission of rainy floods, and the relevant departments cannot obtain the latest data in time to take countermeasures.

(2) Because most of the data transmission uses wired connection, the transmission distance of the system is very limited, and the movement and connection changes between devices will be very difficult [8]. In addition, the transmission rate of serial port connection is low, usually in the range of hundreds to thousands of bits per second, and it is easy to be affected by electromagnetic interference when using cable transmission [9]. These problems have greatly affected the collection of rainfall and flood conditions, and also highlighted that the current system is gradually unable to meet the needs of hydrological development.

In order to solve the above problems, this paper studies the intelligent perception technology of flood and waterlogging prevention. On the premise of maintaining the original equipment, a complete IPv6 gateway system can be introduced, which makes the gateway can realize the application message exchange between the flood sensing equipment with only serial communication and the TCP/IPv6 equipment, and solves the problem of slow, inaccurate and unstable measurement of the original acquisition equipment.

3.3 The IPv6 Gateway System Mainly Has Two Parts, the Front-End Device IoT Router and the Intelligent AI Edge Gateway

(1) Front-end equipment IoT router. Flood sensing IoT devices support IPv6 network protocol in two ways: one is the NB-IoT/4G communication module proposed by Liang, Mou and Guo in "Research on urban underground pipe network monitoring system based on NB-IoT and 4G" [10], which adopts the NB-IoT/4G communication module supporting IPv6 and directly accesses to the Internet; the other is the use of 6LoWPAN protocol (IPv6 over Low-Power Wireless Personal Area Network) proposed by Zheng in the "Design of Industrial Internet of Things Gateway Based on 6LoWPAN" [11], which supports IPv6 and realizes IPv6 single stack operation. Due to the remote location of water conservancy projects such as reservoirs and dams, the cellular network signals in some areas are weak or cannot be fully covered. In view of this situation, the Internet of Things equipment based on the 6LoWPAN protocol can be used. The router is installed in the location where the cellular network signal is better, and the single stack operation of IPv6 network of all monitoring devices is realized by wireless radio frequency ad hoc network router [12]. As shown in Fig. 1, compared with the dual protocol stack, the single stack scheme has less network configuration and maintenance workload, lower operating cost, less risk exposure, lower security risk and higher reliability.

6LoWPAN (IPv6 over Low-Power Wireless Personal Area Network) is a communication protocol that can transmit IPv6 packets over low-power wireless networks (such as IEEE 802.15.4). It aims to provide low-power, low-cost and low-bandwidth solutions for devices in the IoT and other wireless sensor networks [13].

By compressing IPv6 packets to adapt to the small frame size of low-power wireless networks, 6LoWPAN reduces overhead and energy consumption. It also provides address allocation, routing and neighbor discovery mechanisms to achieve seamless communication between devices.

By using 6LoWPAN, devices can connect directly to the Internet and communicate with other IPv6 devices, enabling interoperability and integration with existing IP-based networks. The protocol is popular in various applications such as home automation, smart cities, industrial monitoring, and health care. The core of using 6LoWPAN is to realize the IPv6 Stateless Address Autoconfiguration (SLAAC) of RF child nodes, which is also an important mechanism of IPv6. Compared with the previous DHCP method, its biggest feature is that it can complete the address configuration without DHCP server [14]. Under this mechanism, the node generally first generates the IPv6 link local address through the physical address, and then determines whether the link local address is available through the Duplicate Address Detection (DAD) mechanism. If the local address of the link is available, the autonomous address is configured through the autonomous address prefix carried in the RA message.

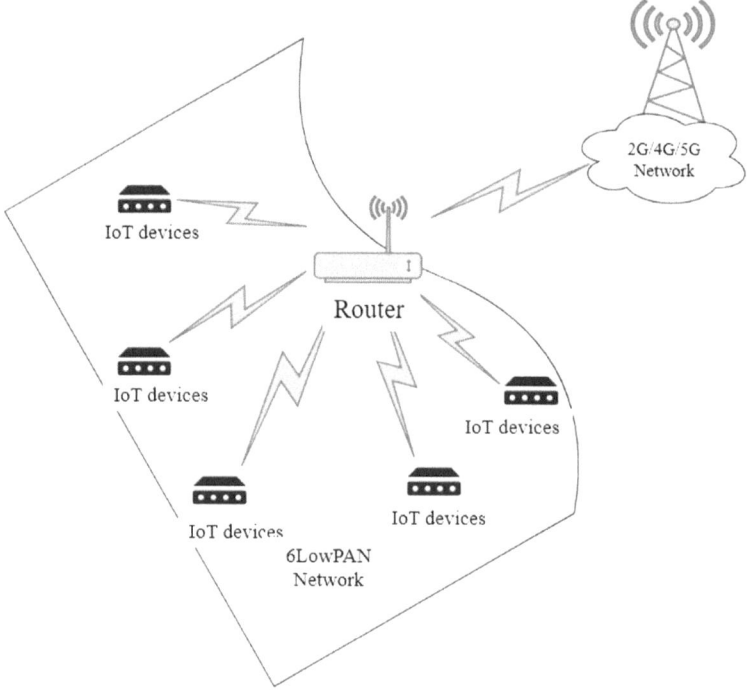

Fig. 1 6LoWPAN network structure diagram

(2) Intelligent AI edge gateway. The intelligent AI edge gateway can connect traditional terminals to the IoT cloud platform and implement edge computing
with cloud services. The emergence of edge computing has gradually sunk the
fields of local intelligent control services, intelligent data acquisition, data analysis, and industrial intelligent manufacturing, and can achieve corresponding
research and development without cloud services [15]. The intelligent AI edge
gateway device applying to flood prevention system uses IPv6 network protocol
to communicate with front-end IoT device, carries out localized intelligent analysis and processing of flood monitoring data in business. Through the built-in
pre-trained AI model, it can carry out image recognition, target detection, face
detection, image segmentation and other applications. And its early warning
information can be released in local through sound, light and other media.

3.4 Application of IPv6 Single Stack End-to-End Network Networking

On the premise of IPv6 protocol, an IPv6 IoT sensing network architecture composed
of front-end sensing equipment, IoT gateway, IoT platform/data receiving platform

and application system platform can be constructed to provide IPv6 network environment for flood sensing data receiving display platform and front-end sensing equipment transmission. As shown in Fig. 2, the IoT sensing network consists of a front-end access area, a business area, an Internet access area, and a terminal area.

(1) Internet access area. The Internet access area is mainly to build a channel for each area to connect with the Internet, and provide boundary access control for each area.
(2) Business area. The business area is mainly to deploy application servers and business application systems, provide data acceptance and storage for front-end sensing devices, and provide flood and waterlogging prevention business applications.
(3) Terminal area. The terminal area is a terminal perception network that integrates IPv6 applications, providing accessing, debugging, testing and researching's capabilities for terminal devices using IPv6 technology.
(4) Front-end access area. The front-end access area is the access Internet area of the front-end sensing device. The IoT sensing device accesses the Internet after adapting the IPv6 protocol, and transmits data back through the operator's IPv6 network, and finally converges to the IoT platform.
(5) Internet users. Internet users can access application system through the support of IPv6 protocol intelligent terminal.

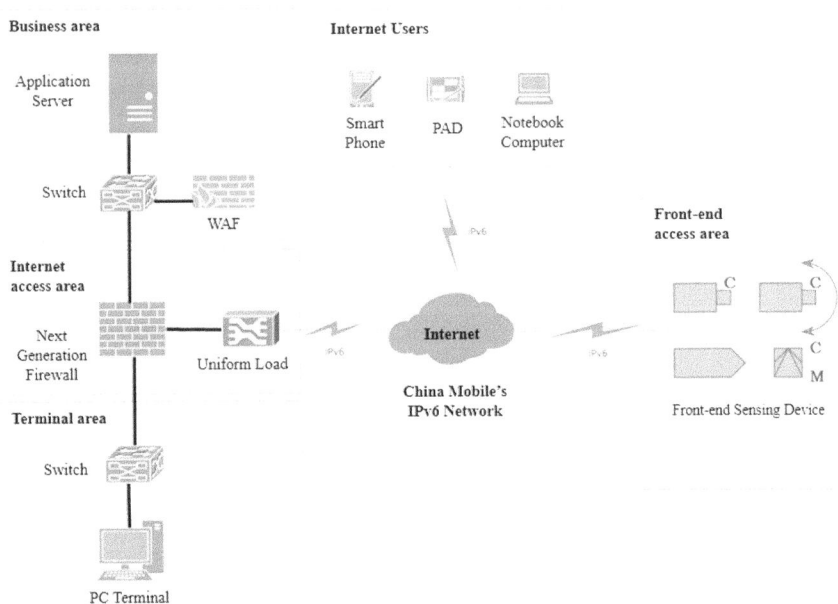

Fig. 2 IoT sensing network architecture diagram

3.5   The Effect of IPv6 Flood Prevention and IoT Perception's Fusion Application

According to the characteristics of flood disasters and the pain points in the defense of the Guangdong-Hong Kong-Macao Greater Bay Area, this project is based on the flood prevention and water project scheduling of the Pearl River Basin, uses the technical advantages of IPv6 single stack transmission, to studied five key technologies, including front-end IoT sensing equipment research and development, 6LoWPAN gateway research and development, network transmission protocol transformation, network architecture construction, and IoT platform development.

(1) The IPv6 single-stack multi-factor telemetry terminal based on 4G network developed supports access to a variety of different types of sensors for estuary storm surge monitoring scenarios. It can realize the functions of acquisition, coding, transmission and storage of sensor data, and directly replace the traditional RTU equipment.

   As shown in Fig. 3, this terminal has been applied in the estuary storm surge observation buoy station in Macao sea area in November 2022. At present, it has received 6480 messages and 810 Mb of storm surge related data. As shown in Fig. 4, the stability of data transmission is effectively improved, and the packet loss rate is increased to one in ten thousand. By monitoring the network, the number of operation and maintenance is effectively reduced.

(2) The IPv6 single stack integrated waterlogging monitoring equipment based on NB network developed, realizes the IPv6 single stack data transmission based on NB network, effectively solves the practical problems such as poor real-time data transmission and network congestion under extreme weather conditions, and provides a solution for the integrated equipment to realize IPv6 upgrade.

Fig. 3 The estuary prototype

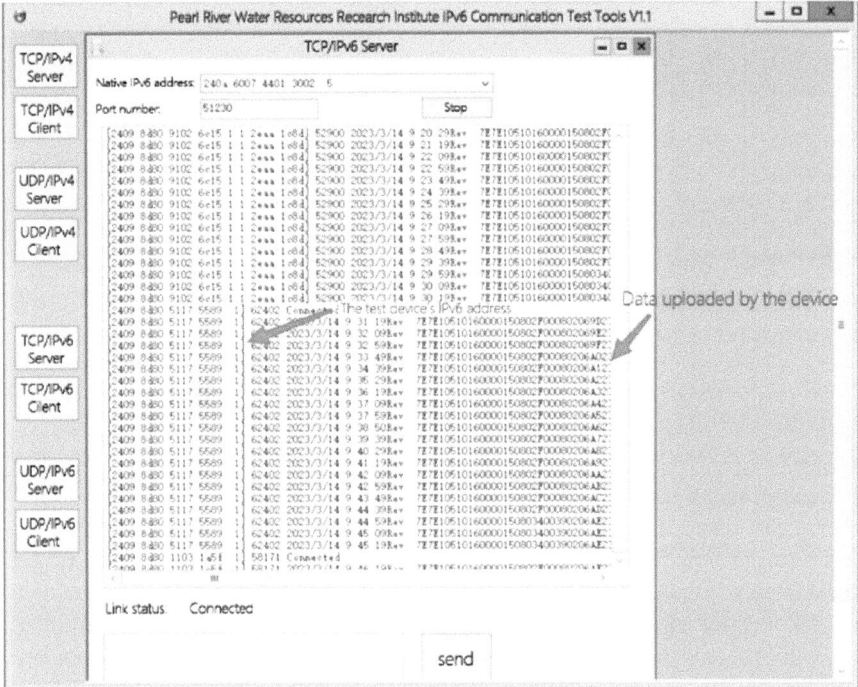

Fig. 4 Server monitoring terminal test observation station spot map

This equipment has be installed in Huachenghui Square, Tianhe District, Guangzhou City in November 2022. It has been running stably for 9 months and has successfully reported water accumulation once. It effectively improves the real-time performance of data transmission, and the average data delay is increased from 60 to 30 ms.

(3) As shown in Fig. 5, the front-end IPv6 Internet of Things router device based on 6LoWPAN can realize the access to LoRA equipment. Its focus is on the scenario where monitoring elements such as reservoirs and dams are widely distributed and local communication networks are used. It can allocate a unique IPv6 address for water level, rainfall, seepage pressure, seepage and other monitoring, so as to realize the interconnection between central station and equipment, equipment and equipment.

This achievement integrates IPv6 smart IoT sensing technology and water conservancy flood tide defense business scenarios, and provides a complete set of IPv6 upgrade solutions for flood disaster IoT sensing equipment, which can effectively solve the innovative integration of smart IoT sensing services in the water conservancy industry and promote the scale deployment and application of IPv6 in the water conservancy industry.

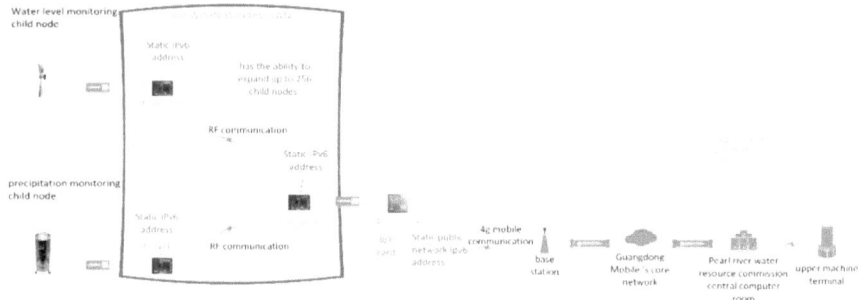

Fig. 5 Front-end sensing device network structure diagram

4 Conclusion

In recent years, the next generation Internet based on IPv6 has accelerated its development and changed the overall pattern of the Internet irreversibly. With the continuous development of China's basic communication facilities construction, IPv6 is the inevitable trend of Internet upgrading and evolution, the important direction of network technology innovation, and the basic support of network power construction. Carrying out IPv6 pilot in the water conservancy industry and promoting IPv6 scale deployment and application are important support for implementing the intelligent water conservancy construction of the Ministry of Water Resources, and also the core task of learning, implementing and implementing the top-level design and planning of water conservancy network information development.

Flood disaster has always been a prominent problem for the safe development of cities. As one of the important strategic areas with the highest degree of openness and the strongest economic vitality in China, the Guangdong-Hong Kong-Macao Greater Bay Area, is located in the Pearl River Delta region, and its social and economic stability and sustainable development need higher requirements for flood control security. Estuary storm surge monitoring, high-density urban waterlogging monitoring and water conservancy project monitoring are the key elements to ensure flood and waterlogging prevention in Guangdong-Hong Kong-Macao Greater Bay Area. In real-time monitoring, it is the most important to ensure the real-time and stable transmission of all kinds of monitoring information. Applying IPv6 related technologies, developing IPv6-based water conservancy collection and communication equipment, and upgrading and expanding communication protocols, can significantly improve the integrity, timeliness, reliability and security of monitoring data transmission. Strengthening the information support capability of flood and drought disaster prevention in the Pearl River Basin is a key measure to adapt to the current situation of frequent flood and waterlogging disasters in the new era and complete the flood and waterlogging disaster prevention tasks with high quality and quantity.

The application of IPv6 in flood prevention can effectively promote the innovation and integration of the sensing business of the water conservancy intelligent IoT, and provide a deeper application integration idea and demonstration of IPv6 in the mobile IoT terminal after the integration application of the information transmission backbone network of the water conservancy industry, so as to promote the scale deployment and application of IPv6 in the water conservancy industry.

References

1. W. Chen, Y. He, Causes and prevention strategies on urban flood disasters in the Guangdong-Hong Kong-Macao Greater Bay Area. China Flood Drought Manage. **31**(03), 14–19 (2021)
2. X. Wang, P. Cong, C. Liu et al., Analysis of vegetation change and its stress in the Pearl River Basin from 2004 to 2013. Acta Ecol. Sin. **37**(19), 6494–6503 (2017)
3. G. Huang, Y. Chen, The variation characteristics of urban flood risk in the Zhusanjiao Area under the background of high urbanization. China Flood Drought Manage. **31**(11), 7–13 (2021)
4. L. Song, X. Hu, L. Tang, Study on flood routing mathematical model for digital watershed, in *Chinese Hydraulic Engineering Society. 2013 CHES Annual Conference Academic Essays—Flood Control and Drought Relief* (2013), p. 6
5. X. Dai, *Research and Design of the Embedded IPv6 Gateway Based on LwIP* (China University of Mining and Technology, 2016)
6. J. Shang, Analysis of IPv6 message format. Wireless Internet Technol. (06), 9+45 (2013)
7. H. Ayman, Implementation and evaluation of IPv6 with compression and fragmentation for throughput improvement of internet of things networks over IEEE 802.15.4. Wireless Pers. Commun. **130**(2) (2023)
8. L. Srikanth, M. Senthilkumar, Investigation of binding update schemes in next generation internet protocol mobility. Int. J. Adv. Intell. Paradigms **24**(1–2) (2023)
9. H. Tian, X. Zhao, K. Zhang et al., Design of data transmission system in coal mine underground based on 6LoWPAN. J. Mine Autom. **45**(08), 6–12 (2019)
10. Y. Liang, Y. Mou, W. Guo, Research on urban underground pipe network monitoring system based on NB-IoT and 4G, in *2022 China Water Conservancy Academic Conference Essay* (Chinese Hydraulic Engineering Society)
11. H. Zheng, *Design of Industrial Internet of Things Gateway Based on 6LoWPAN* (Hainan University, 2022)
12. J. Sun, C. Wang, D. Tang et al., Research and design of 6LoWPAN network multicast communication scheme. Comput. Eng. Sci. (2021)
13. Q. Yang, *Research on Optimization Technology of 6LoWPAN Network* (Harbin Institute of Technology, 2017)
14. F. Yang, Implementation of IPv6 services in GPRS and WCDMA mobile networks. Commun. Today **10**, 42–43 (2004)
15. W. Shen, *Research and Design of an Adaptive Data Acquisition IoT Gateway Based on Edge Computing* (Hunan University of Technology, 2020)

Intelligent Prediction on Cement Take of Dam Foundation Grouting Based on GOA-ELM Model

Yushan Zhu, Zhu Yang, and Jian Huang

Abstract Accurate and reasonable cement take prediction is of great significance for effective control of dam foundation grouting quality and cost. This article combined the previous research results and engineering practice to explore the different influencing factors of cement take, and conducted parameter correlation analysis to determine the input parameters for prediction. Then, an intelligent prediction model for cement take based on improved extreme learning machine (ELM) is proposed, which uses the grasshopper optimization algorithm (GOA) to optimize its input weights w and hidden layer thresholds b. Finally, taking sets of cement take data from a real dam foundation project as an example to verify the performance of the proposed prediction model, the results illustrates that the proposed model has good prediction accuracy and can assist grouting engineers in adjusting grouting construction design and controlling grouting quality, which has a wide application prospect.

Keywords Dam foundation grouting · Cement take prediction · Extreme learning machine · Correlation analysis

1 Introduction

Dam foundation grouting construction is crucial to ensure the stable operation of hydraulic buildings [1]. The cement take can fully characterize the groutability of the rock mass and the cement grout filling situation, which is a key parameter in grouting construction control [2]. Besides, during the construction of the dam, a significant portion of the budget was spent on the cement grouting process [3], which requires determining the amount of cement injection to prepare the corresponding budget work in advance. Therefore, it is of great significance for the quality and cost control of grouting construction to develop the prediction model of cement take and realize the reliable prediction of cement take before or during grouting construction [4].

Y. Zhu (✉) · Z. Yang · J. Huang
PowerChina Northwest Engineering Corporation Limited, Xi'an 710065, Shaanxi, China
e-mail: zhuyushan33@126.com

© The Author(s) 2025
W. Wang et al. (eds.), *Hydraulic Structure and Hydrodynamics*, Lecture Notes in Civil Engineering 608, https://doi.org/10.1007/978-981-97-7251-3_11

121

In recent years, cement take prediction has always been a research focus of grouting scholars. With the development of computer technology, machine learning algorithms are widely applied in research on grouting prediction. Yang [5] introduced the BP neural network model into grouting engineering, established the relationship between grouting physical quantities, and conducted predictive research on grouting volume. Zhang et al. [6] and Xiao et al. [7] constructed an artificial neural network to estimate the grout take under given conditions by analyzing hydrogeological conditions and grouting construction records. Fan et al. [8] established a cement take prediction model using support vector machine based on identification analysis of the main influencing factors. Öge [9] selected drilling record data and grouting data as model input parameters, and used an adaptive fuzzy neural inference system to predict the grout take of mine grouting engineering. Yan et al. [10] developed a genetic neural networks model to realize accurate cement injection prediction. The above studies have provided positive guidance for cement take prediction and have also proven that the use of intelligent prediction models in grouting engineering is an inevitable trend. Besides, due to the concealment of grouting construction, the factors affecting the cement take are complex and numerous, mainly including geological condition, grouting materials, grouting construction methods and so on, Table 1 shows the selection of influencing factors in the studies of grout take prediction. The selection of influencing factors directly affects the accuracy of cement take prediction, which is also an issue that needs attention.

In summary, there is a highly nonlinear relationship between the cement take and its influencing factors, and the cement take prediction is a complex problem. However, the prediction models used in previous studies are still lacking in nonlinear modeling, and the prediction accuracy needs to be improved. Therefore, it is necessary to adopt machine learning methods with better performance to achieve more scientific and accurate prediction of cement take. The extreme learning machine (ELM) [11] is a popular algorithm that has been widely applied in various fields to construct nonlinear relationships between influencing factors and outputs. At the same time, using optimization algorithms to improve ELM to enhance prediction accuracy is also the mainstream application at present. Thus, in this study, according to the grouting engineering situation, the influencing factors of cement take are selected considering whether the parameters could be obtained and quantified, and then the appropriate

Table 1 Selection of influencing factors in grout take prediction studies

Studies	Influencing factors
Sohrabi-Bidar et al. [3]	Q-value, Lugeon value, secondary permeability index (SPI) value, joint apertures
Fan et al. [8]	Rock compressive strength (RCS), joint spacing (JS), acoustic velocity (AV), rock quality designation (RQD), Lugeon value, transmissivity
Yan et al. [10]	Permeability, development of weak interlayers, RQD, degree of fragmentation, grouting pressure, depth of grout section
Feng et al. [12]	Permeability, rock permeability coefficient, resistivity, P-wave velocity

input parameters for prediction are determined by parameter correlation analysis. And a GOA-ELM model with good nonlinear processing ability is proposed to predict the cement take of grouting engineering. The purpose is to provide necessary decision support for grouting engineers to judge the cement take of each grouting borehole.

2 The Prediction Model Based on GOA-ELM

The ELM is a single hidden layer feedforward neural network (SLFNs) that has the advantages of fast training speed, easy implementation, and better generalization performance compared to other machine learning methods. In this study, ELM is adopted to predict cement take, for data set (x_i, y_i), $i = 1, 2, …, N$, where prediction input vector $x_i = [x_{i1}, x_{i2}, …, x_{in}]^T \in \mathbf{R}^n$, and prediction target vector $y_i = [y_{i1}, y_{i2}, …, y_{in}]^T \in \mathbf{R}^m$, the outputs of ELM with L hidden neurons and activation function $g(x)$ are expressed as:

$$\sum_{j=1}^{L} \beta_j g\left(w_j \cdot x_i + b_j\right) = o_i \quad i = 1, 2, …, N \tag{1}$$

where o_i is the ith output of ELM relative to the input x_i, β_j represents the output weight which connect the output neuron and the jth hidden neuron, $w_j = [w_{j1}, w_{j2}, …, w_{jn}]^T$ represents the input weight which connect the jth hidden neuron and the input neuron, b_j represents the jth hidden neuron threshold.

ELM has been widely used in the field of prediction due to its excellent performance, however, its regression accuracy is affected by the parameters w and b, in order to further improve the prediction accuracy of ELM, it is necessary to optimize its parameters.

The grasshopper optimization algorithm (GOA) [13] is a swarm intelligent optimization algorithm proposed by Mirjalili in 2017, which simulates the behaviour of grasshopper swarms. This algorithm has few parameters and easy to implement, it also has strong global and local search ability, and has been successfully applied to various optimization problems. The mathematical expressions of GOA are defined as following:

$$X_i = S_i + G_i + A_i \tag{2}$$

where X_i is ith grasshopper's position, S_i, G_i and A_i are define the social interaction factor, ith grasshopper's gravity force and the wind advection, respectively, and they are calculated as Eqs. (3)–(5):

$$S_i = \sum_{j=1 \neq i}^{N} s\left(d_{ij}\right)\widehat{d_{ij}} \tag{3}$$

$$G_i = -g\widehat{e_g} \tag{4}$$

$$A_i = u\widehat{e_w} \tag{5}$$

where N is the number of grasshoppers, $d_{ij} = |x_j - x_i|$ defines the distance between the ith grasshopper and the jth grasshopper, $s(r) = fe^{\frac{-r}{l}} - e^{-r}$ is the strength function of social forces, $\widehat{d_{ij}} = \frac{x_j - x_i}{d_{ij}}$ defines a unit vector from the ith to the jth grasshopper; g is the gravitational constant and $\widehat{e_g}$ is the unit vector of gravitation; u is the drift constant and $\widehat{e_w}$ is the unit vector of the wind direction.

Substituting Eqs. (3)–(5) into Eq. (2) and modify its global convergence performance to solve optimization problems. The modified equation is shown as:

$$X_i^d = c\left(\sum_{j=1\neq i}^{N} c\frac{ub_d - lb_d}{2}s\left(\left|x_j^d - x_i^d\right|\right)\frac{x_j - x_i}{d_{ij}}\right) + \widehat{T_d} \tag{6}$$

where X_i^d is the component of ith grasshopper in the dth dimension, ub_d and lb_d are the upper bound and lower bound in the dth dimension, $\widehat{T_d}$ is the target location of the grasshopper in the dth dimension so far, c is the convergence factor.

Therefore, a GOA-ELM prediction model for cement take is established in this study, in which the mean squared error (MSE) is taken as the fitness function and the GOA is adopted to optimize the parameters w and b of ELM. The process of GOA-ELM prediction model is shown in Fig. 1.

3   Case Study and Analysis

Taking the dam foundation grouting project of a hydropower station in southwest China as an example, the proposed intelligent prediction model is applied to predict the cement take of a certain grouting unit, and the predictive performance and engineering applicability of the model is analysis and discussed.

3.1   Data Collection

339 grouting sections of 27 grouting boreholes in the grouting unit were selected as the research objects for relevant data collection. According to the actual situation of the project, the data of cement take influencing factors that can be collected at present consist of the following three categories: First, the parameters that characterize the geological conditions, including the rock quality designation (RQD), fracture density (F_ρ), fracture zone range (F_r) which obtained according to the drilling core and

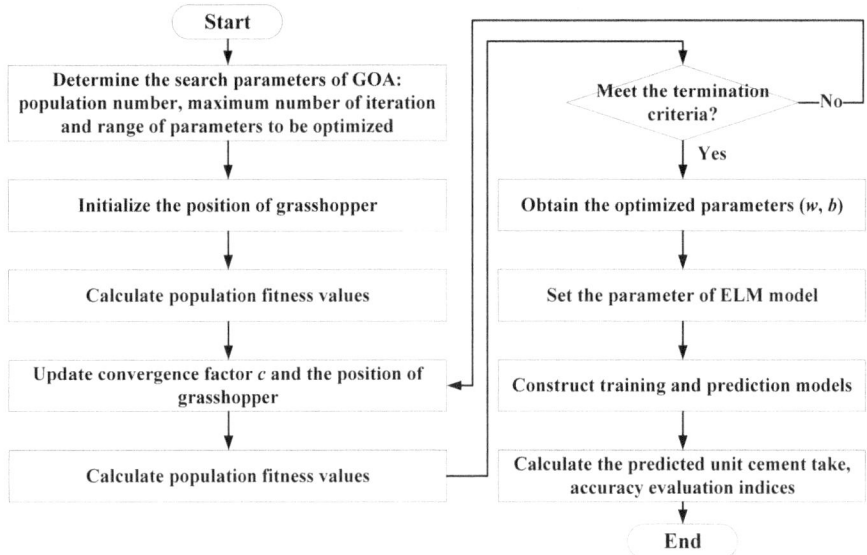

Fig. 1 Flow chart of the proposed GOA-ELM prediction model

its description, as well as the permeability (LU) before grouting obtained through the water pressure test (WPT); The second is the parameters to characterize the grout material, including the initial water-cement ratio (WC_s) and the final water-cement ratio (WC_e); The third is the parameters to characterize grouting construction technology, including the initial grout flow rate (F) and grouting pressure (P) which obtained from the grouting real time monitoring system, as well as the length of borehole section (H), the diameter of grouting borehole (d).

However, due to the fact that drilling core are usually only carried out in specific inspection boreholes, it is not possible to obtain the drilling core results of all grouting boreholes, and the results are always presented in the form of pictures and text descriptions, which leads the data of parameter RQD, F_ρ and F_r are incomplete and cannot be scientifically quantified, and therefore they are not suitable as input parameters for predicting the cement take to avoid adversely affect in the prediction results. These three parameters mainly demonstrate the geological conditions in the grouting area, fortunately, the parameter LU, as the result of the water pressure test on each borehole, is a visual representation of the connectivity of fractures and can fully reflect the geological conditions to a certain extent. Therefore, in order to ensure the integrity of the data and the accuracy of the prediction results, this study selects parameter LU, WC_s, WC_e, H, d, F and P as the influencing factors for cement take (CT) prediction.

3.2 Cement Take Prediction Based on GOA-ELM Model

Parameter Correlation Analysis

According to the data collection results, the unit cement is influenced by many factors, and it is necessary to analyse the correlation between each influencing factors and the unit cement to determine the appropriate input parameters of the prediction model, reduce data redundancy and adverse effects. The Pearson correlation analysis results are shown in Fig. 2.

In the constructed correlation coefficient matrix diagram, the number in the lower left represents the Pearson correlation coefficient between parameters; The upper right corner is a visual display of parameter correlation, the larger the radius of the circle and the darker the color, indicating a higher correlation; the marked * in the circle indicates significant correlation ($p \leq 0.05$).

As can be seen from Fig. 2, the correlation coefficient between LU and CT is 0.81, showing a very significant positive correlation; the correlation coefficient between F and CT is 0.69, showing a significant positive correlation; the correlation coefficient between WC_e and CT is -0.60, between H and CT is -0.47, showing a significant negative correlation; the correlation coefficient between d and CT is -0.18, which has a certain negative correlation; while the Pearson correlation coefficient between CT with WC_s and P is small, indicating the correlation relationship is poor.

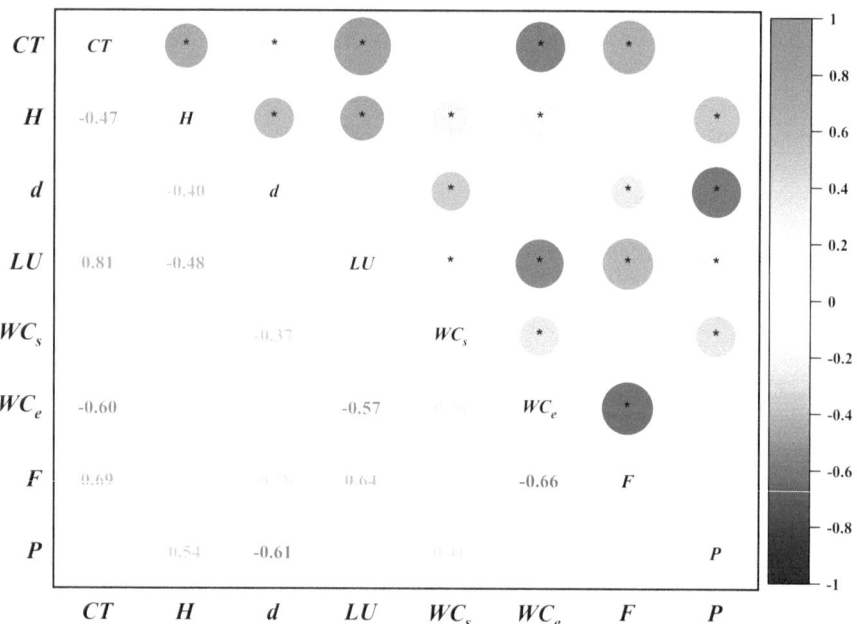

Fig. 2 Correlation coefficient matrix diagram

Based on the correlation analysis results, the length of borehole section (H), the diameter of borehole (d), the permeability (LU), the final water-cement ratio (WC_e) and the initial grout flow rate (F) were selected as the influencing parameters to input into the GOA-ELM based cement take prediction model.

Prediction of Cement Take and Results Analysis

Firstly, the population number of GOA is set to 50, the maximum number of iterations is set to 100, and the number of ELM hidden layer is 20. Then, 339 sets of data are randomly divided into training sets (289 sets of data) and test sets (50 sets of data). Partial sample datasets are shown in Table 2.

After data normalization, GOA is used to optimize the parameters of ELM model. Finally, based on the trained GOA-ELM model, the cement take is predicted and the results as shown in Fig. 3.

From Fig. 3, it can be seen that the predicted cement take value has good consistency with the actual data, and the fitting degree is high ($R^2 = 0.9702$). For several samples with relatively large errors, the possible reasons are systematic errors within the prediction model, and the stratigraphic discontinuity and uncertainty in the grouting area. But overall, the absolute prediction error is small: the maximum error is 33.30 kg/m, the minimum error is 0.18 kg/m, and the average absolute error is 9.94 kg/m, the prediction error is within the acceptable range. This indicates that the selected input parameters are appropriate, the proposed intelligent prediction model has good prediction performance, plays a good role in independent learning and prediction, which can meet the practical needs and can provide technical guidance for grouting engineers.

Table 2 Sample datasets for cement take prediction

Number	Input					Output
	H (m)	d (mm)	LU (Lu)	WC_e (–)[a]	F (L/min)	CT (kg/m)
1	2	110	73.33	1:1	16.8	219.35
2	3	75	36.62	1:1	28	20.60
3	5	56	19.68	0.8:1	63	181.38
4	5	56	42.58	1.5:1	60.8	132.70
5	5	56	21.66	0.8:1	57.8	295.58
6	4	56	2.45	3:1	20.7	15.35
7	5	75	20.09	1.5:1	52.9	165.14
8	3	75	5.71	3:1	14.9	15.833
9	5	56	46.2	0.6:1	74.3	311.34
10	5	75	13.15	0.8:1	47.7	141.44
…	…	…	…	…	…	…
339	5	56	8.84	3:1	30.1	46.12

[a] Denotes dimensionless

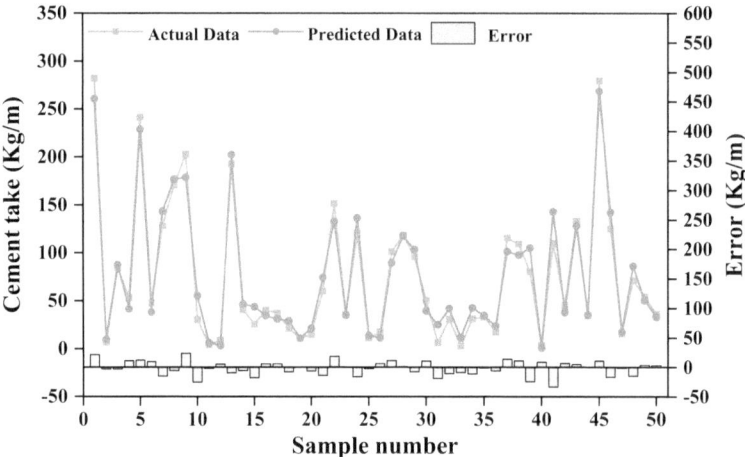

Fig. 3 Comparison of predicted and actual cement take value

Comparison of Different Prediction Models

The proposed GOA-ELM intelligent prediction model is compared with ELM model and BP neural network (BPNN) model. The predicted comparison results are shown in Fig. 4. From Fig. 4, it is obviously that all three models can obtain prediction results that fit the actual data well, and the proposed GOA-ELM model has the best fitting result; In addition, the samples with the highest prediction error among the three models are not quite the same, which is due to the differences in model structure and training learning methods among different models, while the proposed model in this study has advantages over other models in this regard.

Furthermore, taking root mean square error (RMSE), mean absolute percentage error (MAPE) and mean absolute error (MAE) as error evaluation indexes, and the prediction error results of the three methods are shown in Table 3. As can be seen from Table 3, the three models all have good prediction effect on the test data, in which the proposed model has the smallest RMSE, MAPE and MAE, indicating the superiority of the proposed model in prediction accuracy.

Practical Application of the Proposed Prediction Model

The proposed prediction model not only has the advantages of predictive performance mentioned above, but can also be practically applied in grouting construction control. By using the intelligent prediction model of cement take, the consumption of cement grouting is analysed, and the grouting volume is evaluated and judged. Based on the results of the intelligent prediction analysis of cement take and the control standards, the grouting construction parameters are optimized and adjusted to ensure effective grouting of the dam foundation rock mass, thereby achieving feedback control of the grouting construction process. For example, if the predicted cement take into a certain borehole section is less than the standard requirement, grouting engineers can

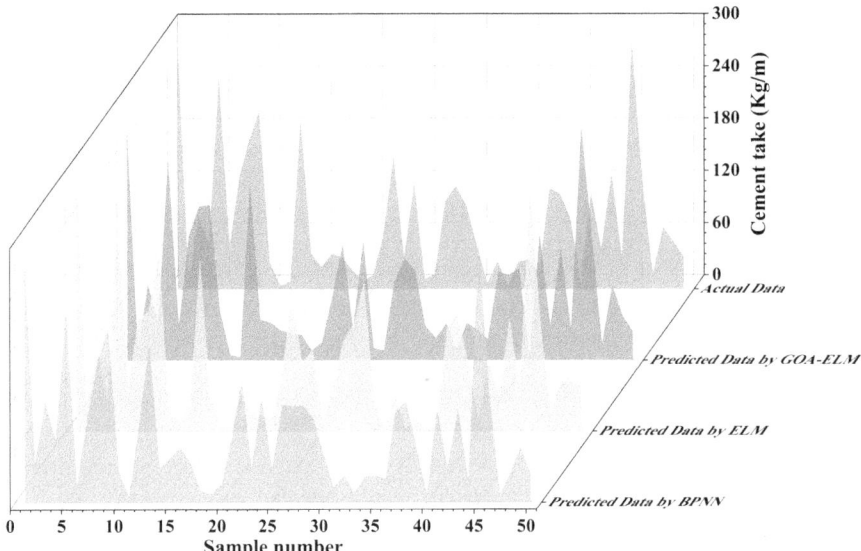

Fig. 4 Comparison of prediction results obtained by three predication models

Table 3 Prediction error analysis results

Prediction model	RMSE (kg/m)	MAPE (%)	MAE (kg/m)
GOA-ELM	12.465	16.93	9.94
ELM	20.230	23.75	13.90
BPNN	25.789	36.09	15.831

increase the grouting pressure to fill the rock mass with more cement grout; If the predicted value is higher than the standard, the grouting pressure can be reduced, or the water cement ratio can be increased to limit grout diffusion and accelerate grout solidification.

4 Conclusion

Accurate prediction of cement take is of great significance to improve grouting design and save grouting cost. This study proposed an intelligent prediction model to forecasting cement take, and carried out the corresponding analysis and discussion, the conclusions are obtained as below:

(1) Considering the nonlinear relationship between cement take and its influencing factors, this study established a GOA-ELM prediction model to solve the cement take prediction problem, which combines the strong non-linear mapping ability

of ELM and the global optimization advantage of GOA. The proposed prediction model has good reliability and high precision.

(2) The cement take is affected by many factors, and the influencing factors should be analysed and selected according to the actual project. It is worth noting that, factors have high correlation with cement take should be selected as input parameters for the prediction model. For some factors that are difficult to obtain or scientifically quantify at present, relevant in-situ test results can be found to supplement and replace, and advanced technologies such as image recognition can be introduced later to solve the above problems.

(3) The proposed prediction model can satisfy the judgment of the cement take in the actual grouting construction. Based on the specific grouting requirements and standards of the project, the grouting engineer can adjust the grouting process based on the predicted value to ensure grouting construction quality and reduce unnecessary material waste, the intelligent prediction model has practical application value.

(4) The cement take prediction problem is complicated and uncertain, so it is still necessary to conduct more in-depth research such as exploring different algorithms or more diverse data sets to achieve accurate control of the whole construction process of grouting engineering.

Acknowledgements The authors are grateful to the editor and the reviewers for useful comments and suggestions that helped to improve the paper. This work was financially supported by the Natural Science Fund of Shaanxi Provincial (2022JC-LHJJ-14) and National Natural Science Foundation of China (52079109).

References

1. D.S. Park, J. Oh, Permeation grouting for remediation of dam cores. Eng. Geol. **233**, 63–75 (2017). https://doi.org/10.1016/j.enggeo.2017.12.011
2. R.M. Taylor, P. Choquet, Automatic monitoring of grouting performance parameters, in *International Conference on Grouting and Deep Mixing* (2012), pp. 1494–1505. https://doi.org/10.1061/9780784412350.0125
3. A. Sohrabi-Bidar, A. Rastegar-Nia, A. Zolfaghari, Estimation of the grout take using empirical relationships (case study: Bakhtiari dam site). Bull. Eng. Geol. Environ. **75**(2), 425–438 (2016). https://doi.org/10.1007/s10064-015-0754-5
4. B.S. Elledge, M. Dubeau, D.M. Heenan, Modeling grout injection volume in fractured rock using borehole imagery, in *Grouting and Deep Mixing* (2012), pp. 1055–1064. https://doi.org/10.1061/9780784412350.0086
5. C. Yang, Estimating cement take and grout efficiency on foundation improvement for Li-Yu-Tan dam. Eng. Geol. **75**(1), 1–14 (2004). https://doi.org/10.1016/j.enggeo.2004.04.005
6. L. Zhang, Q. Li, Y. Song, Neural network-based experimental study on shaft water sealing by grouting, in *IEEE International Geoscience and Remote Sensing Symposium* (2007), pp. 3142–3145. https://doi.org/10.1109/IGARSS.2007.4423511
7. F. Xiao, Z. Zhao, Grouting knowledge discovery based on data mining. Tunn. Undergr. Space. Technol. **95**, 103093 (2019). https://doi.org/10.1016/j.tust.2019.103093

8. G.C. Fan, D.H. Zhong, J.J. Wang, B.Y. Ren, Cement take evaluation and prediction based on empirical relationships and support vector regression, in *International Conference on Energy & Environmental Protection* (2016). https://doi.org/10.2991/iceep-16.2016.120

9. İ.F. Öge, Prediction of cementitious grout take for a mine shaft permeation by adaptive neuro-fuzzy inference system and multiple regression. Eng. Geol. **228**, 238–248 (2017). https://doi.org/10.1016/j.enggeo.2017.08.013

10. F.G. Yan, Z.K. Li, K. Zhong, Prediction method of grouting unit amount of cement injection based on genetic neural networks. Water Resour. Hydropower Eng. **54**(S2), 224–230 (2023) (in Chinese). https://doi.org/10.13928/j.cnki.wrahe.S2.037

11. G.B. Huang, Q.Y. Zhu, C.K. Siew, Extreme learning machine: theory and applications. Neurocomputing **70**(1/3), 489–501 (2006). https://doi.org/10.1016/j.neucom.2005.12.126

12. S.X. Feng, Y.F. Zhao, Y.J. Wang, A comprehensive approach to karst identification and groutability evaluation: a case study of the Dehou reservoir, SW China. Eng. Geol. **269**, 105529 (2020). https://doi.org/10.1016/j.enggeo.2020.105529

13. S. Saremi, S. Mirjalili, A. Lewis, Grasshopper optimisation algorithm: theory and application. Adv. Eng. Softw. **105**, 30–47 (2017). https://doi.org/10.1016/j.advengsoft.2017.01.004

Analysis of the Saltwater Intrusion Prevention Effect of Pump for Ship Locks in the Estuary

He Wang, Chihong Li, Yazhou Hu, Jing Zheng, and Hong Chen

Abstract This article comprehensively considers hydrogeological and meteorological conditions, including topography, tides, runoff, and wind. Integrating factors such as ship lock operation rules and tidal dynamic parameters, a two-dimensional tidal salinity model is established. The study further conducts computational analyses on the desalination effects of various pump operation processes in the newly constructed ship lock on the estuary. The results show that during the ship lock operation, saltwater plumes can enter inland rivers through tidal dynamics and accumulate on one side of the ship lock, leading to upstream propagation. In addition, high saltwater backflow and rapid upstream propagation of saltwater accumulation were observed during high tides. This study also studied the model calculation and analysis of anti-salt pump stations of different scales and operating processes, providing theoretical basis and technical support for engineering design plans.

Keywords Estuary · Ship lock · Saltwater intrusion · Pump · MIKE 21

1 Introduction

Saltwater intrusion in ship lock is an important issue in estuary engineering, which is of great significance for protecting inland water quality and ecological environment [1–3]. The article involves the regulation project, which starts from the river in the west and extends to the south branch of the estuary in the east [4–7]. The project is located in the estuary, which is influenced by tides. Salty water flows upstream along the navigation channel with the rising tide, resulting in saltwater intrusion problems [8–11]. The ship lock and sluice gate designed in this project are located inside the inland water system. During the operation, there is a saltwater-freshwater exchange that causes saltwater intrusion, polluting the freshwater area and seriously affecting industrial and agricultural production and people's lives along the river banks [12, 13].

H. Wang (✉) · C. Li · Y. Hu · J. Zheng · H. Chen
Shanghai Waterway Survey and Design Research Institute Co., Ltd., Shanghai 200120, China
e-mail: wanghesjtu@126.com

W. Wang et al. (eds.), *Hydraulic Structure and Hydrodynamics*, Lecture Notes in Civil Engineering 608, https://doi.org/10.1007/978-981-97-7251-3_12

During the operation of the ship lock, due to the accumulation of mixed brine brought about by ship lock opening and closing and its upstream movement, there is no mature numerical model research method for guiding the design of salt prevention measures for river-sea direct access channel projects. By systematically combining the aforementioned research results with the specific conditions of the regulation project, and fully considering factors such as tidal dynamics, river flow, wind, and ship lock operation rules, a two-dimensional tidal current salinity model is established to systematically study the influence range of saltwater intrusion and upstream movement during ship lock operation. Based on this, suitable pump sizes and operating processes are selected for this region to provide theoretical basis and technical support for the implementation of the project design plan.

2   Tide and Salinity Model

2.1   Numerical Model

In this paper, the two-dimensional model widely used by DHI for the world water environment simulation MIKE 21 FM is selected to establish the two-dimensional tidal current salinity model of the estuary. Equation (1) is the salinity transport equation.

$$\frac{\partial h\bar{s}}{\partial t} + \frac{\partial h\overline{us}}{\partial x} + \frac{\partial h\overline{vs}}{\partial y} = h\left[\frac{\partial}{\partial x}\left(D_h\frac{\partial \bar{s}}{\partial x}\right) + \frac{\partial}{\partial y}\left(D_h\frac{\partial \bar{s}}{\partial y}\right)\right] + hs_sS \tag{1}$$

In where, S is the average salinity in the direction of water depth, S_s is the salinity value of the source, F_s is the horizontal diffusion term, and D_h is the horizontal diffusion coefficient.

2.2   Domain and Meshing

The topographic data in the estuary are based on the measured topography after the construction of the deep-water channel in the estuary, the open sea data is based on the nautical chart data of recent years. The boundary line is comprehensively considered using both Google Maps coastline and the seawall embankment line.

Figure 1 shows the distribution map of the calculation grid of the model, using a triangular grid to divide the model area, about 154,558 grid elements, 79,865 grid nodes, accurately fitting the complex and changeable shoreline and islands, the grid side length is between 500 and 2000 m, and the local encrypted grid in the beach area near the river mouth is about 5 ~ 50 m.

Fig. 1 Computational grid distribution plot of the tidal salinity model

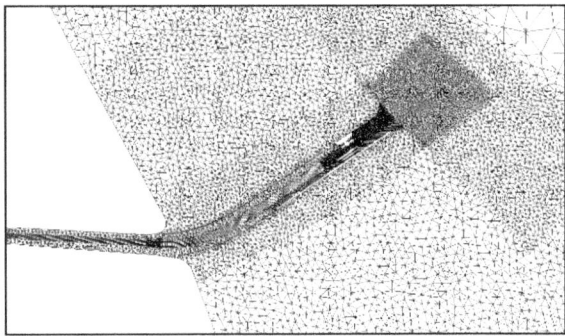

2.3 Domain and Meshing

The runoff of the inland river has little influence on the tide level of the calculated sea area. The tidal level boundaries of the open sea (east, south and north) were generated by a tidal prediction model with a spatial resolution of $1/30°$. The initial conditions of the model are obtained by trial calculations under the given boundary conditions.

The model is given an adaptive calculation time step according to the water depth and flow velocity at different times, the number of CFLs is less than 0.8 to meet the requirements of model stability, and the calculation time step is between 0.01 and 30 s; and the vortex viscosity calculation formula of Smagorinsky formula is $0.28 \text{ m}^2/\text{s}$; the dry and wet dynamic boundary treatment is included; the wind field adopts the National Centers for Environmental Prediction (NCEP), the wind friction empirical coefficient is $0.0015 \sim 0.003$.

The roughness coefficient of the estuary is determined based on the measured data, and gradually changes from the open sea to the estuary, with the roughness of the main trough and the open sea with deep water depth taking $0.011 \sim 0.016$, and the roughness of the shoal land taking $0.020 \sim 0.028$.

3 Model Validation

The saltwater intrusion of the estuary is particularly serious in the winter dry season, and winter is selected as the calibration data of the model. The measured data of two tide level measurement stations and two tidal flow and salinity measurement points (SW1# and SW2#) in the sea area near the estuary were selected to verify the tide level, flow velocity and salinity.

Fig. 2 Tide level calibration
results (two tide stations at
the estuary)

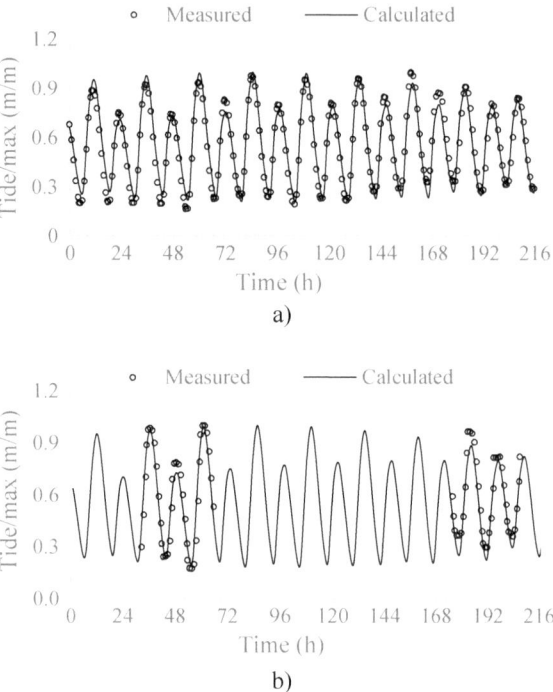

a)

b)

3.1 Tide Current Verification

The verification results of the calculation model on tide level and power flow are
shown in Figs. 2, 3 and 4.

According to the model verification results, the calculated tide level process is
basically consistent with the measured tide level process, and the linear correlation
coefficient between the calculated tide level value and the measured value of each
tide level station is greater than 0.97. The simulated values of the tidal flow process
line during the spring tide and neap tide periods are basically consistent with the
trend and peak value of the measured values. Therefore, the proposed model can
accurately reflect the actual tidal process tide in the estuarine waters, and the tide
level calculation accuracy of the model is high, which can be used for the simulation
of saltwater intrusion prevention measures.

3.2 Salinity Verification

The validation results of the estuary salinity model calculation are shown in Fig. 5.
The salinity verification results show that the salinity simulation results and the

Fig. 3 Tidal flow velocity and flow direction calibration results (SW1#)

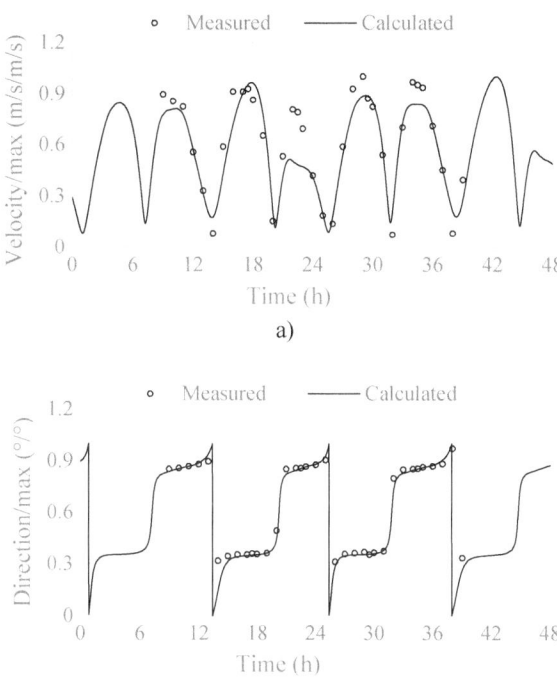

a)

b)

measured data meet the requirements of the model in terms of periodic variation and error, and have extremely high accuracy.

4   Calculation of Salt Discharge Effect of Different Pumping Station Scales

4.1   Pump Operation Process for Saltwater Intrusion Prevention

Under the assumption that the water level in the upper reaches of the river is constant, the navigation of the lock will inevitably bring salt water into the inland waterway due to the tidal movement of the open sea and the inland river water level, which will cause the risk of salinization of the inland water. And the design scheme of operation process of pump for saltwater intrusion prevention is shown in Table 1.

(1) No pump flow

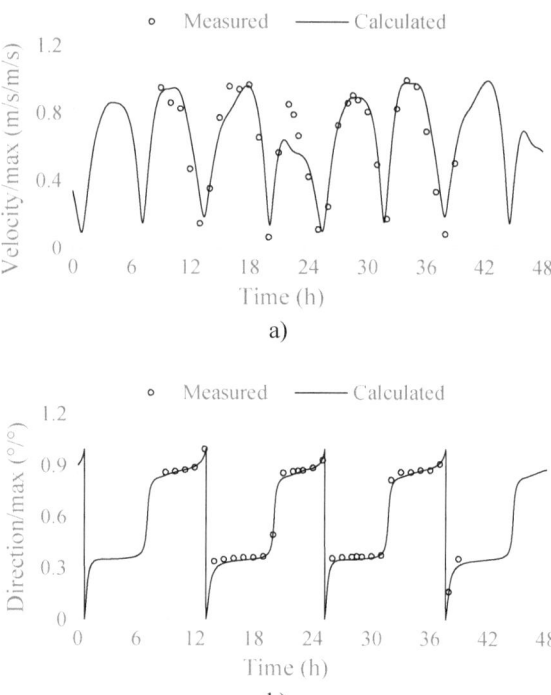

Fig. 4 Tidal flow velocity and flow direction calibration results (SW2#)

As shown in Fig. 6, the area of the 7 days and 15 days envelope area of 0.14 ~ 14 PSU in the freshwater area is 0.581 km^2 and 0.896 km^2, respectively, and the upward distance exceeding the 0.45 PSU threshold is about 2052 m and 2804 m, respectively.

(2) Pump flow: 9 m^3/s

As shown in Fig. 7, the areas of the 7-day and 15-day envelope areas of 0.14 ~ 14 PSU in the flush freshwater area are 0.436 km^2 and 0.737 km^2, respectively, and the upward distance exceeding the 0.45 PSU threshold is 1594 m and 2045 m, respectively.

(3) Pump flow: 18 m^3/s

As shown in Fig. 8, the area of the envelope area of 0.14 ~ 14 PSU in the freshwater area for 7 days and 15 days is 0.354 km^2 and 0.662 km^2, respectively, and the upward distance exceeding the 0.45 PSU threshold is 1198 m and 1865 m, respectively.

4.2 Comparative Analysis of Pump Operation Process

Table 2 shows the area and upstream distance affected by seawater intrusion in the winter high tide prevention programme. According to the calculation results, the

Fig. 5 Salinity calibration results

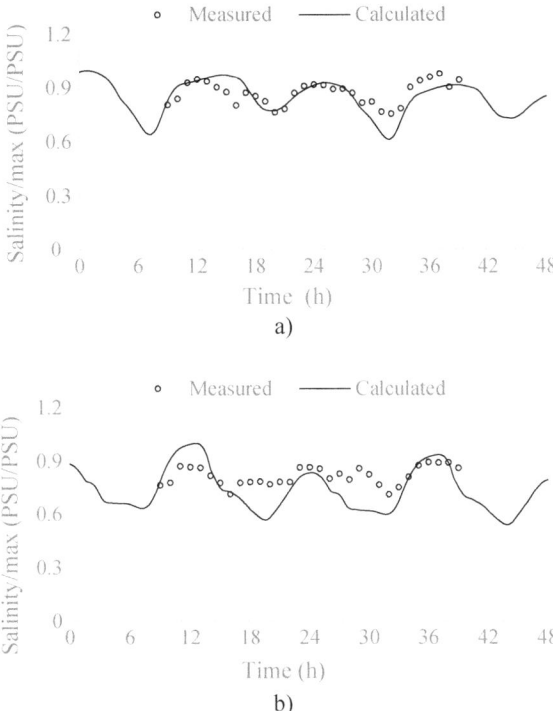

a)

b)

Table 1 Design scheme of operation process of pump for saltwater intrusion prevention

No.	Hydrological conditions	Upper reaches of the river	Operation process	Running time (h/day)
1–0	Winter spring tide	Normal water level Salinity: 0 PSU	No pump flow	0
1–2			Pump flow: 9 m³/s	2
1–3			Pump flow: 18 m³/s	2

most affected area range and the maximum upward distance are basically linearly correlated with the specifications of the pump, and the selection of the pump scale also needs to meet the navigation requirements.

According to the comparative analysis of the salt discharge effect of different specifications of the strong discharge pump, it can be seen that the larger the scale of the pumping station, the effect is about obvious, but the investment and operating expenses of the strong discharge pump should be comprehensively evaluated, and the small size of the strong discharge pump should be selected on the basis of meeting the needs of the operation process. At present, according to the existing comparative analysis results, the following process optimization scheme is given.

The daily operation of the anti-salt pumping station is mainly aimed at the situation that the interval is between 7 days ≤ the interval between the opening and closing of

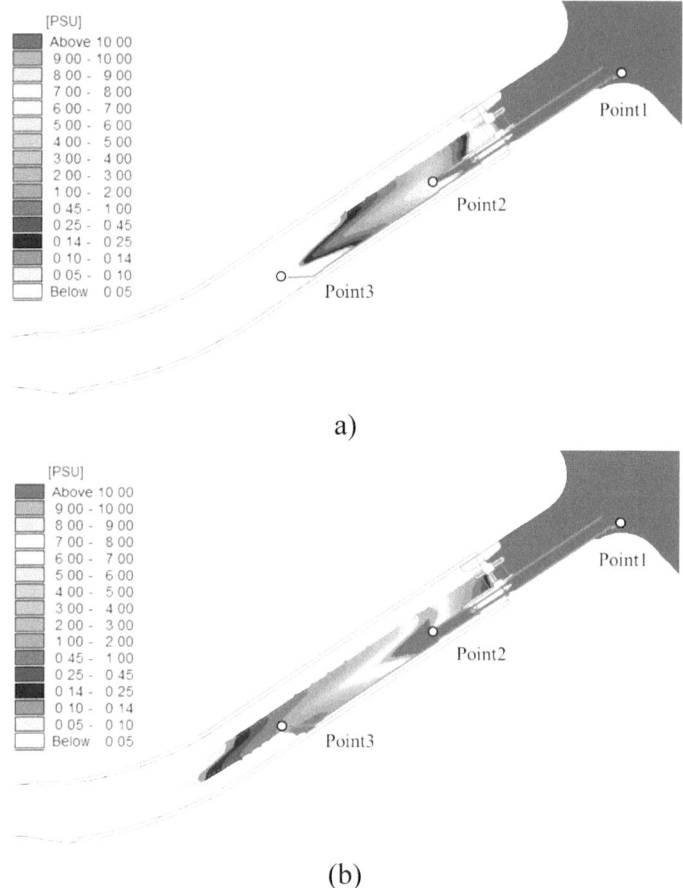

Fig. 6 The 7 days and 15 days maximum envelope ranges of spring tide saltwater intrusion (0 m³/s)

the control gate is ≤ 15 days when there is no ultra-high water level in the river, and the opening and closing rules are: for the pumping station with a scale of 18 m³/s, it is opened for 2 h or 3 h a day, combined with sensitive points 2 and 3 for automatic control or continuous operation until the navigation warning water level is 2.5 m closed, which can effectively deal with the upstream risk of the saline water mass and the scope of invasion.

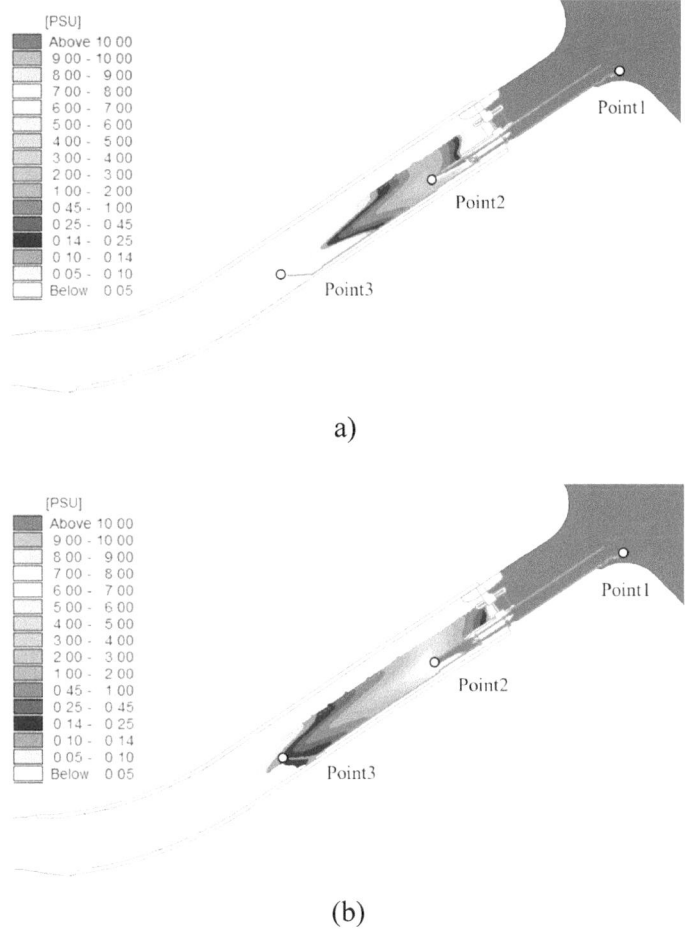

Fig. 7 The 7 days and 15 days maximum envelope ranges of spring tide saltwater intrusion (9 m³/s)

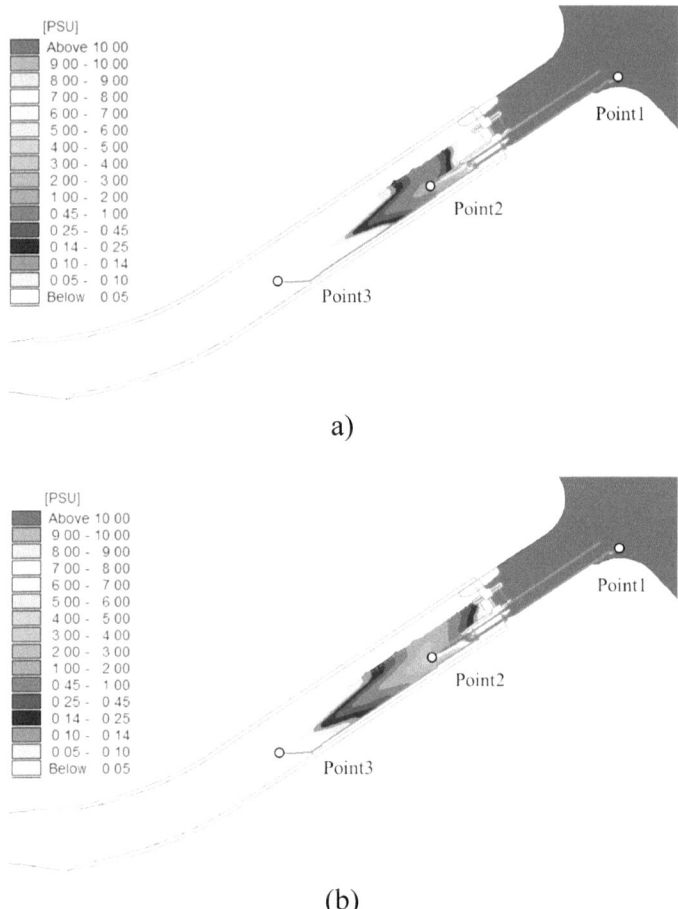

a)

(b)

Fig. 8 The 7 days and 15 days maximum envelope ranges of spring tide saltwater intrusion (18 m³/s)

Table 2 Design scheme of operation process of strong discharge pump

Operation process	7-day flushing freshwater		15-day flushing freshwater		Navigation conditions
	Area (km²)	Distance (m)	Area (km²)	Distance (m)	
No pump flow	0.581	0.896	2052	2804	Yes
Pump flow: 9 m³/s	0.436	0.737	1594	2405	Yes
Pump flow: 18 m³/s	0.354	0.662	1198	1865	Yes

5   Conclusion

Based on the results and discussions presented above, the conclusions are obtained as below:

(1) On the basis of fully considering the tide, runoff, wind and lock operation, a verified and reliable two-dimensional tidal salinity calculation model was established, and the accurate simulation of inland river salt water intrusion during the operation of the lock was realized.
(2) The larger the pumping station, the more significant the salt discharge. The affected area and maximum distance are essentially linearly related to the flow rate of the pump.
(3) The specifications of the strong discharge pump need to be designed in combination with the actual salt discharge process and comprehensive energy consumption, and the different flow specifications and operating hours given in this paper can provide important basic data for the subsequent design of the pump scale.

References

1. L. Li, J. Zhu, H. Wu et al., A numerical study on water diversion ratio of the Changjiang (Yangtze) estuary in dry season. Chin. J. Oceanol. Limnol. **28**(3), 700–712 (2010)
2. W. Gong, J. Shen, The response of salt intrusion to changes in river discharge and tidal mixing during the dry season in the Modaomen Estuary, China. Cont. Shelf Res. **31**(7–8), 769–788 (2011)
3. H.T. Shen, Z.C. Mao, J.R. Zhu, *Salt Water Intrusion in the Yangtze River Estuary* (Ocean Press, Beijing, 2003)
4. X. Xu, G. Xiong, G. Chen et al., Characteristics of coastal aquifer contamination by seawater intrusion and anthropogenic activities in the coastal areas of the Bohai Sea, eastern China. J. Asian Earth Sci. **217**, 104830 (2021)
5. C. Qiu, J. Zhu, Y. Gu, Impact of seasonal tide variation on saltwater intrusion in the Changjiang River estuary. Chin. J. Oceanol. Limnol. **30**(2), 342–351 (2012)
6. T. Cao, D. Han, X. Song, Past, present, and future of global seawater intrusion research: a bibliometric analysis. J. Hydrol. **603**, 126844 (2021)
7. R. Yuan, H. Wu, J. Zhu, The response time of the Changjiang plume to river discharge in summer. J. Mar. Syst. **154**, 82–92 (2016)
8. Z.L. Sun, G.G. Li, D. Xu et al., Predicting the impact of sea level rise on saltwater intrusion in the Qiantang estuary. Zhongguo Huanjing Kexue **37**(10), 3882–3890 (2017)
9. P. Mac Cready, W.R. Geyer, Advance in estuarine physics. Ann. Rev. Mar. Sci. **2**, 35–38 (2010)
10. H. Xiao, W. Huang, E. Johnson et al., Effects of sea level rise on salinity intrusion in St. Marks River estuary, Florida, U.S.A. J. Coast. Res. **68**, 89–96 (2014)
11. P. Xue, C. Chen, P. Ding et al., Saltwater intrusion into the Changjiang River: a model-guided mechanism study. J. Geophys. Res. Oceans **114**(C2) (2009)
12. J. Panthi, S.M. Pradhanang, A. Nolte et al., Saltwater intrusion into coastal aquifers in the contiguous United States—a systematic review of investigation approaches and monitoring networks. Sci. Total Environ. **836**, 155641 (2022)
13. İ. Demirci, N.Y. Gündoğdu, M.E. Candansayar, P. Soupios, A. Vafidis, H. Arslan, Determination and evaluation of saltwater intrusion on Bafra plain: joint interpretation of geophysical, hydrogeological and hydrochemical data. Pure Appl. Geophys. **177**, 5621–5640 (2020)

Study of Hydraulic Soil Stability and Seepage Effects

Research on Groundwater Classification and Grading Method for Underground Space Development

Junhong Zhou and Jiajia Yan

Abstract This research paper focuses on the development of a groundwater classification and grading method specifically tailored for underground space development projects. The study aims to provide a systematic approach to assess the groundwater conditions and associated risks. Based on the principle of multi-objective optimization, groundwater classification and grading method is proposed in this paper. According to the types of confined water in different foundation pits, this method accurately evaluates the impact of the changes in the confined water on the surrounding environment, and comprehensively analyzes other limiting factors such as project investment, project safety, surrounding buildings, and ecological environment, and then controls the confined water.

Keywords Groundwater · Underground space development · Classification · Grading

1 Introduction

In recent years, with the rapid development of China's social economy, the contradiction between the rapidly growing urban population and the limited urban soil resource space has become increasingly prominent. Problems such as urban traffic congestion, air pollution, ground roads occupying limited urban ground space, and urban environmental deterioration have become a serious problem. Sustainable urban development poses serious challenges. The development and utilization of urban underground space has become one of the main methods for the world's major cities to solve the above urban development problems. Urban rail transit is an important means to solve problems such as traffic congestion and land resource shortage in

J. Zhou
Ningbo Rail Transit Group Corporation Limited, Ningbo, China

J. Yan (✉)
Powerchina Huadong Engineering Corporation Limited, Hangzhou, China
e-mail: yan_jj@hdec.com

© The Author(s) 2025

W. Wang et al. (eds.), *Hydraulic Structure and Hydrodynamics*, Lecture Notes in Civil Engineering 608, https://doi.org/10.1007/978-981-97-7251-3_13

most large-sized cities in the world. The development of urban rail transit in China is particularly rapid. According to statistics, 55 cities in mainland China have built a total of 10,291 km of urban rail transit lines by December 2022. In the past five years, the growth trend of China's urban rail transit mileage is as shown in Fig. 1. With the large-scale development of urban construction, a large number of deep foundation pit projects have emerged in urban underground space development projects, and their scale and depth are constantly increasing. And the safety risk of the foundation pit is also getting higher and higher. At the same time, in the development of urban underground space, the environment around foundation pits has become more complex, and the requirements for environmental control of surrounding buildings, pipelines and other structures have also continued to increase. In areas with rich groundwater resource, especially under geological conditions that contain confined water layers, the thickness of the water-resisting layer between the pit bottom and the confined layer will continue to decrease with the excavation of the foundation pit. When it is less than the anti-surge of the critical thickness, foundation pit surge accidents may occur. The forms of foundation pit outburst are mostly water gushing, flowing soil and sand escaping, and the traditional soil mechanics theory is difficult to explain the mechanism of their occurrence [1]. According to statistics, about 20% of underground space disaster accidents are related to groundwater, and groundwater has become an important factor affecting the safe development of underground space, as shown in Fig. 2. The effective control of groundwater is of great significance to reduce the impact on the surrounding environment of the construction. Many researchers have carried out more research on the control of groundwater. Hong et al. [2] and Ding et al. [3] used plasticity theory to analyze the inrush damage of foundation pit when the bottom of the pit is an impermeable layer. Wang et al. [4] and Zhang et al. [5] used seepage theory to analyze the inrush damage of foundation pit when the bottom of the foundation pit is an impermeable layer. Gong et al. [6] studied land subsidence caused by confined water precipitation.

Although many researchers have carried out a lot of research on groundwater precipitation, seepage and so on, but the research results are mainly for a specific project or a specific construction method, the study of quantitative constraints is less [7–10]. In practical engineering, the control method of groundwater is still mainly

Fig. 1 The growth trend of China's urban rail transit during 2018 ~ 2022

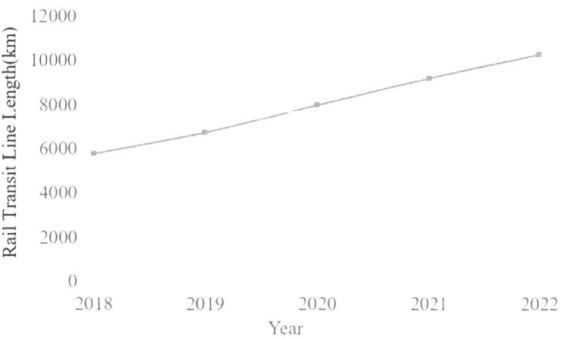

Fig. 2 Proportion of disaster types in underground space during 2016–2020

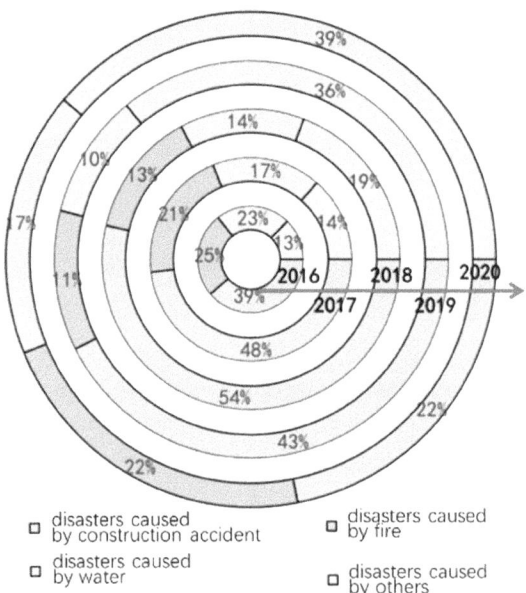

disasters caused by construction accident

disasters caused by water

disasters caused by fire

disasters caused by others

based on engineering experience, and the engineering control method of confined water in foundation pit is mainly based on precipitation and complete isolation. In order to ensure the safety of the project, the method of complete isolation of groundwater is usually adopted. The effect of this method is remarkable, but the construction is difficult and the project cost is high. Especially under the deep conditions of underground aquifer, the difficulty and cost of complete isolation are very high. For the use of precipitation methods to control the impact of groundwater on the foundation pit, there is still a lack of simple and practical methods for predicting the settlement caused by precipitation and its impact on the surrounding environment, often resulting in uncontrollable settlement caused by precipitation during construction. Therefore, in order to formulate more refined groundwater control methods in the process of underground space development and form groundwater classification control technology, it is urgent to develop a multi-objective optimization method that takes into account the safety, economy and feasibility of the project and is used to determine the confined water control scheme of the foundation pit.

2 Groundwater Classification and Grading Method

This paper mainly studies a quantitative, refined and engineering-oriented classification and multi-objective optimization method for determining the confined water control scheme of foundation pit. According to the types of confined water in different foundation pits, this method accurately evaluates the influence of confined water

Table 1 Classification of influence level of confined water on foundation pit

Influence level	Classification standard	
	Thickness of confined water layer (m)	Depth of drawdown of confined water level (m)
Level 1	> 20	> 16
Level 2	13 ~ 20	6 ~ 16
Level 3	6 ~ 13	< 6
Level 4	< 6	–

changes on the surrounding environment, comprehensively analyzes other limiting factors such as engineering investment, engineering safety, surrounding buildings, and ecological environment, and then evaluates the confined water control scheme, which provides a scientific basis for the final determination of the confined water control scheme.

2.1 Classification of Influence Degree of Confined Water on Foundation Pit

Based on the design parameter data, the site hydrogeological parameter model is established to classify the influence degree of confined water on the foundation pit. The required main parameters include: diving and confined water depth, thickness and the influence range of surrounding buildings; soil mechanics index, permeability coefficient, confined water quality, confined water level drawdown value and single well water inflow. Based on the established hydrogeological parameter model, the aquifer with the greatest influence on the foundation pit is selected as the division basis. According to the thickness of the aquifer, the drawdown value of the aquifer water level and the single well water inflow, the influence degree of the aquifer on the foundation pit is divided into four grades from high to low, as shown in Table 1.

2.2 Control Level Classification of Confined Water

In order to guide the design of confined water control measures, according to the safety level of foundation pit and the influence degree of confined water on the safety of foundation pit in Table 1, the control level of confined water is classified. The control level of confined water in foundation pit is divided into four levels from high to low, as shown in Table 2. Among them, the safety level of foundation pit is divided into first, second and third levels according to the complexity of surrounding environment, excavation depth and environmental protection control requirements. The safety level of foundation pit with excavation depth > 10 m is usually set as

Table 2 Control level classification of confined water

Confined water control level	Classification standard	
	Excavation safety grade	Influence level of confined water on foundation pit
Level 1	Grade 1	Level 1
Level 2	Grade 1	Level 2
	Grade 2	Level 2
Level 3	Grade 1	Level 3
	Grade 2	Level 2, 3
	Grade 3	Level 1, 2, 3
Level 4	Grade 1, 2, 3	Level 4

the grade 1, and the safety level of foundation pit with excavation depth < 5 m is set as the grade 3. The safety level of foundation pit is adjusted according to the surrounding environmental conditions, geological conditions and the influence degree of foundation pit damage.

2.3 Determination of Classification Processing Scheme

Specific treatment scheme for confined water control shall be selected based on the refined classification of the above confined water control levels and in combination with the existing construction control technology. The principle of confined water control shall be determined according to Table 3.

According to the control grade of confined water in the foundation pit, the treatment scheme of confined water is selected by level. There are many different construction methods for each type of treatment scheme. A variety of alternative confined water control schemes are proposed, the advantages and disadvantages of different construction methods are analyzed, and the appropriate treatment method is selected for the treatment method of confined water in the foundation pit to determine the alternative confined water control scheme.

The target factors to be considered in determining specific control measures mainly include the maturity of technology, reliability of groundwater control, project

Table 3 Processing scheme of confined water according to the control level

Confined water control level	Processing scheme of confined water
Level 1	Completely cut off confined water
Level 2	Depressurization in pit and recharge outside pit
Level 3	Depressurization in pit
Level 4	Completely cut off or depressurization in pit

investment, environmental impact, construction period, etc. However, there is a mutually exclusive relationship between these control objectives. For example, the most advanced and reliable groundwater control measures may be required to achieve the best safety of the project, but such schemes often bring unbearable cost inputs. Therefore, there is usually no optimal solution for groundwater control that can optimize all objectives. This kind of problem is called multi-objective optimization problem or Pareto frontier problem.

The multi-objective optimization problem can be defined as:

$$\begin{cases} minF(x) = (f_1(x), \ldots f_m(x))^T \\ s.t. \ x \in \Omega \end{cases} \tag{1}$$

where, x is the decision vector and n-dimensional decision space, $F{:}\Omega \in R_m$ contains m objective functions to be optimized simultaneously, and Rm is the objective space.

For solving such multi-objective optimization problems, Schaffer first proposed a multi-objective evolutionary algorithm (MOEA). The main method is to decompose the multi-objective problem into several different single objective problems through the aggregation function, and then optimize all single objective problems at the same time. The multi-objective optimization analysis is carried out in combination with various limiting factors, and the optimization analysis of alternative confined water control schemes is completed to obtain the recommended schemes that meet the selection conditions and their selection opinions. In combination with various limiting factors, multiple alternative treatment schemes can be optimally evaluated, among which the limiting factors include but are not limited to engineering economy, structural durability, engineering importance, long-term adjustment ability, degree of influence of disturbance, degree of difficulty of construction and ecological balance.

According to the above methods, this paper developed a multi-objective optimization algorithm for groundwater control, established a classification and classification control system with "water level control as the premise and settlement control as the core", overcame the problem of extensive selection and design of traditional groundwater control measures, and realized the refined control of multiple types of complex groundwater in urban complex underground space. Take the double objective optimization analysis of control effect project cost as an example to analyze the relationship between the control effect of various alternatives and the project cost, as shown in Fig. 3. The Pareto front can be found through multiple existing codes. It should be noted that the schemes on the Pareto front are superior to other alternatives outside the front, while each scheme on the front has no advantages or disadvantages, so the schemes on the Pareto front can be used as the final scheme. Considering that the cost is one of the main factors affecting the selection of the actual construction scheme, the final scheme can be selected in combination with the allowable value, or further combined with other limiting factors, the hierarchical analysis can be carried out to further screen the schemes on the Pareto front.

Fig. 3 Dual-objective optimization of control effect-project cost

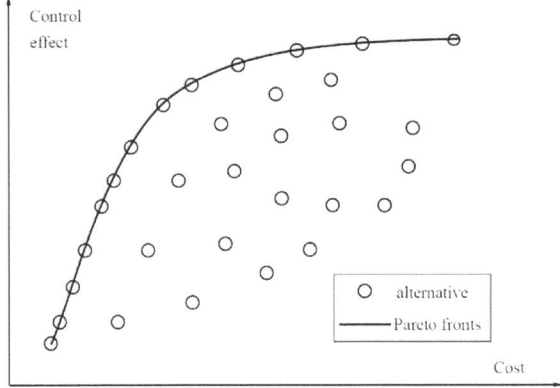

3 Discussion

A classification system to assess the impact of confined water on foundation pits was proposed in this paper. Parameters such as diving and confined water depth, thickness, and influence range of surrounding buildings, soil mechanics index, permeability coefficient, confined water quality, confined water level drawdown value, and single well water inflow are considered. Based on these parameters, the aquifer with the highest impact on the foundation pit is identified, and the influence degree is classified into four grades.

In order to guide the design of confined water control measures, it was suggested that classifying the control levels of confined water based on the safety level of the foundation pit and the influence degree of confined water on its safety. The safety level of the foundation pit is categorized into three levels, primarily based on factors such as surrounding environment complexity, excavation depth, and environmental protection requirements. The control level of confined water is then divided into four levels, aligned with the safety levels of the foundation pit.

It was recommended that the selecting specific treatment schemes for confined water control based on the refined classification of control levels and existing construction control technologies. A processing scheme is determined according to the control level, taking into account factors such as technology maturity, groundwater control reliability, project investment, environmental impact, and construction period. Various alternative schemes are proposed and analyzed to identify the most appropriate treatment method for confined water control in the foundation pit.

The selecting of the optimal confined water control measures was formulated as a multi-objective optimization problem, considering factors such as technology maturity, groundwater control reliability, project investment, environmental impact, construction period, and other limiting factors. The paper employs a multi-objective evolutionary algorithm to analyze and optimize the alternative confined water control schemes. The algorithm decomposes the problem into single objective problems and optimizes them simultaneously. Through Pareto frontier analysis, the paper identifies recommended schemes that meet the selection conditions and provides selection opinions.

Overall, this paper aims to refine the classification and control of confined water in foundation pits by considering multiple factors, including safety, engineering investment, environmental impact, and construction feasibility. The proposed method allows for a comprehensive evaluation of alternative control schemes through multi-objective optimization, ultimately providing scientifically justified recommendations for selecting the most suitable confined water control measures in construction projects.

4   Conclusions

Groundwater has become one of the important factors restricting the development and construction of urban underground space in the southeast coastal areas of China. Since the design method of groundwater control in the design of the target underground project is mainly based on engineering experience, it cannot take into account the control parts of project safety, project cost and environmental protection. Through extensive engineering research and based on the multi-objective optimization algorithm, this paper proposes a quantitative, refined, and engineering-guided classification multi-objective optimization method to determine the foundation pit confined water control scheme. Constructed a refined and quantitative design method for urban groundwater control schemes. Compared with traditional design methods, the control determines the corresponding groundwater level for specific engineering projects by finely decomposing the degree of impact of confined water on foundation pits and the level of confined water control. Control strategy, and then according to the principle of multi-objective marker optimization, groundwater control measures that comprehensively consider control conditions such as project safety, project cost, and environmental protection can provide a tool entity guide for the development of underground space in areas rich in groundwater resources.

References

1. H.W. Ying, D.Y. Xu, D. Wang, L.S. Zhang, Analysis of non-Darcy flow in aquitard at bottom of foundation pit under fluctuation of confined water conditions. J. Shanghai Jiaotong Univ. **54**(12), 1300–1306 (2020)
2. Y. Hong, C.W.W. Ng, L.Z. Wang, Initiation and failure mechanism of base instability of excavations in clay triggered by hydraulic uplift. Can. Geotech. J. **52**(5), 1–10 (2015)
3. C.L. Ding, Summary of study on calculation method of inrushing for confined water foundation pit in soft soil area. J. Undergr. Space Eng. **3**(2), 333–338 (2007)
4. Y.L. Wang, K.H. Xie, M.M. Lu, K. Wang, A method for determining critical thickness of base soil of foundation pit subjected to confined water. Rock Soil Mech. **31**(5), 1539–1544 (2010)

5. L.S. Zhang, H.W. Ying, K.H. Xie, X.G. Wang, C.W. Zhu, Analytical study on exit gradient at base aquitard of deep excavations under dynamic artesian water. Chin. J. Geotech. Eng. **39**(2), 295–300 (2017)
6. X.N. Gong, J. Zhang, Settlement of overlaying soil caused by decompression of confined water. Chin. J. Geotech. Eng. **33**(1), 145–149 (2011)
7. A. Singh, M.K. Hasan, P.K. Srivastava, Groundwater vulnerability assessment: a multi-criteria decision making approach. J. Water Land Dev. **34**(1), 21–29 (2017)
8. Y. Wang, L. Yin, J. Chen, Groundwater quality assessment and classification in the context of water resource and environmental management: a case study in China. Environ. Sci. Pollut. Res. **24**(14), 12910–12919 (2017)
9. C. Xu, L. Xu, F. Sun, J. Yu, D. Yang, Risk control and dealing example of confined water of deep foundation pit. Geotech. Eng. **35**(S1), 353–358 (2014)
10. A. Chowdhury, M.K. Jha, V.M. Chowdary, Groundwater management practices: a review of successful case studies. Water Resour. Manage **24**(10), 2013–2040 (2010)

Study on the Influence of Anchor Reinforcement Methods on the Stability of Slope Under Rainfall Infiltration Condition

Nan Zhang, Zhilin Mao, Shijie Zou, and Yi Wang

Abstract Rainfall infiltration is a major cause of slope instability, and many researchers have conducted extensive studies on the mechanism of slope instability under the influence of rainfall infiltration. The design of slope support currently faces several challenges, including a shortage of appropriate design basis, dependence on imprecise, experience-based, or analogy-based designs, and the absence of focused and optimized design methods. Using numerical simulation, this study aimed to find the impact of various anchor reinforcement schemes on slope stability during heavy rainfall. Several anchoring reinforcement schemes have been proposed, the impact of anchoring angle, anchoring distance, anchoring length, and other parameters on the slope's safety factor has been analyzed and compared, and the optimal solution has been suggested.

Keywords Slope · Anchor reinforcement · Anchorage angle · Spacing · Safety factor

1 Introduction

The stability of slopes is significantly impacted by rainfall, which raises the water table and weakens the stability of slopes [1]. In this study, the effect of rainfall infiltration on a high-risk slope in Fengshun County, Meizhou City, Guangdong Province was analyzed. Using the Midas-GTS finite element strength reduction method, a numerical simulation was conducted to examine the impact of various anchor reinforcement schemes on the stability of the slope. The study considered different anchor spacing, anchorage angles, and anchors' design lengths to understand these factors' effect on slope stability. The results of this study provide valuable insights for enhancing the scientific prevention and control of high-risk slopes, reducing landslide risks, and improving local disaster prevention, mitigation, and relief capabilities.

N. Zhang · Z. Mao (✉) · S. Zou · Y. Wang
School of Civil Engineering, Jiaying University, Meizhou, Guangdong, China
e-mail: mzl13822538227@163.com

157

W. Wang et al. (eds.), *Hydraulic Structure and Hydrodynamics*, Lecture Notes in Civil Engineering 608, https://doi.org/10.1007/978-981-97-7251-3_14

2 Finite Element Strength Reduction Method and Numerical Simulation

The finite element strength reduction method approach does not require assumptions regarding the shape, location, or conditions of the sliding surface. It is a widely used approach in geotechnical engineering for evaluating the stability of slopes [2–4]. In this method, the ratio of the maximum shear strength produced by the geotechnical body and the actual shear stress produced by the external load in the geotechnical body of the slope as the discount factor F_s. The discount factor F_s was applied to the shear strength parameters C and φ of the soil body in order to account for the reduction in shear strength caused by external loads. Then use the discounted shear strength index C_F and φ_F to replace the original shear strength index C_F and φ_F. When the two values are calculated until convergence is no longer achieved, the slope will reach the limit state and destabilization will occur. And the last calculated discount factor F_s serves as the safety factor of the slope [5–7].

The principle of the strength reduction method can be expressed as follows:

$$C_F = C/F_S \tag{1}$$

$$\varphi_F = tan^{-1}(tan\,\varphi)/F_S \tag{2}$$

$$\tau_{fF} = C_F + \sigma\,tan\,\varphi_F \tag{3}$$

where C_F is the cohesive force of the soil; φ_F is the internal friction angle of the soil; τ_{fF} is the shear strength of the slope material.

In this study, the geotechnical body is modeled as an isotropic and homogeneous medium, represented by 2D solid units. The diagram of the slope is presented in Fig. 1, characterized by a height of 24 m and a dip angle of 40°. The anchor reinforcement is represented as a 1D implanted truss unit. The strength of the geotechnical body is described using the Mohr–Coulomb strength theory. Boundary conditions are established through the use of static constraints on the sides and bottom of the model, with the maximum rainfall being simulated as a line flow on the upper part of the model. The head on the left side of the model is set at 20 m, while the head on the right side is set at 10 m. To better understand the infiltration pattern of rainfall and the impact of anchor reinforcement on the stress, displacement, and stability of the slope, monitoring points are placed on the slope surface. The upper layer of the slope is soil, and the bottom layer is bedrock. The physical and mechanical properties of the slope and anchor can be found in Tables 1 and 2, respectively.

The numerical simulation of slope stability was performed to evaluate the effect of heavy rainfall. The simulation conditions included a surface flow rate of 3.47e−6 m³/s/m² and a rainfall duration of 10 h. The relationship between the soil permeability coefficient and matrix suction, as described by Zhao [8], was considered in the

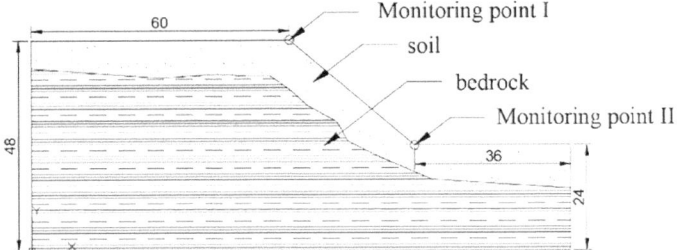

Fig. 1 Schematic diagram of the slope

Table 1 Physical and mechanical properties of the soil mass and bedrock

Material	Soil mass	Bedrock
Elastic modulus (MPa)	35	900
Poisson's ratio	0.27	0.23
Unit weight (kN/m³)	21	24
Initial void ratio	1	0.5
Cohesion (kPa)	14.6	120
Frictional angle (°)	20.5	40

Table 2 Physical and mechanical parameters of anchor

Material	Anchor
Elastic modulus (MPa)	200
Poisson's ratio	0.3
Length (m)	10
Initial void ratio	1
Diameter (m)	0.25
Angle (°)	20

simulation. The close correlation between unsaturated permeability and the soil–water characteristic curve has been utilized by numerous researchers to calculate the unsaturated permeability coefficient.

By utilizing the established soil–water characteristic curve, the unsaturated permeability coefficient can be estimated and the unsaturated permeability characteristics of the soil can be determined. The unsaturated infiltration characteristics of the soil can also be calculated by using the established soil–water characteristic curve to estimate the unsaturated infiltration coefficient. In this study, the Van Genuchten model was used to fit the soil–water characteristic curves and calculate the unsaturated permeability coefficients.

Table 3 Slope stability classification standard [9]

Slope stability factor-Fs	Slope stability state
Fs < 1.00	Unstable
1.00 ≤ Fs < 1.05	Less stable
1.05 ≤ Fs < Fst	Basic stability
Fs ≥ Fst	Stable

2.1 Effect of Anchor Arrangement Spacing on the Slope the Safety Factor

To investigate the influence of anchor arrangement spacing on the safety factor of the slope, the initial anchorage angle is set as $20°$, the anchor length is 10 m, and three different spacing arrangements have been designed for the slope with the spacing of 2.5, 3, and 3.5 m.

The simulation results indicate that without anchor support, the slope's safety factor is 1.01 and the maximum total displacement is 8.85 m under storm conditions. An evaluation based on the slope stability classification standard (as outlined in Table 3) shows that the slope is in a less stable condition. The simulation results demonstrate that the safety factor of the slope increases as the spacing between the anchors increases, and the maximum total displacement of the slope decreases, indicating that a more spaced-out anchor arrangement leads to a more stable slope. Additionally, the slope stability state classification criteria, presented in Table 3, show that the slope is in a relatively stable state for all three anchor arrangements.

In the case, the slope is reinforced by anchors, the safety coefficient increases with different spacing of the anchors. The simulation results indicate that as the spacing between anchors increases, the safety factor of the slope initially rises, reaches its peak at a spacing of 3 m, and then decreases as shown in Fig. 2. The total displacement within the slope reaches its minimum value when the anchors are spaced 3 m apart (as shown in Fig. 3). When compared to the conditions without anchor support, the total displacement of the slope is reduced by 79.26%, 90.23%, and 89.38% respectively. It is clear that the optimal arrangement of anchors is achieved when the spacing is 3 m. These findings suggest that anchor reinforcement plays a crucial role in improving the stability of slopes under rainfall conditions, and the spacing of the anchors is a significant factor that influences the safety factor of the slope.

2.2 Effect of Anchorage Angle on the Safety Factor of Slope

The angle between the anchor and the horizontal slope surface is referred to as the anchorage angle. This study focuses on examining the impact of anchorage angles ranging from 15 to $25°$ on the safety factor of an unsaturated slope, using the anchor

Fig. 2 Total displacement under each arrangement spacing of anchors

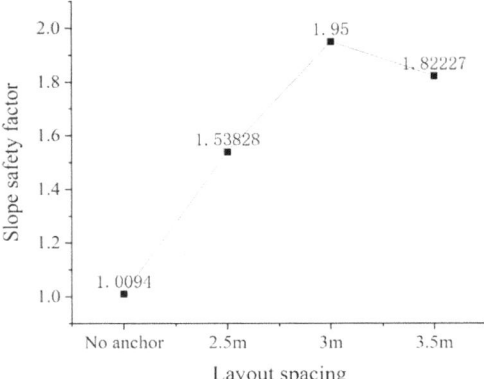

Fig. 3 Safety factor under each arrangement spacing of anchors

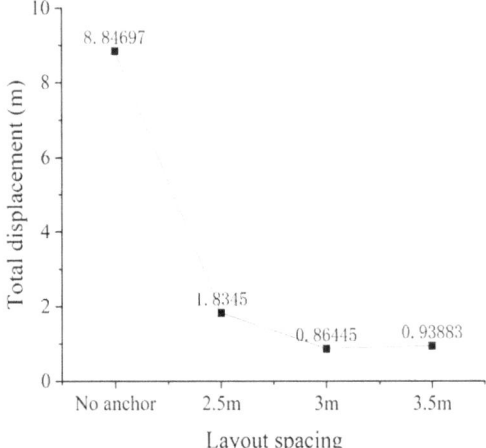

arrangement with a spacing of 3 m and a length of 10 m. The numerical simulation results are presented in Figs. 4 and 5.

The results show that the safety factor of the slope increases and then decreases with increasing anchorage angle, suggesting that there exists an optimal anchorage angle. There are two viewpoints on the optimal anchoring angle: one is that the anchor should be arranged horizontally along the slope, and the other is that the anchor should be arranged along the vertical sliding surface [10]. In this study, it was found that when the anchorage angle is greater than 20°, the safety factor of the slope decreases. The slope has the highest safety factor at an anchorage angle of approximately 20° (Fig. 4). A summary of the total movement of the slope at two monitoring points is presented in Fig. 5, and the detailed contour of the total displacement can be found in Figs. 6, 7, 8 and 9. The results of the anchor reinforcement study indicated that as the anchorage angle increases, the sliding surface position shifts inward initially and then outward, this indicates that the stability of the slope initially improves, but

Fig. 4 Safety factor of the slope under each anchorage angle

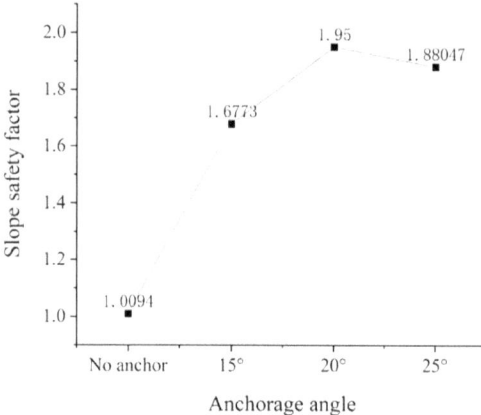

Fig. 5 Total displacement of the slope at two monitoring points

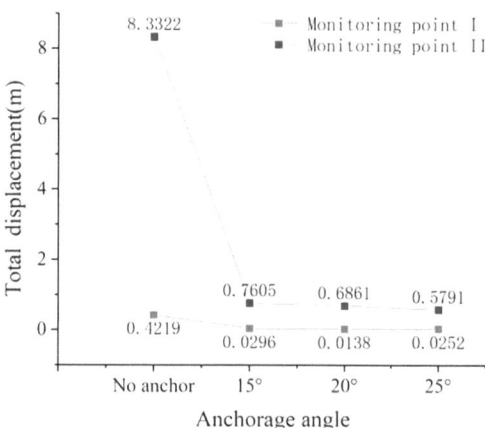

then decreases. After anchor reinforcement, it shows that with increasing anchorage angle, the sliding surface position first moves inward and then outward, indicating that the stability of the slope increases and then decreases (see Figs. 6, 7, 8 and 9).

As the anchor length increases, the slope's safety factor improves, resulting in a decrease in total displacement. The minimum displacement value achieved is 0.477 m, a 94.66% reduction compared to prior to anchor support, demonstrating the anchoring function's effectiveness in enhancing slope stability.

2.3 Effect of Anchor Length on the Slope Safety Factor

In this case study, seven anchors were strategically placed from the bottom to the top of the slope at 3 m intervals, each with a 20° anchoring angle. The impact of

Fig. 6 Total displacement of the slope without anchor reinforcement

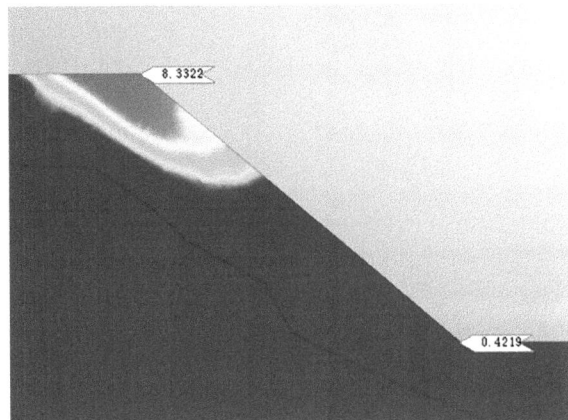

Fig. 7 Total displacement of the slope with 15° anchorage angle

Fig. 8 Total displacement of the slope with 20° anchorage angle

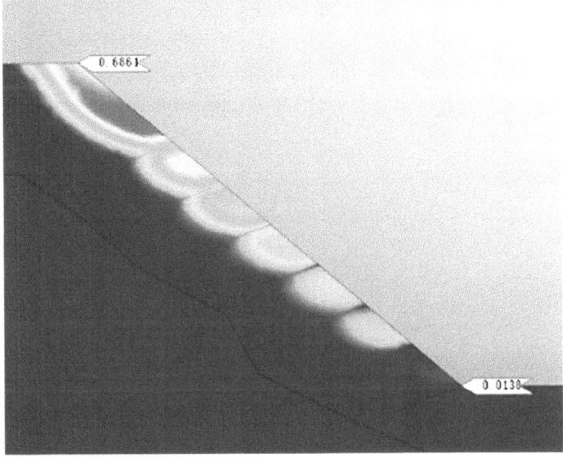

Fig. 9 Total displacement of the slope with 25° anchorage angle

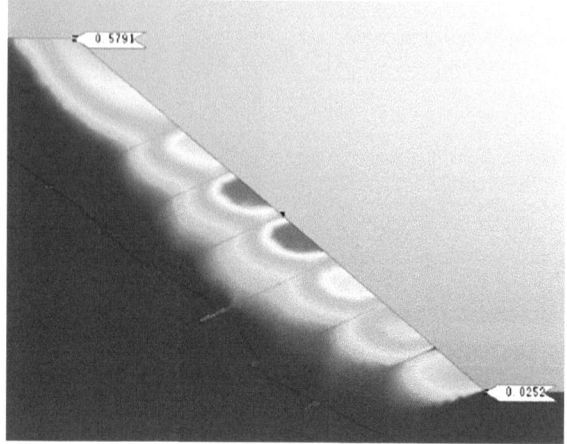

Fig. 10 Safety factor of the slope for each anchor length

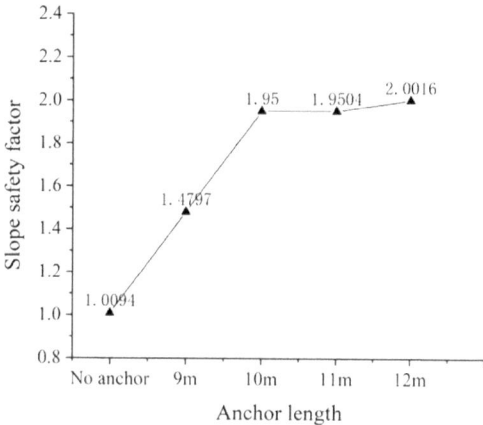

anchor lengths of 9, 10, 11, and 12 m on the safety factor of the slope was determined through numerical simulations, and the results were illustrated in Figs. 10 and 11, which demonstrate the correlation between the anchor length and the safety factor of the slope.

Fig. 11 Total displacement of the slope for each anchor length

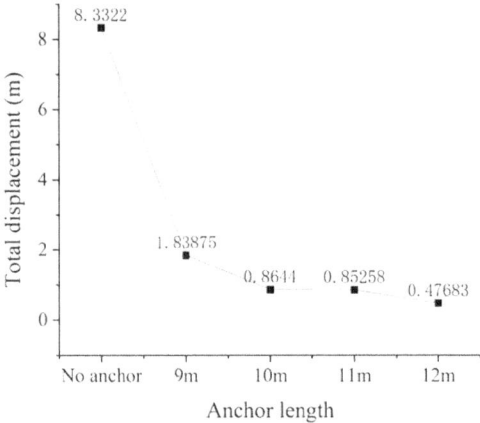

3 Conclusion

In this study, the stability of an anchor-reinforced slope under rainfall infiltration was analyzed, and the impact of various reinforcement schemes was evaluated. The following conclusions were drawn.

The anchorage angle plays a crucial role in determining the safety factor of a slope. It is necessary to consider various factors affecting slope stability and determine the optimal anchorage angle.

The spacing between anchors can have a significant impact on the distribution of reinforcement forces and ultimately, the overall stability of the slope. Too much spacing can result in insufficient reinforcement, while too little spacing can result in over-reinforcement and waste of materials. It is important to carefully consider the optimal anchor spacing that balances cost-effectiveness and stability in the design of a reinforced slope.

An increase in anchor length results in a noticeable improvement in the safety factor of the slope. Optimal anchor length should provide the necessary stability and safety for the slope while being economically and practically feasible for implementation.

Acknowledgements This work was financially supported by Research Startup Project of Jiaying University (Grant No. 2023RC69) and the Research Project (Grant No. 2023KJY01).

References

1. J. Li, N. Zhang, W. Wang, J. Hong, W. Chen, Z. Mao, Hazards, causes and control countermeasures of rainfall-induced landslide in Meizhou. Constr. Des. Eng. **1**, 47–49 (2023) (in Chinese)

2. E. Dawson, W. Roth, A. Drescher, Slope stability analysis by strength reduction. Geotechnique **49**, 835–840 (1999)
3. J. Chen, F. Dang, W. Tian, Discussion on analysis of slope instability criterion by strength reduction FEM. Northwest Hydropower **2**, 14–18 (2009) (in Chinese)
4. H. Chi, Z. Wu, X. Liu, Application of three constitutive models of soil based on Midas GTS NX. Geotech. Eng. Techn. **3**, 143–149 (2020) (in Chinese)
5. S. Wan, T. Nian, J. Jiang, M. Luan, Discussion on several issues in slope stability analysis based on shear strength reduction finite element methods (SSR-FEM). Rock Soil Mech. **7**, 2283–2288, 2316 (2010) (in Chinese)
6. Y. Li, C. Zhou, H. Zhang, Stability analysis of slope by using strength reduction FEM based on the criterion of uncontrolled displacement at three characteristic nodes. Yellow River **2**, 146–148 (2012) (in Chinese)
7. X. Bi, T. Yan, J. Lu, Application of Midas-GTS (SRM) to 2D stability analysis of slope. J. Nat. Disasters **1**, 170–176 (2015) (in Chinese)
8. P. Zhao, Analysis of unsaturated expansive soil slopes under rainfall infiltration in Zhen Jiang. Master thesis, Jiangsu University of Science and Technology, 2015 (in Chinese)
9. Ministry of Housing and Urban-Rural Development of the People's Republic of China, *Technical Code for Building Slope Engineering GB/50330—2013* (China Architecture & Building Press, Beijing, 2013) (in Chinese)
10. W. Wei, Y. Cheng, Soil nailed slope by strength reduction and limit equilibrium methods. Comput. Geotech. **37**, 602–618 (2010)

Seismic Response Analysis of Subsea Tunnels Under the Influence of Overlying Seawater

Weizhen Huang, Cong He, and Guoyuan Xu

Abstract The safety of subsea tunnels during earthquakes is a crucial factor in the design process. It is appropriate to consider the impact of the seawater above when assessing the seismic resistance of a subsea tunnel. In this paper, finite element models of the seawater-seabed-tunnel structure system are established using ABAQUS software to investigate the fluid–structure interaction and plastic damage of the lining. The seismic response and damage characteristics of the subsea tunnel are compared at various seawater depths. The study found that the overlying seawater inhibits the vertical movement of the seabed surface, but has minimal effect on the horizontal movement. With the increase in seawater depth, the stress on the tunnel structure will noticeably increase, with the spandrel and arch foot becoming the weak points in seismic design. Overlying seawater will increase the dissipation energy of the tunnel system, leading to tunnel damage. When the seawater depth exceeds 20 m, the tunnel system's dissipation energy at varying depths of seawater will significantly increase.

Keywords Finite element calculation · Subsea tunnel · Seismic response · Seawater-seabed-tunnel interaction · Damage evolution

1 Introduction

Subsea tunnels have become prevalent in the construction of river and sea crossings in recent decades, due to their convenience, high traffic efficiency and low susceptibility to climate conditions. In contrast to land tunnels, subsea tunnels are mostly built in areas that are vulnerable to earthquakes [1], and the construction sites in the sea are challenged by the overlying layer of infinite seawater. Seismic waves can travel through seawater, resulting in intricate transmission and reflection at the interface between the seabed and seawater, which can significantly affect the seismic response of the subsea tunnel structure. Thus, comprehending the seismic functionality of

W. Huang · C. He · G. Xu (✉)
School of Civil Engineering and Transportation, South China University of Technology, Guangzhou, Guangdong, China
e-mail: 430855690@qq.com

© The Author(s) 2025
W. Wang et al. (eds.), *Hydraulic Structure and Hydrodynamics*, Lecture Notes in Civil Engineering 608, https://doi.org/10.1007/978-981-97-7251-3_15

subsea tunnels under the sway of overlying seawater is fundamental to ensure the secure design of tunnels.

The influence of overlying seawater on site and structure dynamic response is researched using theoretical derivation, shaking table tests, and numerical simulation. A vertical incidence-based P-wave model [2] showed that P-waves transmit and reflect in seawater. The vertical ground motion of the seafloor decreases due to destructive interference from saltwater layer waves approaching its natural frequency. Furthermore, Boore and Smith [3] analyzed strong-motion records from the seafloor seismic observation system near the Southern California coast and used theoretical calculations to determine how seawater affects ground motion. As seawater cannot transmit shear waves, their findings show that seawater has no impact on horizontal ground motion. Chen [4] conducted seawater-seabed models. During the P wave occurrence, the seabed model's displacement time history reveals sub-peak spots and travels right on the time axis with increasing seawater depth. Cui et al. [5] examined how overlying seawater affects the reaction of an immersed tunnel using shaking table testing and ADINA finite element software. Based on Biot's theory of a saturated porous medium and inviscid fluid wave theory, Zhu et al. [6, 7] used the Hankel function integral transformation method to derive the analytical solution for the scattering problem of plane P waves and plane SV waves by subsea tunnel lining. It was concluded that the largest site displacement and tunnel stress occurred when the water depth was ten times the tunnel radius.

In brief, most subsea tunnel seismic behavior investigations have concentrated on a single seismic wave type. However, subsea tunnels are affected by both the early-arriving P wave and the later-arriving SV wave. Elastic constitutive models were used to simulate subsea tunnel lining, which made it difficult to adequately capture damage evolution and energy dissipation during seismic events.

This paper addresses an offshore subsea tunnel case study. The ABAQUS finite element program simulates seawater using an acoustic fluid medium. A coupling model is used to examine seawater-seabed-tunnel interaction at different depths. This model uses a viscous-spring artificial boundary and equivalent load input. The objective is to analyze the seabed site's reaction characteristics, tunnel lining stress distribution, and damage progression under P-SV wave delay incidence.

2   Numerical Analysis Method

2.1   Seawater Simulation

Since the viscosity of common fluids such as air and water is very low, the fluid can be treated as an acoustic medium without affecting the substance.

Taking seawater as the acoustic fluid, the wave equation for an ideal inviscid, linear, low-disturbance fluid is

$$\frac{1}{K_f}\ddot{p} + \frac{\partial}{\partial x} \cdot \left(\frac{1}{\rho_f}\frac{\partial p}{\partial x}\right) = 0 \tag{1}$$

where, p is the acoustic pressure; ρ is the density of the acoustic medium; K_f is the bulk modulus.

The truncated boundary of the fluid domain adopts the non-reflecting boundary condition built into ABAQUS to simulate the infinite domain of the fluid domain. The radiation condition of the non-reflecting boundary can be expressed as

$$\nabla p \cdot \mathbf{n} = -\frac{1}{c}\dot{p} \tag{2}$$

where $\mathbf{n}$ is the inner normal direction of the boundary of the truncated fluid domain.

2.2 Viscous-Spring Artificial Boundary and Equivalent Seismic Load Input

The viscous-spring artificial boundary effectively replicates the radiation damping effect of the infinite domain foundation, so that the scattered waves generated by the structure and the surface can pass through the infinite domain or be absorbed by the artificial boundary, which has the characteristics of accurate and stable computation. Tangential and normal springs and dampers are applied on both sides and bottom nodes of the seabed soil. The calculation parameters outlined below:

$$K_N = A_l \alpha_N \frac{G}{R}, \quad C_N = A_l \rho c_p \tag{3}$$

$$K_T = A_l \alpha_T \frac{G}{R}, \quad C_T = A_l \rho c_s \tag{4}$$

where, G and ρ are the shear modulus and density of the medium, respectively; R is the distance from the wave source to the artificial boundary; A_l is the equivalent area of node l; α_N and α_T are the correction coefficients of normal and tangential viscous-spring artificial boundary, respectively, and the recommended values for 3D problems are 1.33 and 0.67 [8]. The subscript N represents the normal direction, and the subscript T represents the tangential direction.

For seismic waves, it is necessary to decompose the motion of the artificial boundary into the inner field and outer field. The viscous-spring boundary is able to simulate the outer field, while the method proposed by Liu et al. [9] is able to simulate the inner field. The input problem of the inner field is transformed into a wave source problem, and the input formula for the inner field on the viscous spring artificial boundary is derived as follows:

$$F_l(t) = A_l\left[\sigma_0(x_l, y_l, t) + K_l u_0(x_l, y_l, t) + C_l \dot{u}_0(x_l, y_l, t)\right] \tag{5}$$

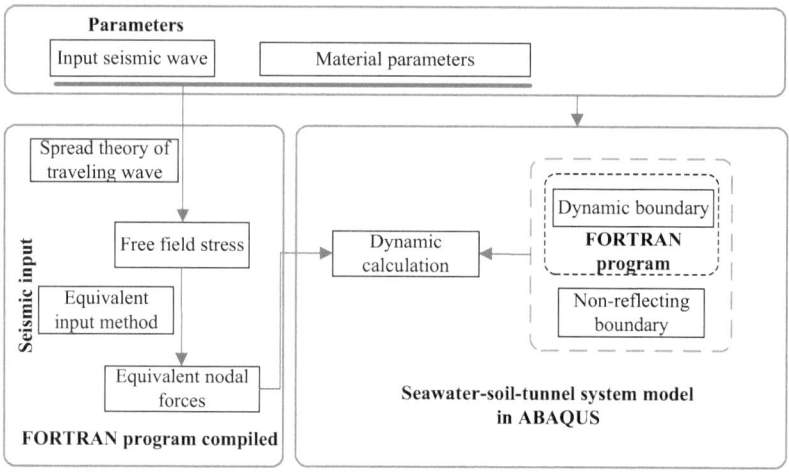

Fig. 1 Flowchart of numerical simulation process

where $u_0(x_l, y_l, t)$ is the displacement of node l; $\sigma_0(x_l, y_l, t)$ is the stress of the inner wave field at the truncated boundary of node l.

This study utilises a self-developed Fortran programme to compute the viscous artificial boundary parameters and equivalent node loads for large-scale finite element simulations, owing to the significant number of boundary nodes involved. The calculation results are exported in the inp file format and subsequently fed into the ABAQUS software in batches for computational analysis. The technique of computation is illustrated in Fig. 1.

In order to illustrate the veracity of the seismic analysis model of the seabed proposed in this study, we use the P-wave vertical incidence single-covered seabed site model established by Crouse et al. as an example, depicted in Fig. 2. The vertical transfer function is defined as the proportion of the vertical movement at the ground surface under seawater to that at the free surface without seawater. The P-wave vertical incidence model presented in this paper is compared to previous studies [2, 10]. Figure 3 illustrates the computational results of the proposed submarine ground vibration analysis model, which are in good agreement with the solutions of other researchers. This indicates the accuracy of the proposed submarine site model.

Fig. 2 Vertical incidence model case

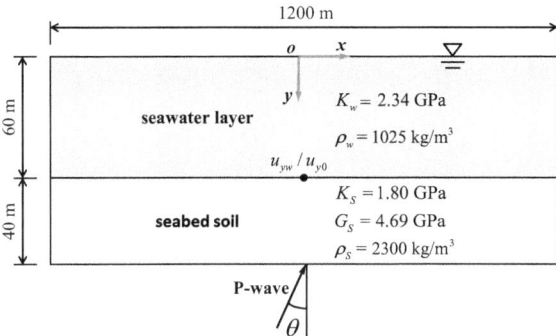

Fig. 3 Comparison of the proposed vertical transfer function model for the underwater site with those proposed in the previous studies

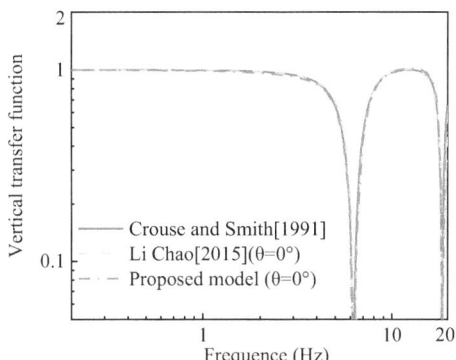

3 Establishment of Numerical Model of Subsea Tunnel

3.1 Finite Element Model and Parameters

In this paper, a numerical model is established based on an offshore subsea tunnel project. The tunnel is buried at a depth of 10 m, while the overlying seawater ranges from 0 to 40 m. The tunnel ring has an outer diameter of 14 m, an inner diameter of 12.8 m, and a lining thickness of 0.6 m. Moreover, the concrete strength grade is C60. Figure 4 presents the overall finite element model with tunnel monitoring points. The lining adopts the CDP plastic damage constitutive, which can account for damage energy dissipation and plastic energy dissipation. The damage parameters are listed in Table 1. The tunnel is situated in the silty clay stratum. The structure and soil elements are of the CPE4 type. The overlying seawater adopts the AC2D4 acoustic element type. The parameters for the soil and seawater are listed within Table 2.

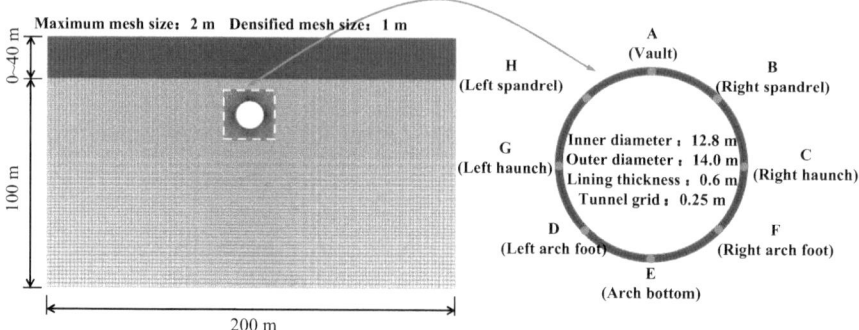

Fig. 4 Finite element meshing of seawater-seabed-tunnel coupling model and schematic diagram of tunnel monitoring points

Table 1 Damage parameters of concrete

Dilation angle	Eccentricity	F_{b0}/f_{c0}	K	Viscosity parameter
36	0.1	1.16	0.6667	0.0005

Table 2 Site material parameters

Type	Density ρ (kg/m^3)	Elastic modulus E (MPa)	Poisson ratio μ	Shear wave velocity v_s (m/s)
Soil	2000	320	0.48	232
Seawater	1000	2340	0.33	–

3.2 Seismic Wave Selection

When the seismic source is far away from the structure, the seismic wave may be nearly vertical incidence. There is a difference in propagation velocity between P wave and SV wave, and the propagation velocity of P wave is about 2 ~ 3 times that of SV wave. To achieve a more accurate simulation of earthquake damage to the structure, 15 s acceleration time history data were intercepted from the 1995 Kobe earthquake (Fig. 5). To comply with the Ministry of Housing and Urban–Rural Development's standards for analysing rare earthquakes in areas with an eight-degree seismic fortification intensity, the original records were subjected to equal-scale amplitude modulation. It is assumed that the arrival time of the P wave is six seconds prior to the S wave. Furthermore, the only external load considered is the incident seismic wave.

Fig. 5 Kobe seismic wave acceleration time history (0.3 g)

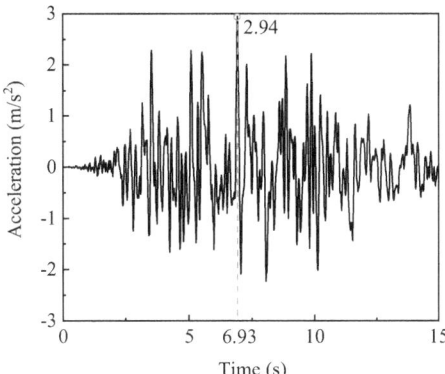

4   Results and Discussion

4.1   Seismic Response of Seabed Surface

Formula 6 defines the peak acceleration, velocity, and displacement ratio (Da, Dv, and Du) as metrics to quantify the impact of water depth on the peak ground motion of the seabed surface.

$$
\begin{cases} Da_x(x) = \frac{\max(|a_x|)}{\max(|a_0|)} \\ Da_y(x) = \frac{\max(|a_y|)}{\max(|a_0|)} \end{cases}, \quad
\begin{cases} Dv_x(x) = \frac{\max(|v_x|)}{\max(|v_0|)} \\ Dv_y(x) = \frac{\max(|v_y|)}{\max(|v_0|)} \end{cases}, \quad
\begin{cases} Du_x(x) = \frac{\max(|u_x|)}{\max(|u_0|)} \\ Du_y(x) = \frac{\max(|u_y|)}{\max(|u_0|)} \end{cases}, \quad (6)
$$

where, a, v and u represent acceleration, velocity and displacement respectively; x is the horizontal coordinate of the monitoring point on the seabed surface; the subscript 0 represents the seismic response of the seabed surface when the seawater depth is 0 m. The vertical incidence of P wave mainly affects the vertical movement of the site (y direction), and the vertical incidence of SV wave mainly affects the horizontal movement of the site (x direction).

Figure 6 illustrates the distribution of seismic response peak ratio in both the vertical and horizontal directions of the seabed surface, considering various seawater depths.

Firstly, as seawater depth increases, the seabed surface accelerates less vertically. This is because seawater propagates compression waves. Seismic waves, specifically P waves, from the seabed can penetrate the seawater and bounce back to the bottom at sea level. The phase change will lower vertical ground motion near seawater frequency. The bottom surface above the subsea tunnel has a lower vertical velocity than the surrounding land site. The bottom surface's vertical seismic response alternates as distance from the subsea tunnel increases. The range of change increases with seawater depth. The incident and scattered seismic waves propagate together to the seabed surface and enter the seawater for back-and-forth transmission and reflection as the submarine tunnels disperse them, resulting in different seismic responses at

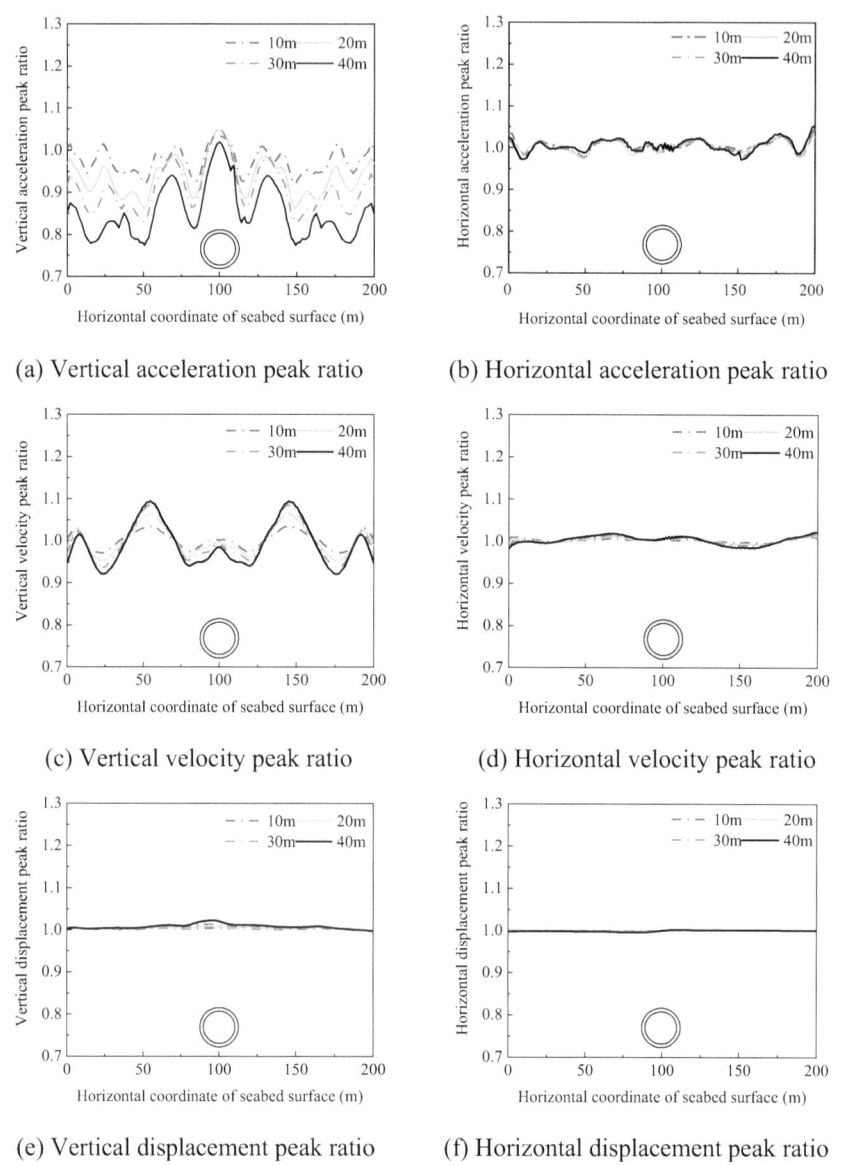

(a) Vertical acceleration peak ratio

(b) Horizontal acceleration peak ratio

(c) Vertical velocity peak ratio

(d) Horizontal velocity peak ratio

(e) Vertical displacement peak ratio

(f) Horizontal displacement peak ratio

Fig. 6 Spatial distribution of peak ratio of seabed surface motion

different locations. Seawater depth has little effect on vertical displacement. Seabed vertical displacement above the tunnel increases little with seawater depth. Additionally, the ratio of horizontal ground motion on the bottom surface remains generally constant as water depth increases. This is due to seawater cannot transmit SV waves.

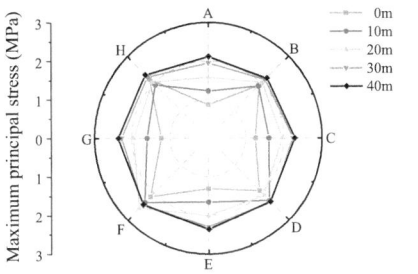 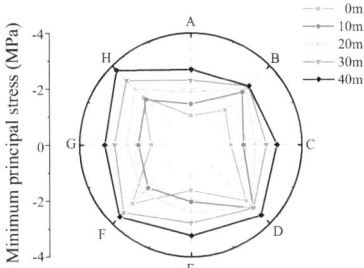

(a) Maximum principal stress peak   (b) Minimum principal stress peak

Fig. 7 Principal stress peak envelope diagram of subsea tunnel

4.2 Stress Response Pattern of Subsea Tunnel

The dynamic stress response pattern of the monitoring points in the subsea tunnel is analysed first. An envelope diagram in Fig. 7 illustrates the maximum and minimum principal stress at each monitoring point. The stress change is considerably apparent at various seawater depths in the subsea tunnel.

Overall, as the depth of seawater increases, the maximum principal stress and minimum principal stress of each monitoring point in the subsea tunnel also increase. Notably, the vault experiences the greatest increase in both the maximum principal stress (146.50%) and the minimum principal stress (159.62%). The larger stress values occurred at the foot and spandrel of the tunnel, with the maximum principal stress extreme value of 2.56 MPa and the minimum principal stress extreme value of − 3.77 MPa, which are the weak points in the seismic design and should be given great attention and strengthening measures.

4.3 Energy Dissipation and Damage Evolution in Subsea Tunnel

Under the impact of earthquakes, the damage and failure of materials in subsea tunnels will accumulate as energy dissipation. The process of damage evolution in the tunnel lining is closely linked to the evolution of damage dissipation energy and plastic dissipation energy. It is also directly connected to the development of cracks. The greater the number of cracks, the larger the damage area. Figure 8 indicates that lining energy dissipation increases prematurely with seawater depth.

The dissipation of damage energy and plastic loss energy displays a tendency towards stability at time t = 15 s. This time period coincides with the end of P wave incidence and the subsiding of SV wave. The total dissipated energy of the tunnel system shows varying extreme values at different seawater depths. These

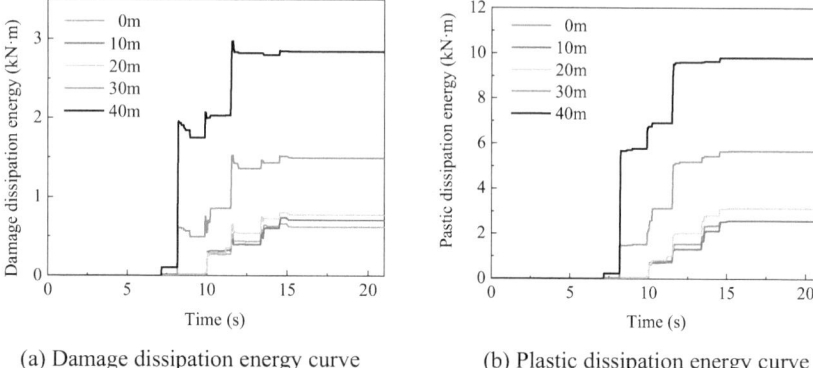

(a) Damage dissipation energy curve   (b) Plastic dissipation energy curve

Fig. 8 Dissipation energy curve

extreme values have been recorded as 2.48 kN m, 2.56 kN m, 3.33 kN m, 7.14 kN m, and 12.64 kN m, respectively. The dissipation energy of nearby seawater depths exhibits a percentage rise of 3.22%, 30.08%, 114.41%, and 77.03%, respectively. It is evident that while the seawater depth is taken in equal proportion, the accompanying alteration in total dissipation energy does not follow a linear pattern, instead, it escalates as the seawater depth increases. Significant increases in dissipation energy can be observed when the depth of seawater exceeds 20 m.

Taking the example of the conditions at a water depth of 40 m, Fig. 9 explains the evolution of the damage area of the tunnel. The concrete stiffness degradation rate index has been chosen to evaluate the damage state of the concrete lining structure, with values ranging from 0 (no damage) to 1 (complete damage).

The process of damage development in the lining under the influence of P waves and SV waves can be described as follows: At t = 8.2 s, there is a limited extent of

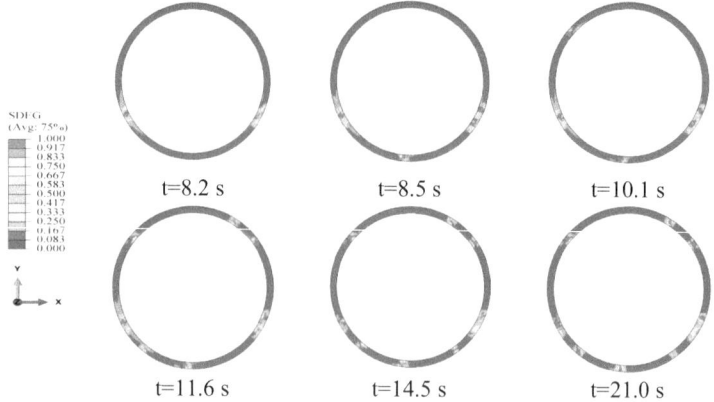

Fig. 9 Damage evolution process of subsea tunnel

cumulative damage observed in the vicinity of both the left and right arch feet. In particular, the left arch foot damage area has a higher concentration and manifests as internal and external penetration damage. At t = 8.5 s, significant deterioration of the lining arch base was observed, with the damage at the right arch foot intensifying and the affected area further expanding. The left spandrel showed structural degradation at t = 10.1 s. At 11.6 s, two significant instances of cumulative damage are observed in close proximity to the right spandrel, as well as a notable occurrence of cumulative damage between the arch bottom and the left arch foot. At 14.5 s, a large range of cumulative damage is observed between the right arch foot and the arch bottom. Damage to the lining remained relatively stable from 15.0 s until the end of the earthquake. The timing of damage at various locations in the lining correlates with the moment of sudden dissipation energy curve alteration.

5 Conclusions

In this paper, the seawater-seabed-tunnel coupling model is established by ABAQUS software. The simulation of the delayed incident process of P-SV waves is conducted. The dynamic response law of the subsea tunnel and the motion law of seabed surface under different seawater depth conditions are discussed. The main conclusions are as follows:

(1) The existence of overlying seawater will inhibit the whole vertical movement of the seabed surface. Affected by the tunnel structure, the vertical seismic response of the seabed surface changes alternately with the increase of the distance from the subsea tunnel.

(2) With increasing seawater depth, the stress experienced by tunnel structure increases. The stress level around the arch foot and spandrel is obviously higher than that of other parts. The weakness in seismic design necessitates more attention and the implementation of reinforcing measures.

(3) This investigation evenly chooses seawater depth, yet tunnel dissipation energy cumulatively does not demonstrate a linear relationship. Tunnel dissipation energy increases as seawater depth exceeds 20 m. Thus, the seawater is crucial to practical engineering.

(4) During the successive incidence of vertical P waves and SV waves, damage to the lining coincides with rapid changes in the dissipation energy curve. The left and right arch feet of the tunnel first show little damage. Thereafter, the cumulative injury between the arch bottom and the arch foot increases significantly. Finally, the spandrel sustains localized damage.

Acknowledgements The authors gratefully acknowledge the support from Natural Science Foundation of Guangdong Province, China (grant number 2023A1515012630).

References

1. W. Zhao, H. Gao, W. Chen, P. Xie, Analytical study on seismic response of subsea tunnels in a multi-layered seabed subjected to P- and SV-waves. Tunn. Undergr. Space Technol. **134**, 105015 (2023)
2. C.B. Crouse, J. Quilter, Seismic hazard analysis and development of design spectra for Maul A platform, in *Proceedings of the Pacific Conference on Earthquake Engineering*, vol. 3 (1991), pp. 137–148
3. D.M. Boore, C.E. Smith, Analysis of earthquake recordings obtained from the Seafloor Earthquake Measurement System (SEMS) instruments deployed off the coast of southern California. Bull. Seismol. Soc. Am. **89**, 260–274 (1999)
4. B. Chen, *Characteristics of Offshore Ground Motions and Seismic Response Analysis of Sea-Crossing Bridges* (Dalian University of Technology, Dalian, 2016)
5. J. Cui, Y. Lu, J. Qu, Y. Li, Influencing factors analysis of seismic responses of water immersed tunnel. J. Southwest Jiaotong Univ. **55**, 1224–1230 (2020)
6. S. Zhu, W. Li, V.W. Lee, C. Zhao, Seismic response of undersea lining tunnels under incident plane P waves. Chin. J. Geotech. Eng. **42**, 1418–1427 (2020)
7. S. Zhu, K. Chen, W. Li, Analytical solution to the scattering of plane SV waves in an underwater double-lined tunnel. J. Xi'an Univ. Archit. Technol. (Nat. Sci. Ed.) **55**, 217–226 (2023)
8. J. Liu, Y. Du, X. Du, Z. Wang, J. Wu, 3D viscous-spring artificial boundary in time domain. Earthq. Eng. Eng. Vib. **5**, 93–102 (2006)
9. J. Liu, Y. Lu, A direct method for analysis of dynamic soil-structure interaction based on interface idea. Dev. Geotech. Eng. **83**, 261–276 (1998)
10. C. Li, H. Hao, H. Li, K. Bi, B. Chen, Modeling and simulation of spatially correlated ground motions at multiple onshore and offshore sites. J. Earthq. Eng. **21**, 359–383 (2016)

Research on Excavation Deformation of Foundation Pit Based on Seepage and Stress Coupling

Linhao Hu, Xuejin Zhao, Zihao Li, and Kunyong Zhang

Abstract In order to study the effect of seepage and stress coupling on excavation deformation of foundation pit, finite element numerical analysis method is used in this paper, taking a hanging water curtain foundation pit project under construction in the rich water stratum area of the Yangtze River floodplain as an example. By comparing the deformation index of the surrounding surface settlement foundation pit with different models, the law of the influence of seepage on the deformation of foundation pit excavation is obtained. The results show that when there is no foundation pit dewatering requirement and the support stiffness is large enough, the settlement caused by the release of ground loss stress is small, and the surface settlement curve around the foundation pit is parabolic. When there is precipitation in the pit and considering the support stiffness is large enough, the surrounding surface settlement caused by foundation pit construction is triangular in distribution. The calculation and analysis of foundation pit excavation considering the coupling of seepage and stress in actual excavation construction show that the surrounding surface settlement curve is a combination of the above two settlement characteristics, and the numerical and curve characteristics are coupled.

Keywords Foundation pit deformation · Seepage and stress coupling · Numerical simulation

1 Introduction

It is expected that by the middle of this century, China will continue to carry out large-scale underground engineering, so a large number of geotechnical engineering and academic problems will emerge, especially about the deep foundation pit of subway

L. Hu · X. Zhao
China Communications Construction Co., Ltd., No. 3 Engineering Co., Ltd. of CCCC First Highway Engineering Co., Ltd., Beijing, China

Z. Li (✉) · K. Zhang
Institute of Geotechnical Engineering Science, Hohai University, Nanjing, China
e-mail: 315433990@qq.com

© The Author(s) 2025
W. Wang et al. (eds.), *Hydraulic Structure and Hydrodynamics*, Lecture Notes in Civil Engineering 608, https://doi.org/10.1007/978-981-97-7251-3_16

stations. Subway stations are often located in busy areas of large and medium-sized cities, surrounded by a wide variety of existing structures, roads and pipelines and other facilities interleaving complex, it is necessary to strictly control the impact of foundation pit excavation on the surrounding environment. At present, underground engineering shows the development trend of "deep, large and long", and multi-layer underground stations are more common, which puts forward higher requirements for subway foundation pit engineering at the technical level. The excavation of foundation pit will inevitably disturb the soil mass, and then cause deformation of the supporting structure, which brings risks to the construction, and also has adverse effects on the surrounding environment [1]. In recent years, foundation pit engineering accidents have occurred frequently, which have brought loss of life and property, and also produced adverse social impacts. Therefore, it is necessary to design the foundation pit support more reasonably, analyze and predict the deformation of the supporting structure more accurately, and ensure the safe construction of the foundation pit, which is always a hot issue in the subway deep foundation pit engineering [2].

Based on the theory of random medium, Yang et al. [3] believed that the surrounding surface settlement caused by foundation pit dewatering was a random process, and established a corresponding random medium model to calculate the final settlement of the surrounding surface caused by foundation pit dewatering, and verified it by engineering examples, demonstrating the rationality and effectiveness of this method Kanghe et al. [4]. assumed the seepage problem in soil caused by foundation pit dewatering as a one-dimensional seepage problem, derived the calculation formula of surrounding surface settlement and stress change in soil layer caused by foundation pit dewatering, and found that the root cause of surface settlement and stress change in soil layer was seepage, and found that the calculation effect was good through application in a foundation pit engineering example. Based on the principle of effective stress of unsaturated soil and the Dupuit hypothesis, Yiqian et al. [5, 6]. Obtained the precipitation curve equation formed by the stability of groundwater level after foundation pit dewatering without considering the lateral deformation of soil caused by precipitation and the influence of group well effect. The ground settlement of unsaturated soil on the upper part of the precipitation curve and saturated soil on the lower part of the precipitation curve is calculated by the layer-summation method. The correctness and effectiveness of the above calculation method are verified by comparing with numerical simulation and field measurement data.

Taking deep foundation pit engineering as the object, Guihong et al. [7]. established a three-dimensional mathematical model of unsteady flow, considered the influence of the construction process and the disturbance of construction personnel on the stability of deep foundation pit, and deduced the soil deformation equation around the foundation pit according to the elastic–plastic constitutive relationship, small deformation hypothesis and Terzaghi's [8–10] effective stress principle. Through the numerical simulation analysis of groundwater seepage and soil stress coupling, the change law of groundwater seepage field around foundation pit under the condition of group well dewatering is studied, and the change characteristics of stress field

Table 1 Physical indicators of soil layer

Soil layer	Elastic modulus (MPa)	Internal friction angle (°)	Permeability coefficient (m/s)	Weight (kN/m³)
Sandy soil	100	27	1.38E-05	18.2
Round gravel	100	36	6.48E-05	20.5

around foundation pit are given considering the influence of foundation pit support. The influence of groundwater seepage on the stability of foundation pit excavation was studied by numerical simulation analysis of the coupling of groundwater seepage and soil stress during excavation of a foundation pit project in Wuhan.

In this paper, GeoStudio finite element calculation software is used to study the deformation and seepage analysis of foundation pit during excavation, which can ensure the stability of foundation pit during excavation construction, so as to ensure the safety of underground engineering itself, and ensure the stability of surrounding buildings and environment.

2 Establishment of Finite Element Model

2.1 Parameter Value

The strata in this area are simplified. The thickness of the soil layer is 50 m and the upper layer is sandy soil with a thickness of 20 m. The lower stratum is a boulder with a thickness of 30 m. The half width of the foundation pit is 20 m, the influence range is 300 m, the excavation depth is 16 m, and the water curtain is inserted 30 m deep. The model adopts a structural beam instead of a suspended water curtain. The length of the structural beam is 29 m, the stiffness is $1e + 10$ kPa, the cross-sectional area is 1 m², and the moment of inertia is 0.083 m⁴. Table 1 shows the parameters of each soil layer.

2.2 Establishment of Finite Element Model

Half of the excavation of the model is taken on the principle of symmetry. The length of the whole model is 320m and the thickness is 50 m. The SIGMA/W module in GeoStudio software was used to numerically simulate the foundation pit model. The whole model was divided into a 1.1 m quadrilateral grid with a total of 17,048 nodes and 13,739 units. Set horizontal constraints on the left and right sides of the model, and set horizontal and vertical constraints on the bottom of the model. The whole construction process is divided into four steps, and the water level is lowered to 1 m at the bottom of the excavation before each excavation. Each time 4 m is excavated,

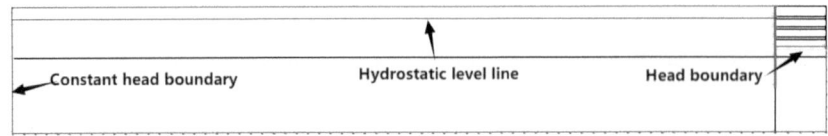

Fig. 1 Model diagram of seepage stress coupling foundation pit excavation

support is set (the support has met the requirements, that is, horizontal constraints are set on the beam).

In order to compare the effects of excavation, dewatering and excavation-dewatering coupling on soil deformation under different working conditions, three calculation conditions are set up respectively: In the excavation model of excavation-precipitation coupling foundation pit, the initial water level line is set two meters below the surface before excavation to simulate the diving level. In order to consider the incomplete well flow effect during excavation, the constant water head boundary is set at the left boundary of the model from two meters below the surface to the bottom of the model, as shown in Fig. 1. As the excavation model only considers the influence of the excavation foundation pit on the soil, no water level line and water head boundary are set, as shown in Fig. 2. In the precipitation model, the clear water level line, the constant water head boundary from two meters below the surface on the left to the bottom of the model, and the constant water head boundary 17 m below the surface on the right side of the water seal curtain are set to simulate the precipitation in the foundation pit, as shown in Fig. 3.

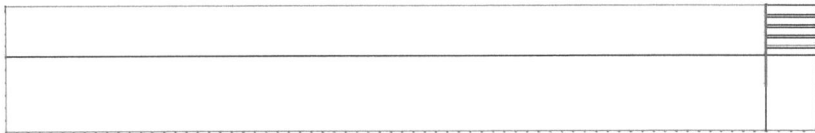

Fig. 2 Only considering excavation foundation pit model diagram

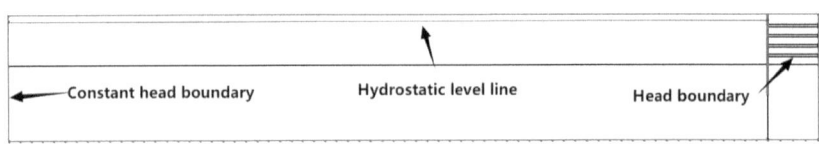

Fig. 3 Only considering the dewatering foundation pit model diagram

3 Analysis of Calculation Results

3.1 Only the Excavation Foundation Pit Model is Considered

When the excavation of the foundation pit is completed, the vertical displacement cloud map of the surrounding strata outside the foundation pit is calculated, as shown in Fig. 4.

According to Fig. 4, the vertical displacement curve of the surface outside the foundation pit is obtained, as shown in Fig. 5.

From Fig. 5, it can be seen that the maximum settlement of soil outside the foundation pit is 7.5 mm, and the further away from the foundation pit, the smaller the settlement value, but the maximum settlement is not in the support position, which is because the top of the support structure is supported by concrete, that is, the so-called parabolic settlement curve.

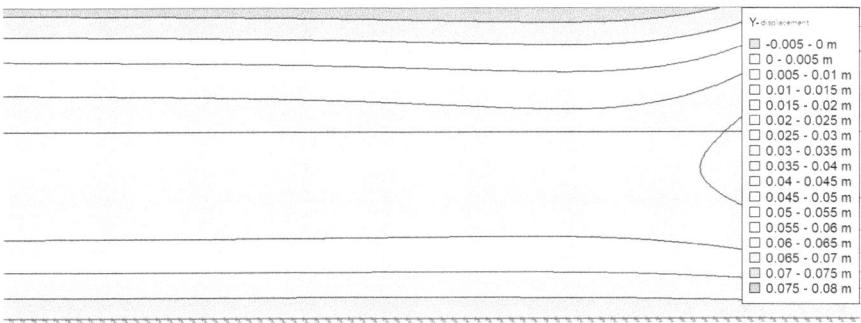

Fig. 4 Only considering the vertical displacement cloud map of the excavation model surface (m)

Fig. 5 Only surface settlement curves around the excavation model are considered

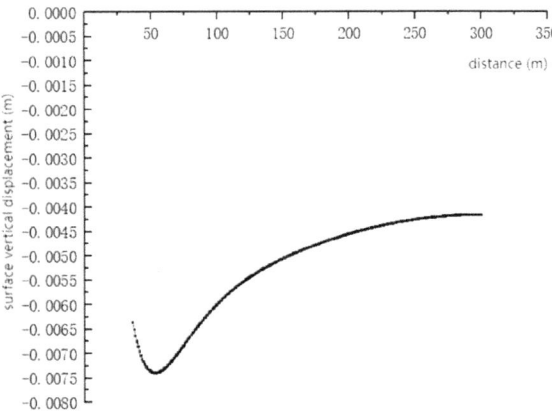

3.2 Only Dewatering Foundation Pit Model is Considered

When the foundation pit dewatering is completed, the vertical displacement cloud map of the surrounding strata outside the foundation pit is calculated, as shown in Fig. 6.

According to Fig. 6, the vertical displacement curve of the surface outside the foundation pit is obtained, as shown in Fig. 7.

From Fig. 7, it can be seen that the maximum settlement of the surface outside the pit is 0.24 m, and the settlement curve is approximate to a straight line. The maximum settlement is at the nearest point, because the precipitation at this location is the most severe, independent of the constraint conditions of the support structure, so it is called a triangular distribution, similar to the settlement of cantilever support excavation.

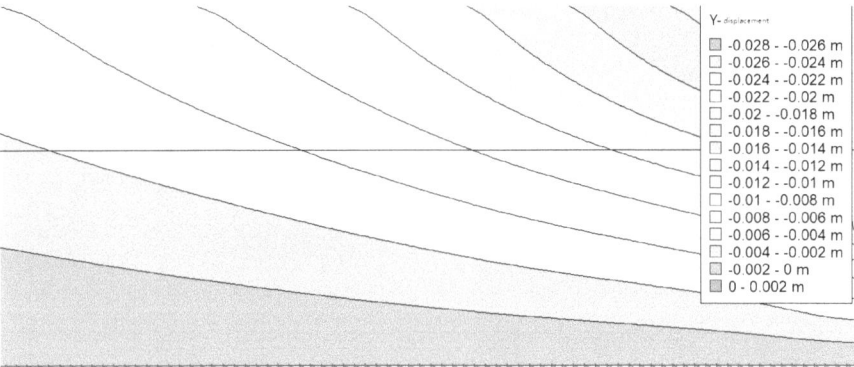

Fig. 6 Only considering the vertical displacement cloud map of the precipitation model surface (m)

Fig. 7 Only the surface displacement curve around the precipitation model is considered

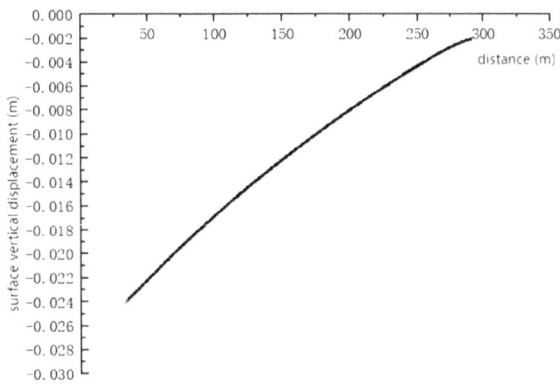

3.3 Seepage Stress Coupling Foundation Pit Excavation Model

When both excavation and dewatering are completed, the vertical displacement cloud map of the surrounding strata outside the foundation pit is shown in Fig. 8.

According to Fig. 8, the vertical displacement curve of the surface outside the foundation pit is obtained, as shown in Fig. 9.

From Fig. 9, It can be seen that the maximum surface settlement outside the pit is 0.25 m, and the settlement curve is approximately straight and the maximum settlement is at the nearest place. This is because the precipitation is the most intense at this position, which has nothing to do with the constraints of the retaining structure, so it is the so-called triangular distribution, similar to the settlement of cantilever support excavation.

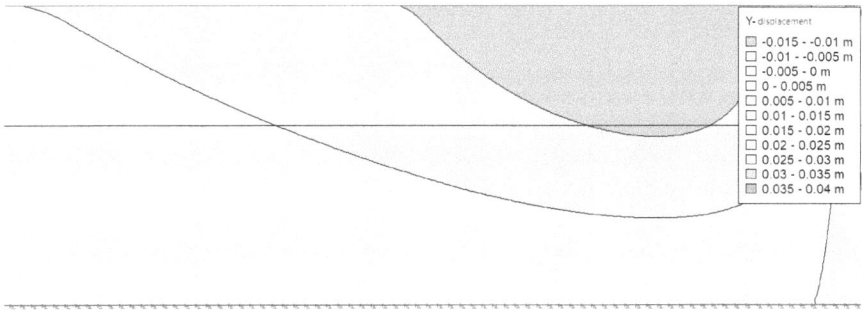

Fig. 8 Surface vertical displacement cloud map considering seepage stress coupling model (m)

Fig. 9 Consider the surface displacement cloud map around the coupled seepage stress model

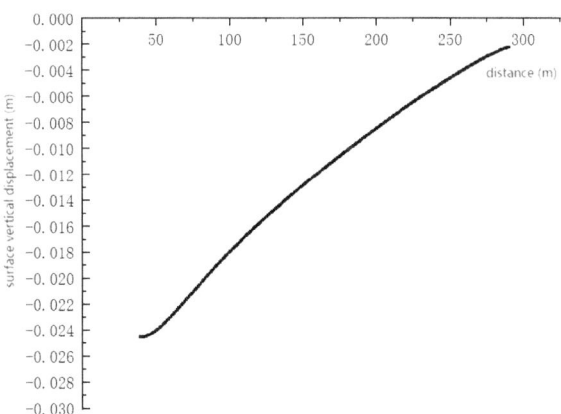

Fig. 10 Surface displacement curves around different models

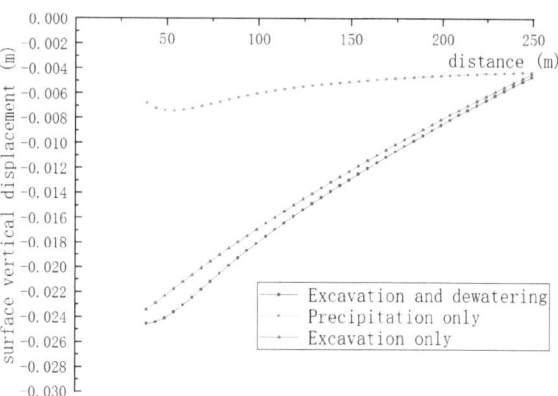

3.4 Comparison of Surface Subsidence Around Different Models

By organizing, the comparison curves of surface subsidence around different models can be obtained, as shown in Fig. 10.

From Fig. 10, it can be seen that the foundation pit excavation model without considering seepage has the smallest surrounding surface settlement, while the foundation pit excavation model with considering seepage stress coupling has the largest surrounding surface settlement. When approaching the edge of the foundation pit, only the slope of the surrounding surface settlement curve of the precipitation model increases, while the slope of the surrounding surface settlement curve of the foundation pit excavation model considering seepage stress coupling decreases.

4 Conclusions

Through the analysis of the calculation results of the surrounding surface settlement of the foundation pit model under different conditions, the conclusions are as follows:

(1) Through the above comparative analysis, it is concluded that the precipitation has a great influence on the deformation around the foundation pit. In the actual construction process, when the support stiffness is large enough and no precipitation is needed, the excavation of foundation pit has little influence on the surrounding surface settlement, that is, the settlement caused by the release of ground loss stress caused by excavation is small. In the water-rich stratum area, even if the support structure stiffness is very large, the precipitation will cause the surrounding stratum to settle, resulting in the foundation pit deformation.

(2) In the construction process of the Yangtze River floodplain rich water formation area, the dewatering process should be strictly controlled, according to demand dewatering, graded dewatering to avoid excessive dewatering which will cause

excessive deformation of foundation pit and cause construction danger. When the ground settlement caused by the decrease of the ground water level has an impact on the surrounding environmental security, the method of recharge can be considered to control the groundwater level of the aquifer near the building that is sensitive to the ground settlement, so as to avoid the influence of construction precipitation on the sensitive surrounding buildings.

References

1. X. Hongzhong, C. Wensen, H. Wenjie, Deformation behavior analysis of deep foundation pit in Nanjing Metro station [J]. J. Disaster Prev. Reduction Eng. **38**(04), 599–607 (2018)
2. X. Xirong, W. Lifeng, W. Kang et al., Analysis and prediction of underground deep foundation pit excavation behavior in soil-rock composite strata [J]. J. Undergr. Space Eng. **16**(S1), 247–254 (2020)
3. Y. Junsheng, L. Baochen, *Surface movement and deformation caused by urban tunnel construction [M]* (China Railway Press, Beijing, 2002)
4. L. Yuqi, Z. Jian, X. Kanghe, The time effect of foundation pit engineering considering the coupling effect of seepage and deformation. J. Hydraul. Eng. **37**(6), 694–698 (2006)
5. W. Yiqian, *Study on Deformation Caused by Dewatering of Deep Foundation Pit of Subway Station in Diving Area [D]* (Lanzhou University of Technology, Lanzhou, 2016)
6. W. Yi-Qian, Z. Yan-Peng, An improved algorithm of ground settlement induced by foundation pit dewatering in diving area [J]. J. Zhejiang Univ. (Eng. Technol.) **50**(11), 2188–2197 (2016)
7. P. Guihong, W. Jun, L. Jianjun et al., Numerical simulation of seepage stress combination in deep foundation pit excavation [J]. Chin. J. Rock Mech. Eng. **23**(2), 4975–4978 (2004)
8. K. Terzaghi, Settlement and consolidation of clay [J]. Edbaummechanic **95**(3), 1–10 (1925)
9. K. Terzaghi, *Theoretical Soil Mechanics M* (Wiley, New York, 1943)
10. C.-Y. Ou, J.-T. Liao, W-L. Cheng, Building response and ground movements induced by a deep excavation **50**(3), 209–220 (2000)

Study on Urban Flood Control Strategies in River Valley Plain Areas Under Multiple Factors Constraints

Ziqian Huang and Tugui Fan

Abstract The valley plain has the characteristics of both mountainous hills and plains, with special terrain conditions and frequent floods. In order to improve the drainage capacity of the river valley plain, this article takes the Bihu Plain in the Oujiang River Basin as an example, uses MIKE11 one-dimensional hydrodynamic software to establish a mathematical model of the river network, analyzed the flood and drainage in the river valley plain, studied the impact of Traditional Flood control mode and Flood control mode based on sponge cities concept in regional drainage, analyzes the advantages and disadvantages of different flood control strategies in practical cases, and provides solutions for regional flood control, Summarized and clarified the relationship between urban flood control and watershed flood control.

Keywords River valley city · Flooding and waterlogging control · Sponge city · MIKE11

1 Introduction

Valley is a trough-shaped zone caused by geological processes of the river on the surface [1]. The topographical characteristics of valley are mainly high on both sides and low in the middle, with a relatively flat terrain and minimal slope. Both sides of valley are surrounded by mountains, and the river enters the valley from the mountainous section, with the river slope decreasing sharply. The Bihu Plain is located in the middle mountainous area of southern Zhejiang. The Bihu Plain is the location of Bihu New Town in Lishui City, and is the future sub-center of Lishui City.

At present, the research on urban flood control mainly focuses on the combination of traditional water conservancy engineering methods, the transformation of urban undersurface under the guidance of the sponge city concept, and the traditional way of municipal storm sewer system transformation, Fajun et al. [2] used traditional

Z. Huang (✉) · T. Fan
Zhejiang Guangchuan Engineering Consulting Co., Ltd., Hangzhou, China
e-mail: 283345127@qq.com

© The Author(s) 2025
W. Wang et al. (eds.), *Hydraulic Structure and Hydrodynamics*, Lecture Notes in Civil Engineering 608, https://doi.org/10.1007/978-981-97-7251-3_17

189

methods such as newly opened river channels, set up control gates, and river drainage pumping stations to build an urban flood control and drainage system in Yuyao City with river drainage as the main body; Lin et al. [3] proposed to divert rainwater from the urban drainage system of Wenshan City, Yunnan Province to increase the drainage capacity of rainwater drainage pipes; Jinhui et al. [4] adopted the low-impact development concept to simulate the impact on surface runoff under different LID facilities. The research of Xiang et al. [5] shown that the essence of urban flood control problem is that urban drainage and flood control of outer rivers are contradictory, in most cases, urban waterlogging and outer river floods rise and fall at the same time, and it is difficult to take into account both flood control and drainage. This article takes the Bihu Plain as an example, constructed a one-dimensional hydrodynamic model, simulates the water flow of urban drainage in river valley plains under different flood control modes, and gives flood control solutions that met with various external constraints.

2   Construction of Research Area Model

2.1   Model Generalization

The interior of the plain is a spindle-shaped river network system, with major rivers including the Xinzhi River, Dalongyuan River, Ruoxi River, Jincunxi River, Cangkengxi River, Gaoxi River, Zhuxi River, Shangenxi River, Langqixi River, and Tongji Canal system. The Xinzhi River is located in the northeastern part of the plain and is the backbone drainage channel of the plain, receiving all floods and water from the plain. There are currently 2 reservoirs upstream of the Bihu Plain, namely Gaoxi Reservoir and Langqi Reservoir, both of which have no flood control function. According to the distribution characteristics of the river system, flood source and hydraulic structures in the Bihu Plain, author generalized 28 rivers 13 dams 4 regulating gates and 26 bridges within the calculation scope, the generalized river is totally 88.7 km length, 689 calculation sections is included. The calculation sections, bridges, dams, etc. were all based on actual measurement data, and the upper boundary of the model was designed using corresponding frequency flood processes. After considering the moderating effect of the reservoir, the discharge flow of the reservoir was used as the inflow boundary, and the lower boundary of the model was designed using the corresponding frequency flood process of Daxi River.

2.2   Model Construction

This study uses the Mike11 model for numerical simulation. The Mike11 model can simulate the water flow situation in subgroups and supercritical conditions under the

Table 1 Calculate the range of the regional rough rate (after calibrated)	Serial No.	River section and characteristics	Rough rate range
	1	Straight	0.025 ~ 0.033
	2	Bending	0.033 ~ 0.038
	3	Hetan beach	0.038 ~ 0.045

condition of numeral water flow. From river channels to the estuary affected by the tidal, the models have good adaptability and have been widely used worldwide. The model is based on the limited difference method, and the hydration model of the continuous equation and the dynamic volume equation. The model uses the six point Abbott Ionescu finite difference scheme for the calculation of unsteady flow in rivers and estuaries, One dimensional open channel unsteady flow differential equation (that is Saint Venant's equation).

MIKE Abbott Ionescu finite difference scheme calculates the water level or flow alternately in sequence at each grid point, sets the water level point h at the location of the river cross-section, and automatically insert the flow point Q between the two h points. This calculation model has a total of 1737 h points and 1326 Q points in this study.

2.3 Model Parameter Rates Calibrated

Due to historical reasons, the Bihu Plain has few records of floods, with only a small amount of pictures and video materials, and there is no historical flood data and flood records in the local area. After the analysis of hydroponics designers, the large floods in the Bihu Plain in recent years include "2014-08-20" floods, "2016-09-28" floods, and "2019-08-09" floods. Therefore, this study collected historical flood data through on-site visits and measurements. Through on-site surveys, a total of 18 flood points were obtained, and the historical flood fields were "2016-09-28" floods. This study gives different rough rates on different rivers, and adjusts on the basis of the initial value (0.035). The rough rate used in the models sees Table 1.

2.4 Rate Calibration Results

Comparing the surveyed "2016-09-28" flood levels with the model-calculated water levels, by adjusting the model parameters appropriately, the model's calculated values are close to the surveyed values. The calibration error is between 0.01 m and 0.08 m, meeting with the design requirements.

2.5   Current Flood Prevention and Drainage Capacity Analysis

Author used a calibrated model to analyze the current flood control and drainage capacity of the Bihu Plain. The simulation results show that the range of the mountain flood control area in the study area is near the banks and on the north of Xinzhi River, and the range of the waterlogging dominant zone is south of Xinzhi River. Villages are scattered throughout the Bihu Plain, and Bihu Town is completly located in the midlands and southern part of the Bihu Plain, belonging to the waterlogging dominant zone. And there is insufficient drainage capacity of Tongji irrigation canal in the midlands and southern plains. The river drainage capacity in Bihu Town are 10–15 years recurrence. Villages is 5 ~ 8 years recurrence. It is a large gap between the standard of drainage in the planned of town in 20 years and villages in 10 years.

3   Flood Control Strategies and Effect Evaluation

3.1   Strategies and Ideas for Flood Control in Waterlogged Areas

The main water system in the BiHu plain is the canal system of Tongjiyan Irrigation District. Tongjiyan Irrigation District is a national important historical relic protection unit and World irrigation engineering heritage site. The project was established during the Southern Dynasty (502–519 AD). It is one of the famous large-scale water conservancy projects in ancient China, In China, the historical relic protection unit shall designate a protection zone and a construction control zone. The boundary of the protection zone of the Tongjiyan Canal System shall be extended by 10 ~ 30 m from the location of the historical relic [6]. In addition, except the urban construction and development land, a large amount of permanent basic farmland is also involved in the plain area. In summary, the contradiction between regional urban waterlogging control, historical relic protection, and land use is very prominent. Jameson [7] He suggests that flood management should focus on the synergy between rainwater harvesting, rainwater drainage networks, and land use planning. Based this reality, propose 2 different flood control modes.

(1)   **Traditional flood control mode (Mode 1).**

The formulation of flood control strategies follows the traditional water control approach. The study bases on the characteristics of the water system in the Bihu Plain, it is proposed to excavating new river channels, water system connectivity, excavating XinBi River and Weicun Canal as drainage channels, and connecting rivers along the line. Xinbi River and Xinzhi River form a plain drainage system with the "Xinbi River-Xinzhi River" as the main river channel; The Weicun Canal is located in the southern part of the Bihu Plain, extending from north to southeast. It starts from a branch canal near the south of Lianhe Street in Bihu Town in the north and ends at a large stream southeast of Sanfeng Village in the south. The engineering layout of mode 1 is shown in Fig. 1.

(2) **Flood Control Mode Based on Sponge City Concept (Mode 2).**

Based on the theory of the "nature-society" binary water cycle system in river basins, ecological sponge basin construction is carried out to promote the harmonious complementarity of natural and social water cycles. This is not only a need for comprehensive river basin management, but also an inevitable trend of human civilization development [8]. Since the rise of the concept of sponge cities in 2011, scholars have shown an exponential growth in their research on this topic. Hamidi [9] conducted statistics on sponge city publications after 2015 by country, and found that due to the rapid development of cities in China, the number of related studies worldwide reached 178. In the new period, the construction of water conservancy infrastructure needs to continuously expand its thinking, actively learn advanced concepts, and integrate with other subject.

Researchers are exploring new urban drainage methods, such as Abhinav [10] in India, which uses green roofs, infiltration ditches, roof disconnection, and permeable paving to compare and analyze the drainage effects. Author proposes a waterlogging control mode with the concept of sponge city to study the feasibility of integrating

Fig. 1 Layout of new excavated river channel scheme

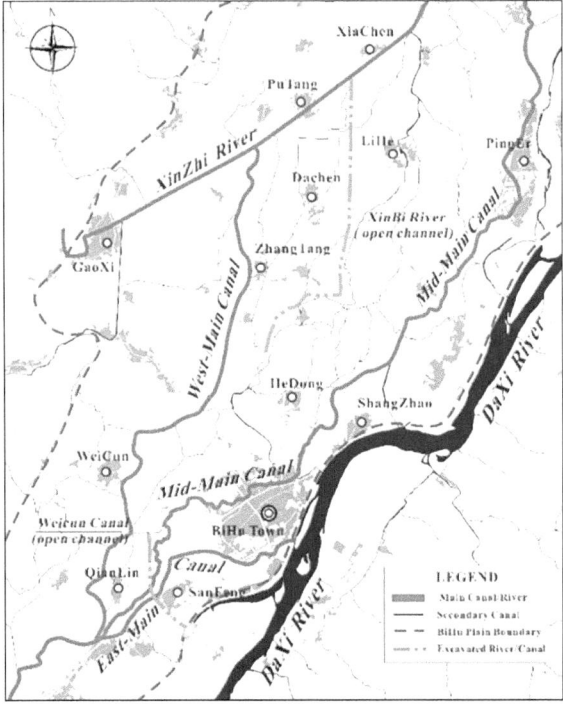

Fig. 2 Layout of
underground drainage
pipeline scheme

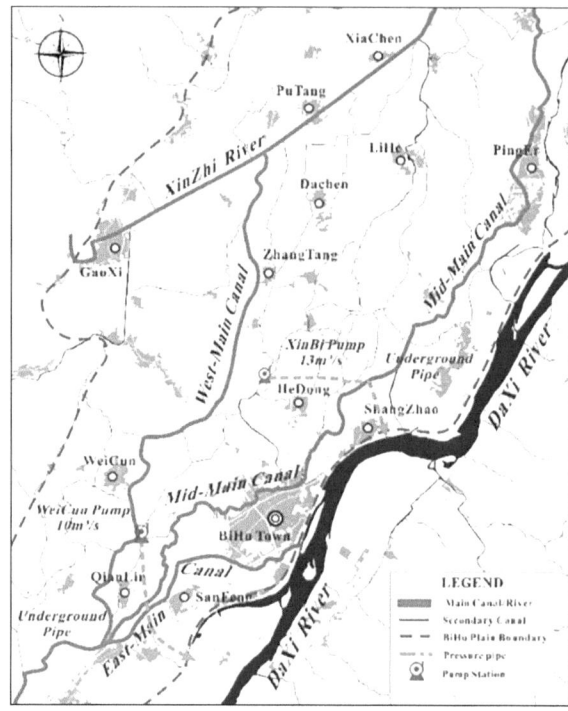

waterlogging control with historical relic protection and farmland protection. The
engineering plan is as follows: an underground forced drainage pump station is set up
in Zhoucun and Weicun in the north and west of Bihu Town. The drainage pipeline
is set at 8–10 m underground. During the flood season, waterlogging water enters
the underground drainage system directly from the river, and is discharged directly
to the outer river through the pump station and pipeline. This utilizes the advantages
of modern engineering technology, while avoiding the protection of permanent basic
farmland and historical relic. The engineering layout of mode 2 is shown in Fig. 2.

3.2 Assessment of Renovate Effectiveness

(1) Evaluation of Flood Control and Drainage Effects.

From the water level calculation results, it can be seen that the traditional flood
control mode of the newly opened river can reduce the water level of the river by 0.2
~ 0.6 m. By adopting the sponge flood control mode of the drainage pump station
on the side of the river, the water level of the river can be reduced by 0.1 ~ 1.1 m.
The engineering effect is good. When the region encounters a 20-year recurrence
rainstorm, the project drainage water under the traditional flood control mode is

3.24 million m³, and the project drainage water under the new water control mode is 4.37 million m³, which is higher than the traditional water control mode. Secondly, after the implementation of the project, the drainage water volume of Weicun Pump Station and Xinbi Pump Station accounted for about 62% of the total drainage water volume in the drainage system, indicating their significant contribution to regional drainage.

(2) **Comparative Evaluation of Advantages and Disadvantages**.

The comparison mainly focuses on construction investment, engineering technical difficulty, policy handling difficulty, operations management, and integration with urban construction. Mode 2 occupies only 0.33 ha (Mode 1 is 4 ha) of Permanent basic farmland, Involves 0.01 km (Mode 1 is 0.4 km) of historical relic. Mode 2 doesn't change the basic style of urban water system, organically integrates with the construction of sponge cities with good compatibility. The disadvantages of Mode 2 are it has difficult managements and more construction investment.

4   Conclusion and Prospect

The traditional flood control mode has disadvantages such as large land area, unfriendly urban construction, and vulgar management methods. The flood control mode based on sponge city concept has advantages such as small land area and good urban integration. Setting up drainage channels underground is a new idea for urban three-dimensional water control, and is also an important manifestation of sponge city construction.

For mountainous and valley cities, plains are still a part of the river basin. As a part of the "natural-society" binary water circulation system, cities are inseparable. It is necessary to follow the flood control and waterlogging mode of large-scale water engineering. Innovation in the flood control mode should be carried out within the overall framework of large-scale water engineering control in the river basin.

This study lacks comprehensive indicators. In the future, it is necessary to explore the establishment of a evaluation index system in combination with regional water ecological governance and urban overall planning.

References

1. Hohai University, *ShuiLi Dacidian*. Shanghai Century Publishing (Group) Co., Ltd., Shanghai (2015)
2. Z. Fajun, W. Lingmin, From the comparison of fite and fireworks two typhoons to summarize the Yuyao model of urban flood control. Flood Drought Disaster Prev. **21**, 41–43 (2021)
3. C. Lin, H. Kun, S. LiHua, C. Yang, K. Zhong Gui, (2020) Problem in comprehensive drainage and waterlogging planning of mountain cities in southeast China discussion on the countermeasure. Water Wastewater Eng. **46**, 687–691

4. H. JinHui, P. AoXuan, S. YingNa, Optimization of low-impact development scheme for communities. Pearl River. **44**(1), 116–122 (2023)
5. L. Xiang, J. Qiu, Z. SiYuan, L. ZhenYu, L. XiChun, Research on urban waterlogging problem intypical city of plain area: case of Huarong County, Hunan province. Yangtze River **51**(9), 22–27 (2020)
6. Z. Lu, et al., *LiShui TongJiYan Portecting Planning*. ZheJiang Historical Architecture Design & Research Institute. Hangzhou (2018)
7. S. Jameson, I.S. Baud, Varieties of knowledge for assembling an urban flood management governance configuration in Chennai. Habitat Int. **54**, 112–123 (2016)
8. W. Hao, J. YunZhong, Concept and thinking of ecological sponge basin construction. Yellow River News **06**(1), 25 (2016)
9. H. Ali, R. Bahman, S.A. George, Sponge City—an emerging concept in sustainable water resource management: a scientometric analysis. Resour., Environ. Sustain. **5**, 100028 (2021)
10. W. Abhinav, et al., Low-impact development (LID) control feasibility in a small-scale urban catchment for altered climate change scenarios. Hydrol. Sci. J. **68**, 1881–1894 (2023)

Study on Progressive Failure Mechanism of High and Steep Overburden Slope Under Rainfall

Xing Zhai, Chenxi Li, Yu Liu, and Hui Li

Abstract This paper summarizes the typical landslide characteristics and main triggering factors by statistically analyzing the detailed investigation data of 51 landslides in Hebei Province, among which rainfall-induced cover-type landslides are the most typical form of landslides. In order to study the destabilizing and destructive mechanism of high steep cover type slope under the action of rainfall infiltration, taking a cover type slope as an example, based on the theory of unsaturated seepage and shear strength, the changing rules of seepage field, displacement field, plasticity zone, and sliding cracking surface of the cover type slope under the condition of intermittent rainfall were analyzed by using the slip surface stress method, and the safety coefficient of the slope under the action of rainfall was calculated. The results show that: after the rainfall starts, because the shallow soil layer of the slope is mainly coarse gravel soil and gravel soil, with relatively strong permeability, the rainwater on the slope surface infiltrates into the interior of the slope body and softens the soil body along the soil-rock interface, and the soil body of the slope body produces displacement at the leading and trailing edges, and shallow sliding is formed in the slope; with the continuation of the rainfall, the slope produces a through plastic zone, and the soil body of the leading edge pulls the central pile body to produce traction slip, and the trailing edge of the soil body produces traction slip, and the trailing edge of the soil body produces traction slip. As the rainfall continues, the slope produces through plastic zone, the leading edge soil body pulls the central pile body to produce traction slip, the trailing edge soil body pushes the central pile body to produce nudging slip, and the deformation gradually develops from the surface to the depth, from the leading edge to the trailing edge.

Keywords Intermittent rainfall · Analysis of stability · Progressive destruction · Safety factor

X. Zhai (✉) · C. Li · Y. Liu · H. Li
Hebei Key Laboratory of Geological Resources and Environment Monitoring and Protection, Hebei Geo-Environment Monitoring, Shijiazhuang, China
e-mail: zhaib8188@163.com

C. Li
School of Civil Enginering, Shijiazhuang Tiedao University, Shijiazhuang, China

© The Author(s) 2025
W. Wang et al. (eds.), *Hydraulic Structure and Hydrodynamics*, Lecture Notes in Civil Engineering 608, https://doi.org/10.1007/978-981-97-7251-3_18

1 Introduction

Hebei Province is a province in China where geological disasters occur frequently on the side (slippery) slope. Under the action of heavy rainfall in flood season, a large number of slopes in stable state under natural conditions are easy to produce geological disasters such as slide and instability, which poses a great threat to people's property safety and surrounding environment. Therefore, it is of great significance to study the progressive failure mechanism of slope under rainfall conditions for the prevention and control of slope geological hazards.

Slope stability analysis under rainfall conditions has always been one of the hot and difficult problems in the field of geotechnical engineering, and many scholars at home and abroad have conducted a lot of discussions and studies, and achieved certain results. Li et al. [1] made a statistical analysis of the survey results of several geological disasters across the country and found that about 90% of landslides were caused by heavy rains. Guangcai et al. [2] combined limit equilibrium method and finite element analysis to find that heavy rainfall has a significant impact on slope stability. Xiaojie et al. [3] analyzed the rainwater infiltration rule and slope deformation characteristics during heavy rainfall by using real-time slope monitoring combined with finite element numerical simulation, and pointed out that the occurrence of landslide has a lag and is related to the intensity and frequency of heavy rainfall in 1 to 3 days. Zhongming et al. [4] used FLAC3D software to compile the FISH function of slope rainfall infiltration and post-rain outflow, and completed the numerical simulation of rainfall infiltration. Yi [5], Yabing [6], Jiancong [7], Bin [8], Kailun [9], Kaihuan [10], Hua [11] et al. studied the deformation rule of slope under the action of rainfall and carried out stability analysis through numerical simulation. In addition, the slope model test is also one of the research methods commonly used by many scholars. Longqi et al. [12] analyzed the influence of rainfall infiltration on slope by conducting slope mechanical model tests with different rainfall intensity and different supporting conditions, and pointed out that the influence factor of slope stability under short-term rainstorm is the accumulation and dissipation of excess pore pressure, while the influence factor of rain infiltration to soften the weak interlayer is the influence factor of slope stability under long-term light rain. Zhenming et al. [13] studied the deformation and failure mechanism of accumulation slope under the action of rainfall through laboratory model tests, and found that the key factor affecting the stability of slope under the action of rainfall was to change the pore water pressure inside the slope. Through the indoor artificial rainfall model test, Hui et al. [14] found that gentle slope is easier to start sliding than steep slope, and the infiltration depth of slope foot is the largest, followed by slope waist, and slope foot is the smallest.

At present, many scholars at home and abroad have used numerical simulation and physical model tests to study slope stability analysis, influence mechanism and prevention measures of unsaturated soil under the action of rainfall, but most of the research is limited to homogeneous single-layer soil slope, and there are few studies on the progressive deformation and failure mechanism of multi-formation slope under the action of rainfall infiltration. Based on a covered slope in Hebei Province, this paper established a two-dimensional numerical model considering the influence of rainfall infiltration on slope stability, studied the changes of pore water pressure, displacement field, slip surface, plastic zone and safety factor of the covered slope under intermittent rainfall conditions, and then analyzed the progressive failure mechanism of the slope under rainfall conditions. It provides theoretical reference for the support of high and steep slope and the prevention and control of geological hazards.

2   Main Characteristics and Cause Analysis of Typical Landslides in Hebei Province

2.1   Statistics of Typical Landslide Characteristics

In 2020, a total of 51 hidden landslide points in the province have been surveyed. According to statistics, the slide body thickness of the major landslide disasters in this province is mostly shallow layer within 10 m, and most of them are old landslides formed since the Holocene, which are stable or basically stable under natural conditions. The main features are as follows:

Slope Height Feature

The height difference is mainly between 20 and 40 m, among which, 1 place with a relative height difference < 10 m, 5 places with a relative height difference between 10 and 20 m, 9 places with a relative height difference between 20 and 30 m, 20 places with a relative height difference between 30 and 40 m, 3 places with a relative height difference between 40 and 50 m, 3 places with a relative height difference between 50 and 60 m, 5 places with a relative height difference between 60 and 70 m, and 3 places with a relative height difference between 70 and 80 m. >80 m in 3 places.

Lithologic Characteristics

The number of soil landslides is large and widely distributed, and the lithology of the sliding body is mainly composed of debris, slope sediment and alluvial sediment, accounting for about 81%, and a small amount is loess accumulation layer landslide. The sliding surface is mostly the base covering interface, and the lithology of the sliding zone is the bottom viscous material in the gravel soil, usually silty clay, clay and silty soil. There is no obvious sliding zone, and the lithology of the sliding zone is the same as that of the underlying overlying bedrock.

Deformation Movement Characteristic

Most of the sliding modes are the whole traction type of sliding body, with a total of 29; Push type followed, a total of 24; It basically slides along the slope in the same direction as the slope, mostly in the southeast to southwest direction. The three landslides are in the creep stage, and the deformation characteristics are mainly cracks at the back edge or front edge, while a few landslides have cracks in the middle.

2.2 Cause Analysis of Landslide

Most of the key landslide points in Hebei Province occur in flood season, and rainfall is the main adverse factor of landslide. Loose slope deposits are conducive to rainfall infiltration. Rainfall infiltrates and migrates through the pores of the loose layer, filling the slope with water, increasing the body weight of the soil, softening the sliding body, and reducing the mechanical strength of the soil in the sliding zone. Cause the sliding body to deform. According to the deformation signs, there is no deformation sign in the natural state of the slope. In the case of heavy rain or continuous rainfall, the erosion of surface runoff on the slope leads to the collapse of local small rock and soil mass or even the overall slide.

3 Numerical Simulation

3.1 The Establishment of Computational Model

In this paper, a landslide in the northwest of Hebei Province is selected as the research object. The landslide is a covered slope with loose accumulation layer on the upper part and bedrock on the lower part. The landslide height difference is about 50 m and the axial length is about 115 m. Under the influence of heavy rainfall, partial collapse of the landslide front occurs, and dense cracks appear in the upper part of the landslide front, which then drags the whole slope to slide. The simplified two-dimensional geological model is shown in Fig. 1.

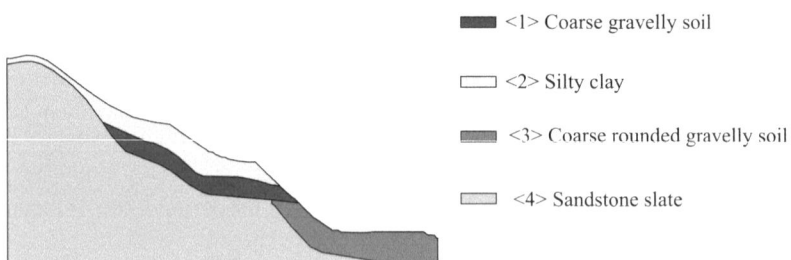

<1> Coarse gravelly soil

<2> Silty clay

<3> Coarse rounded gravelly soil

<4> Sandstone slate

Fig. 1 Numerical model diagram of slope

Considering the rainfall infiltration conditions, SEEP/W, SLOPE/W and SLOPE/W modules of Geostudio, a two-dimensional unsaturated soil seepage analysis software, are used in this paper to study the overburden slope. SEEP/W module mainly performs saturation-unsaturated seepage analysis under rainfall conditions, and automatically saves pore water pressure distribution at each transient moment according to the preset time step size. The pore water pressure at different time steps was imported into SIGMA/W module for pore-stress coupling analysis, and then the analyzed stress and pore water pressure were coupled into SLOPE/W module for stability analysis, and the corresponding safety factor was calculated.

3.2 Rock and Soil Mass Parameters and Boundary Conditions

Physical and mechanical parameters of each rock and soil layer required for this numerical analysis were determined through site investigation and reference to relevant data (Table 1), and the residual water content was 10% of the saturated water content. According to the saturated volume water content and permeability coefficient of soil layer, the saturation permeability coefficient and saturated volume water content of rock and soil mass on slope are estimated by Van Genuchten method in Geostudio software.

Boundary conditions: a fixed water head boundary is applied on the left and right sides of the slope, and an impervious boundary is set at the bottom of the slope. When the saturation permeability coefficient of soil is greater than the rainfall intensity, the slope surface is the flow boundary, and when the saturation permeability coefficient is less than the rainfall intensity, the slope surface is the pressure head boundary.

Rainfall condition: Considering the infiltration condition of the slope under continuous rainfall, the numerical simulation was divided into 8 intermittent rainfall tests. Each time the rain fell for 18 h, and the rain stopped for 6 h. Combined with the local meteorological observation data, 60/() is selected as the rainfall condition in this paper. After the completion of the 8 rainfall tests, the rain stop observation was continued. The entire rainfall simulation lasted for 9 days.

4 Analysis of Calculation Results

4.1 Pore Water Pressure Field Analysis

The SEEP/W module was used to simulate transient seepage under intermittent heavy rainfall conditions, with rainfall intensity of 60/(). Because the saturation permeability coefficient of shallow soil is less than the rainfall intensity, the pressure

Table 1 Physical and mechanical parameters of each soil layer

Floor number	Geotechnical type	Severe/ ($kN \cdot m^{-3}$)	Cohesive force/ kPa	Friction Angle/ (°)	Poisson's ratio	Modulus of elasticity/kPa	Saturation permeability coefficient/(m/s)	Saturated moisture content/%
1	Coarse breccia soil	21	18	25	0.38	35,600	1.08e-8	0.25
2	Silty clay	18	18	25	0.32	31,500	5.4e-7	0.30
3	Coarse boulder soil	21	20	25	0.31	46,700	1.14e-8	0.28
4	Sandstone SLATE	25	630	55	0.25	8,560,000	5.7e-11	0.457

water head boundary is used on the slope. Due to limited space, this paper only selected part of the transient slope pore water pressure field for analysis (Fig. 2).

As can be seen from Fig. 2b–e, since the surface soil of the slope is mainly silty clay and coarse breccia soil, the rainfall intensity is greater than its infiltration intensity. With the rainfall infiltration, the rainwater forms surface runoff along the slope, and the rainwater gradually penetrates from the surface of the slope to the interior of the slope, while the free water entering the interior of the slope moves towards the foot of the slope under the action of gravity. Therefore, the pore water pressure at the foot of the slope is concentrated, and the matric suction quickly dissipates, reaching the saturation state first. At the same time, the matric suction of the surface soil gradually dissipates, and the transient saturation zone is gradually formed. The longer the rainfall duration, the transient saturation area on the slope surface continues to expand to the interior of the slope body, and the groundwater level also shows a trend of increasing at the foot of the slope. However, because the permeability coefficient of bedrock is very small and the matrix suction is very large, the change of the groundwater level is not obvious.

As shown in Fig. 2d and f, when the rainfall stops, pore water in the slope continues to move down under the action of gravity with the increase of duration. However, since there is no rainwater recharge, the infiltrated rainwater gradually drains from inside the slope, the positive pore water pressure begins to decrease, and the range of transient saturation area gradually shrinks.

4.2 Plastic Zone Analysis

The pore water pressure field of the slope at different transient time steps analyzed by the SEEP/W module was coupled into the SIGMA/W module, and the plastic zone of the slope was calculated (Fig. 3).

Due to the large mechanical parameters of the bedrock part, it will not be damaged even if it is subjected to rainfall infiltration. After the first rainfall, the posterior edge, front edge and shoulder of the slope are plastic distribution areas of the slope, but no through plastic zone has been formed, and the slope is relatively stable. With the increase of rainfall duration, the slope produces hydrostatic pressure, the sliding force increases, and the anti-sliding force decreases. The plastic zone at the rear edge and foot of the slope expands to the interior of the slope, and the plastic zone gradually develops into the middle of the slope. According to the occurrence sequence of plastic zone, the most likely instability mode of slope is the deformation and failure of the central deposit pulled by the front deposit and the deformation and failure of central deposit pushed by the rear deposit.

Some plastic areas also appear at the junction between coarse breccia soil and gravel soil and bedrock. This is because the mechanical properties of coarse breccia soil and gravel soil are relatively small, which is greatly influenced by rainfall infiltration. At the same time, the permeability coefficient of the bedrock part is much smaller than that of the coarse breccia soil, so the bedrock part is equivalent to the

Fig. 2 Distribution of slope
pore water pressure under
different rainfall durations

(a) Distribution of the initial slope pore water pressure

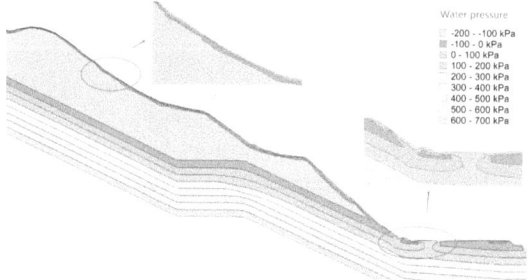

(b) Distribution of pore water pressure after the first rainfall

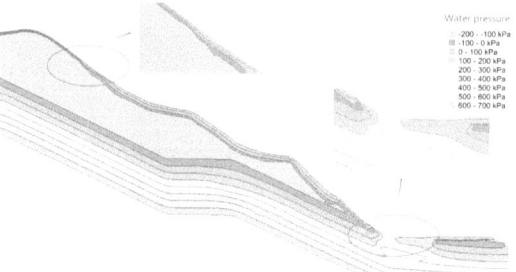

(c) Distribution of pore water pressure after the fifth rainfall

(d) Distribution of pore water pressure after
the rain stopped for the fifth time

Fig. 2 (continued)

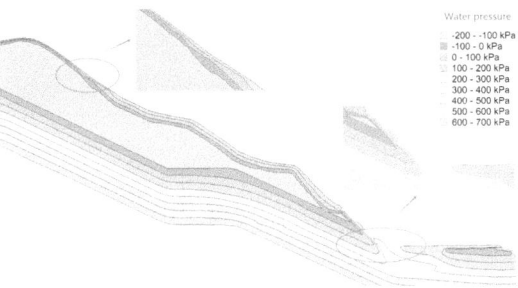

(e) Distribution of pore water pressure after the eighth rainfall

(f) Distribution of pore water pressure after
rain stopped for the eighth time

water-barrier layer, and water will form at the soil-rock interface after rainwater infiltration. Thus, the shear strength of the soil is reduced, and some plastic regions are formed. With the increase of rainfall duration, the range of plastic zone at the soil-rock interface expands continuously.

4.3 Displacement Field Analysis

The displacement of the overburden slope is mainly concentrated at the back edge and foot of the slope (Fig. 4). After the first rainfall, rainwater infiltration causes the surface soil of the slope to soften, and rainwater enters the slope along the coarse breccia soil layer, and deformation occurs at the rear edge of the slope, reaching a maximum of 0.75 m. At the foot of the slope, due to the small matric suction and the greater influence of rainfall, the groundwater level rises rapidly and the shear strength of the soil decreases, so the displacement and deformation occur. With the increase of rainfall duration, rain softened soil continuously, the infiltration range of rain expanded, the shear strength of soil decreased continuously, and the deformation of the rear edge increased significantly, with the maximum displacement reaching 1.6 m. The shallow soil in the middle part of the slope is deformed and destroyed by the pushing action of the soil in the back edge and the pulling action of the soil

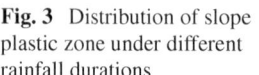

Fig. 3 Distribution of slope
plastic zone under different
rainfall durations

(a) Distribution of plastic zone after the first rainfall

(b) Distribution of plastic zone after the fifth rainfall

(c) Distribution of plastic zone after the eighth rainfall

in the front edge. Runoff is generated along the slope and accumulates at the foot of the slope, resulting in a significant increase in the deformation rate of the slope foot and a continuous expansion of the infiltration depth. The longer the rainfall lasts, the more prone the slope is to deformation and instability.

4.4 Slip Surface Analysis

The slip surface of slope is an important basis to reflect the potential instability position of slope. In SLOPE/W module, pore water pressure force field obtained by

Fig. 4 Displacement
variation diagram under
different rainfall durations

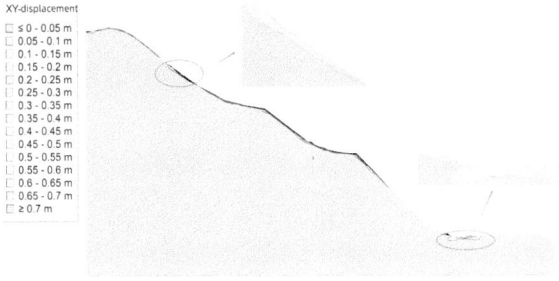

(a) Displacement diagram after the first rainfall

(b) Displacement diagram after the fifth rainfall

(c) Displacement diagram after the eighth rainfall

analysis is introduced, stability analysis is carried out in each transient seepage time step, and potential slip plane is searched automatically (Fig. 5).

After the rainfall began, the surface layer of the slope gradually formed a transient saturation zone, and the soil in the shallow layer of the slope softened. At the same time, the rain penetrated into the slope along the soil-along interface, and the shear strength of the soil decreased, and the slope showed an overall instability state. With the continuous rainfall, the range of transient saturation area continues to expand, the sliding force increases, the anti-sliding force decreases, and the slope changes from the overall instability state to the local instability state. At this time, the most dangerous sliding surface changes and develops from the inside of the slope to the surface of the slope. After the rain stopped, the rain was discharged continuously

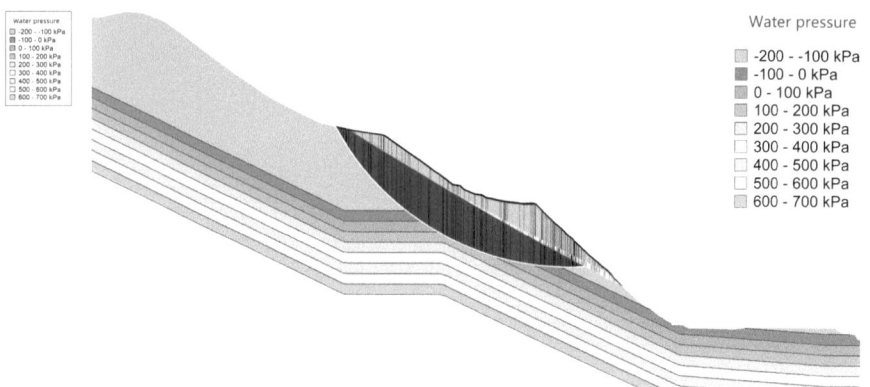

Fig. 5 Variation diagram of slope slip surface under different rainfall durations

under the action of gravity, and the matrix suction began to recover, but the overall stability of the slope was strong, so the change was not large.

4.5 Slope Stability Analysis

Slope safety factor is an important index to evaluate slope stability. With the increase of rainfall duration, the change of safety factor of the overburden slope is shown in Fig. 6. With the continuous rainfall, the rainwater continuously penetrated into the slope, which led to the expansion of the transient saturation area and the dissipation of the matrix suction, and the safety factor of the slope gradually decreased with the increase of time. After the rain stopped, the matric suction gradually recovered and the safety factor gradually increased.

Fig. 6 Variation diagram of slope safety factor

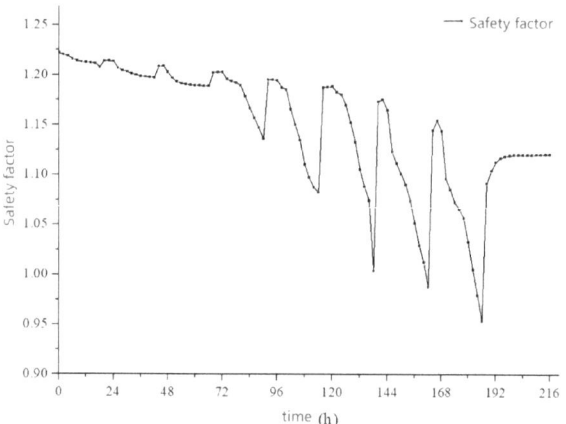

Combined with the change law of slip surface, pore water pressure field, plastic zone and displacement field, the progressive failure mechanism of the overburden slope is as follows: the surface soil of the slope is mainly composed of coarse breccia soil and gravel soil, which has relatively good permeability and is easy to be softened by rain. In the early stage of rainfall, the transient saturation zone formed by rainfall infiltration is relatively shallow, and the shallow soil mass at the front and back edge of the slope is deformed and destroyed. At this time, the most dangerous sliding surface area of the slope is large, and the slope is in the overall instability state. Therefore, during the first four rainfall-stop periods, the slope safety factor changed little. With the increase of rainfall duration, the range of transient saturation area of surface layer expands continuously, which reduces the shear strength of soil mass. The sliding force of slope increases, the anti-sliding force decreases, and the groundwater level rises. As a result, the traction deformation and failure of the central deposit body pulled by the soil mass at the front of slope occurs, and the push deformation and failure of the central deposit body pushed by the soil mass at the back edge occurs. At this time, the area of the most dangerous sliding surface of the slope decreases, mainly in the silty clay layer and coarse breccia soil layer on the surface of the slope body, and the slope is in a local instability state. Therefore, in the last four rainfall-stop periods, the slope safety factor changes more and more. After the rainfall stops, the matrix suction gradually recovers, the range of transient saturation becomes smaller and smaller, and the rainwater gradually drains under the action of gravity, and the slope sliding force gradually decreases, so the slope safety factor gradually rises. When the total sliding force of the potential sliding plane is greater than the total anti-sliding force, the slope will slide along the sliding plane and lead to the instability of the slope.

5 Concluding Discussion

In this paper, a two-dimensional numerical model of a high-steep overburden slope was established by using the finite element software Geostudio. Saturation-unsaturated seepage flow analysis, pore pressure-stress coupling analysis and slope stability analysis were carried out by using the sliding surface stress method, and the changing laws of pore water pressure force field, displacement field, plastic zone, slip surface and safety factor were revealed. Then the progressive failure mechanism of slope under intermittent rainfall condition is analyzed. The main conclusions are as follows:

(1) Under the action of intermittent rainfall infiltration, the matrix suction of the soil mass of the high steep cover slope gradually dissipates, and the soil mass near the slope surface dissipates faster. The rain softens the soil mass and reduces the shear strength of the soil mass. As a result, the groundwater level of the slope increases, the sliding force increases, the anti-sliding force decreases, and the safety factor decreases continuously. After the rain stops, the matric

suction recovers, and the rain water continues to move inside the slope and gradually drains from the slope under the action of gravity, the groundwater level decreases, the sliding force decreases, and the safety factor gradually rises.

(2) In the initial period of rainfall, the rainwater infiltration is shallow, the range of transient saturation zone is small, the area of the most dangerous slip surface is large, and the slope is in the overall instability state; With the continuous rainfall and continuous infiltration of rain, the area of the most dangerous slip surface decreases and the slope becomes local instability. After the rain stopped, the area of the most dangerous slip surface increased slightly.

(3) The slope displacement is mainly concentrated in the silty clay layer, coarse breccia soil layer and gravel soil layer on the slope surface, and the maximum displacement reaches 1.6 m; High-intensity rainfall quickly saturates the shallow surface soil, resulting in shallow sliding. With the increase of rainfall duration, surface water penetrates into the interior of the slope to form groundwater, softens the soil along the soil-rock interface, and gradually forms runoff. The matrix suction of slope foot is small, which is greatly affected by rainfall, and the soil is constantly softened and deformed under the action of rain.

Acknowledgements The authors gratefully acknowledge the support of Hebei Province natural resources Department project (NO. 13000022P00F2D4101293).

References

1. W. Li, S. Jiusheng, Z. Yizhi et al., The formative mechanism and the research idea of Prediction method about coast disaster of rainstorm model [J]. Jiangxi Meteorol. Sci. Technol. **28**(3), 17–22 (2005)

2. L. Guangcai, C. Miaoquan, W. Chunhua, et al., Stability evaluation of weak slopes under heavy precipitation [J]. J. Geotech. Found. **32**(5), 532–535+538 (2018)

3. Y. Xiaoguang, H. Dinggui, H. Zhenli et al., Research on correlations between slope stability and rainfall of high steep slope on Nanfen open-pit iron ore [J]. Chin. J. Rock Mech. Eng. **35**(1), 3232–3240 (2016)

4. J. Zhongming, X. Xiaohu, Z. Ling, Unsaturated seepage analysis of slope under rainfall condition based on FLAC3D [J]. Rock Soil Mech. **35**(3), 81–84 (2014). ((in Chinese))

5. H. Yi, J. Bijiang, L. Jun et al., Numerical simulation of stability of high fill embankment under heavy rainfall [J]. J. Build. Struct. **51**(2), 1768–1776 (2021)

6. H. Ya, C. Xuan, G. Haikuo et al., Seepage analysis and stability study of high and steep slope dump under long duration rainfall [J]. Mod. Min. **11**, 190–194 (2018)

7. X. Jian-Cong, S. Yue-quan, Study on mechanism of disintegration deformation and failure of debris landslide under rainfall [J]. Rock Soil Mech. **29**(1), 106–112+118 (2008)

8. G. Bin, J. Yan, Y. Liang et al., Analysis on process of rainfall-induced landslide in Baolun service area, Sichuan Province [J]. Chin. J. Geol. Hazards Control **31**(4), 45–51 (2020)

9. D. Kailun, L. Hong, L. Xingxing et al., Seepage and stability analysis of high fill embankment with debris under rainfall events [J]. Geotechn. Found. **32**(6), 650–653 (2018)

10. Z. Kaihuan, L. Quanzhou, L. Chengyan et al., Layered gravel soil slope stability of a waste dump considering long-term hard rain [J]. Chin. J. Eng. Sci. **38**(9), 1204–1211 (2016)

11. L. Hua, S. Wenbing, Z. Huaxiang, Study on the influence of rainfall infiltration on the stability of artificial fill slope [J]. J. Guizhou Univ. **37**(3), 30–35 (2020)
12. L. Longqi, L. Shuxue, W. Yunchao et al., Model tests for mechanical response of bedding rock slope under different rainfall conditions [J]. Chin. J. Rock Mech. Eng. **33**(4), 755–762 (2014)
13. S. Zhenming, Z. Siyi, S. Yue, An experimental study of the deposit slope failure caused by rainfall [J]. Hydrogeol. Eng. Geol. **43**(4), 135–140 (2016). ((in Chinese))
14. D. Hui, L. Xiufei, J. Xiuzi et al., Model test research on the gravel-cluttered soil slope under the heavy artificial rainfall condition [J]. J. Saf. Environ.Saf. Environ. **16**(4), 236–240 (2016)

Study on Joint Regulation of Runoff and Sediment of Water Supply Reservoirs in Sediment-Laden River Dongzhuang Reservoir of Jinghe River

Cuixia Chen, Haixia Wang, Qiushi Luo, Zhichao Wen, Junxiu Liu, and Tiange Wang

Abstract There is independence and integration between water supply and sediment reduction in sediment-laden river reservoirs. How to maintain effective storage capacity and meet water supply requirements for a long time is one of the problems to be solved in the efficient operation of sediment-laden river reservoirs. In this paper, Dongzhuang Reservoir of Jinghe River is taken as the research object. Through the analysis of measured data and mathematical model calculation, the measured hydrological sediment and cross-section erosion and deposition in the lower reaches of Jinghe River and Weihe River are analyzed. The reservoir sediment discharge flow index which is beneficial to reduce the sediment deposition in the lower reaches of Weihe River and maintain the effective reservoir capacity for a long time is studied, and the joint regulation mode of reservoir runoff and sediment is put forward. The results show that the sediment regulation of Dongzhuang Reservoir during the sediment retention period is mainly to reduce the deposition of the lower reaches of the Weihe River, and the sediment regulation during the normal operation period of the reservoir is mainly to maintain the effective storage capacity of the reservoir for a long time. The joint regulation of the Dongzhuang Reservoir and the surrounding four storage reservoirs, such as the Ganhe Reservoir, the Xijiao Reservoir, the Longtan Reservoir, and the Helan Reservoir, can reduce the deposition of the lower reaches of the Weihe River by 11 million tons per year, increase the guarantee rate of agricultural irrigation from 30 to 50%, and increase the guarantee rate of industrial water supply from 57 to 95%. And the outflow of the reservoir meets the requirements of the ecological base flow of the river. The research results provide technical support for the optimal operation of reservoirs in sediment-laden rivers.

C. Chen · Q. Luo (✉) · Z. Wen · J. Liu · T. Wang
Key Laboratory of Water Management and Water Security for Yellow River Basin, Ministry of Water Resources, Yellow River Engineering Consulting Co., Ltd., Zhengzhou, China
e-mail: 18655093@qq.com

Z. Wen
e-mail: When012103@tju.edu.cn

H. Wang
Xinxiang Dagong Yellow River Water Supply Co., Ltd., Xinxiang, China

© The Author(s) 2025
W. Wang et al. (eds.), *Hydraulic Structure and Hydrodynamics*, Lecture Notes in Civil Engineering 608, https://doi.org/10.1007/978-981-97-7251-3_19

213

Keywords Sediment-laden River · Dongzhuang reservoir · Reservoir capacity maintenance · Jinghe River · Lower Weihe River

1 Introduction

Reservoir engineering development tasks generally include flood control, water supply, power generation, irrigation, etc. [1, 2]. Reservoir projects developed on sediment-laden rivers must also undertake the task of reducing siltation in downstream rivers. Different from the clear water river, the problem of sediment deposition in the sediment-laden river reservoir often restricts the efficient use of the comprehensive function of the reservoir [3, 4]. To maintain the effective storage capacity for a long time, the sediment-laden river reservoir needs to have a certain scale of sediment discharge [5]. In the flood season, not only the sediment is discharged from the reservoir, but also the sediment deposited during the non-flood season or the flood season is discharged from the reservoir [6]. However, the reduction of water level and sediment discharge restricts the function of water supply and power generation. If the reservoir is stored during the flood season, it is easy to cause serious siltation in the reservoir area, and the contradiction between water storage and sediment discharge is prominent. This problem is more prominent in reservoirs with sediment concentration higher than 100 kg/m^3 [7].

How to deal with the relationship between reservoir storage and sediment discharge in sediment-laden rivers, especially in extra-high sediment-laden rivers, is necessary to reasonably determine the way of runoff and sediment regulation. The established reservoirs consider the synergy between sediment discharge and profit generation when formulating the scheduling rules, and the commonly used scheduling planning methods include linear planning, nonlinear planning [8], and dynamic planning [9]. Wang et al. [10] developed a flood season regulation model for the sediment-laden river reservoir based on the differential response theory with the combined benefits of sediment discharge and water supply as the performance indicator function, which optimizes the combined benefits of water supply and sediment discharge. Jinming [11] constructed calculation methods for flood control benefit, water supply benefit, and power generation benefit respectively, which were used to quickly determine the reservoir benefit corresponding to a given capacity. Li [12] used BP artificial neural network technology to predict the sediment deposition trend in the reservoir area and used the coordinated weight method of multi-criteria group decision-making with preferences to determine the weight between sediment discharge and power generation. Xinhong et al. [13] took Wangyao Reservoir as an example to study the impact of the way of using reservoirs for the water supply in extra-high sediment-laden reservoirs. Existing studies on reservoir sediment discharge and profit generation scheduling methods are mainly in lower sediment river reservoirs or relatively low sediment content of sediment-laden river reservoirs, while there are fewer studies on extra-high sediment-laden reservoirs with sediment content of 100 kg/m^3 or more.

Reservoir regulation mode is related to the conditions of incoming water and sediment, project development tasks, and so on. To achieve the multiple objectives of river siltation reduction, water supply and profitability, reservoir capacity mainte- nance, and ecological improvement, it is necessary to clarify the sediment discharge flow and water supply mode of the reservoir.

2 Study Area, Data, and Method

2.1 Study Area

Dongzhuang Reservoir is constructed in the downstream canyon section of Jinghe River, a secondary tributary of the Yellow River, with a controlled basin area of 4.31×10^4 km^2, accounting for 95% of the basin area of Jinghe River, the location of which is shown in Fig. 1. Jinghe River is characterized by a large amount of sediment transport, high sediment concentraction in the water flow, etc. The average sediment discharge of Dongzhuang Reservoir for many years is 2.37×10^8 tons, with an average sediment concentration of 140 kg/m^3, which is 5 times the average sediment concentration of the Yellow River. The average sediment concentration during the flood season is 206 kg/m^3, with a maximum sediment concentration of 1428 kg/ m^3. The total design capacity of Dongzhuang Reservoir is 3.276×10^9 m^3, which the sediment storage capacity is 2.053×10^9 m^3. The project development task is to focus on flood control and silt reduction, taking into account the water supply and power generation and improving the ecology, i.e., controlling the floods of the Jinghe River and reducing the siltation of the downstream channel of the Weihe River through sediment storage, water and sediment regulation to solve the problem of water resource shortages in the north of Weihe River, as well as to achieve the functions of power generation and improving the ecological environment. Due to the large amount of sediment entering Dongzhuang Reservoir during the flood season, the reservoir storage during the flood season will lead to sediment deposition. If the sediment discharge during the flood season is difficult to meet the requirements of water supply and profitability, and the reservoir regulation is facing the problem of how to coordinate the function of sediment reduction and benefit.

2.2 Study Data

(1) Hydrological data

The design representative station of Dongzhuang Reservoir is Zhangjiashan Hydro- logical Station of Jinghe River, and there is Xianyang Hydrological Station above the confluence of the Jinghe River in the lower reaches of the Weihe River, and Huaxian Hydrological Station below the confluence of Jinghe River. The multi-year average

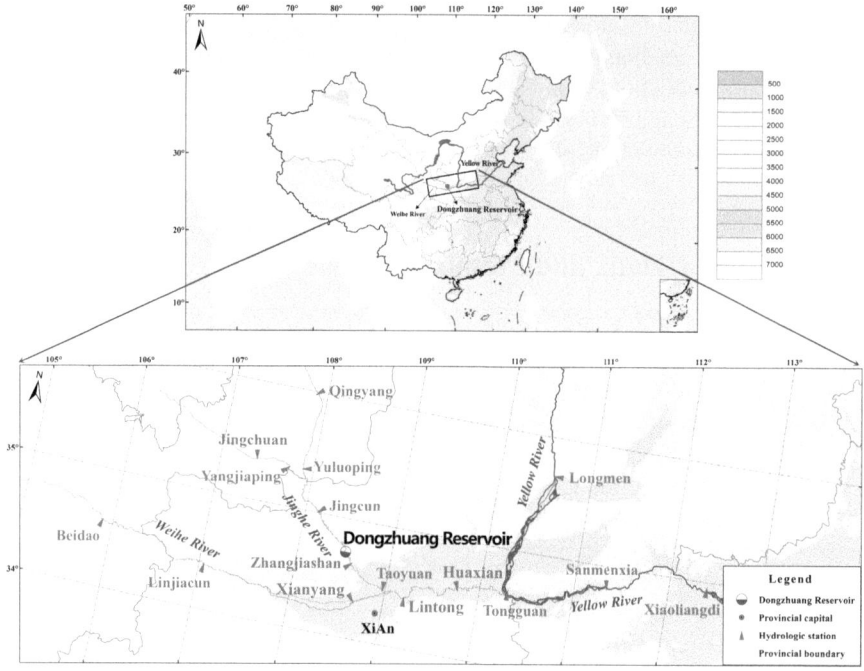

Fig. 1 Location map of the study area

water and sediment quantity of each station is shown in Table 1. Since 2000, the amount of incoming water and sediment at Zhangjiashan Hydrological Station has significantly decreased, with runoff decreasing from 2.195×10^9 m^3 in the 1960s to 1.064×10^9 m^3, and the amount of incoming sediment decreasing from 2.73×10^8 tons to 8.8×10^8 tons. When studying the development plan and operation regulation of Dongzhuang Reservoir in this article, the water and sediment series since 2000 are mainly used for calculation.

(2) Water demand of water supply area

The water supply objects of the Dongzhuang Reservoir are Jinghuiqu Irrigation District, Tongchuan New District, Xixian New District, Fuping and Industrial Park, and Sanyuan County in Shaanxi Province. The irrigation water demand in the Jinghuiqu irrigation area is 3.28×10^8 m^3, and the other areas are 1.65×10^8 m^3 of urban domestic and industrial water.

(3) Ecological flow requirements of rivers

According to the statistics of the measured data of Zhangjiashan hydrological station over the years, the average annual flow of Jinghe River is 53.3 m^3/s. During the sensitive period of fish reproduction from April 15 to June 15 every year, the discharge of Dongzhuang Reservoir should not be less than 20% of the average annual flow. From June 16 to August 31, the discharge should not be less than 30% of the average

Table 1 Year average amount of water and sediment at hydrological gauging stations on the Jinghe and Weihe Rivers

Hydrological station	Series	Amount of water/10^8 m³		
		July–Oct	Nov–Jun	Full year
Jinghe River ZhangJiashan	1960 ~ 2021	8.91	6.20	15.11
	2000 ~ 2021	6.21	4.43	10.64
Weihe River Xianyang	1960 ~ 2021	21.30	14.66	35.96
	2000 ~ 2021	15.87	9.90	25.77
Weihe River Huaxian	1960 ~ 2021	39.09	25.98	65.07
	2000 ~ 2021	34.22	22.62	56.84
Hydrological station	Series	Amount of sediment/10^8 t		
		July–Oct	Nov–Jun	Full year
Jinghe River ZhangJiashan	1960 ~ 2021	1.69	0.18	1.87
	2000 ~ 2021	0.83	0.05	0.88
Weihe River Xianyang	1960 ~ 2021	0.70	0.12	0.82
	2000 ~ 2021	0.20	0.01	0.21
Weihe River Huaxian	1960 ~ 2021	2.30	0.26	2.56
	2000 ~ 2021	0.89	0.07	0.96

annual flow. From September 1 to April 14 of the next year, the discharge should not be less than 10% of the average annual flow.

2.3 Method

The development and utilization of water resources in high sediment-laden rivers need to adjust runoff and sediment. For the ultra-high sediment-laden rivers with sediment concentration above 100 kg/m³, it is necessary to adopt the parallel reservoir development mode of main stream reservoir sediment regulation and storage reservoir water supply [1]. During the flood season, when the main stream reservoir is used for open discharge of sediment during the high sediment-laden flood period, it is supplied by the regulating reservoir. When the main stream reservoir is used for water storage and utilization, the main stream reservoir fills the storage reservoirs with water and supplies water to the water users. The discharge flow of the main stream reservoir is calculated as follows.

Application of sediment discharge:

$$Q_{inflow} = Q_{outflow} + \frac{V_{adjust}}{\Delta t} \tag{1}$$

Water storage benefits:

$$Q_{outflow} = Q_{in-storage} + Q_{ecology} \tag{2}$$

where, Q_{inflow} and $Q_{outflow}$ are the inflow and outflow of the main stream reservoir, m³/s; V_{adjust} is the water storage capacity above the sediment discharge level of the main stream reservoir, 108 m³; Δt is the calculation time step, s; $Q_{in-storage}$ is the flow from the mainstream reservoir to the storage reservoir, m³/s; $Q_{ecology}$ is the ecological flow of the main stream, m³/s.

The inflow of the storage reservoir is the flow $Q_{in-storage}$ introduced from the main stream reservoir, which is generally calculated by the flow capacity of the connecting pipeline between the main stream reservoir and the storage reservoir. The outflow flow $Q_{out-storage}$ is the flow of water supply to the water users. If the storage reservoir is set on a tributary, the flow into the reservoir should be added to the water flow from the tributary, and the flow out of the reservoir should take into account the ecological flow requirements of the tributary below the dam site of the storage reservoir.

Mathematical modeling of water and sediment [14] was used to calculate the flow along the reservoir area, water level, water depth, area, and flow velocity along the reservoir area of the main stream reservoir. The main goal of the calculation of the hydraulic elements of the reservoir is to obtain the change of the reservoir water level and water storage. The reservoir water storage is calculated according to the principle of water balance:

$$V_{t-storage} - V_{t-1-storage} = (Q_{in-storage} - Q_{out-storage})\Delta t \tag{3}$$

Mathematical modeling of water and sediment [14] was used to calculate the sediment concentration of each section of the main stream reservoir, the depth of erosion and deposition of the section, the amount of sedimentation in the reservoir area, etc. The purpose of the calculation of sediment erosion and deposition in the storage reservoir is to obtain the amount of reservoir deposition, which mainly comes from the amount of sediment diversion from the main stream reservoir. If the storage reservoir is set on the tributary, the amount of sediment from the tributary should also be considered. There are generally no sediment discharge facilities in the reservoir, and all the sediment entering the reservoir area is deposited in the reservoir.

3   Results and Analyses

3.1   Reservoir Sediment Regulation Mode

The primary purpose of reservoir regulation is to reduce the deposition in the lower reaches of the Weihe River before the sediment storage capacity of the Dongzhuang Reservoir is full. After the sediment storage capacity of the reservoir is silted up, it enters the normal operation period. The primary purpose of reservoir operation is becoming to maintain the effective storage capacity of the reservoir and give full play to the comprehensive benefits in the long run. Therefore, from the perspective of sediment reduction in the lower reaches of the Weihe River and the long-term storage capacity of the reservoir, the sediment regulation methods of Dongzhuang Reservoir during the sediment retention period and normal operation period are proposed respectively.

(1) Reservoir sediment retention period is conducive to the sediment regulation mode of sediment reduction in the lower reaches of Weihe River

Since 1974, the erosion and deposition of the lower reaches of the Weihe River have been mainly affected by the conditions of incoming water and sediment. A total of 170 non-floodplain floods occurred in the Jinghe River, which lasted 934 days. During this period, a total of 2.98×10^8 tons of sediment was washed out in the lower reaches of the Weihe River. According to the different flow levels and sediment concentration levels of the flood at Zhangjiashan Station of the Jinghe River, the influence of the flood on the erosion and deposition of the lower reaches of the Weihe River is analyzed, as shown in Fig. 2. It can be seen that when the sediment concentration level of the Jinghe River flood is below 200 kg/m³, the lower reaches of the Weihe River are all scoured. The flood sediment concentration of the river is 200 ~ 300 kg/m³, and the lower reaches of the Weihe River are silted up. When the sediment concentration of the Jinghe River flood is greater than 300 kg/m³ and the flow rate is less than 300 m³/s, the deposition in the lower reaches of the Weihe River is serious. As the flow rate increases to more than 300 m³/s, the downstream channel of the Weihe River changes from siltation to erosion, especially, when the flow rate is more than 600 m³/s, erosion of the downstream channel of the Weihe River increases significantly. Therefore, the deposition in the lower reaches of the Weihe River is mainly concentrated in the water and sediment process with sediment concentration greater than 300 kg/m³ and flow level less than 300 m³/s at Zhangjiashan Station.

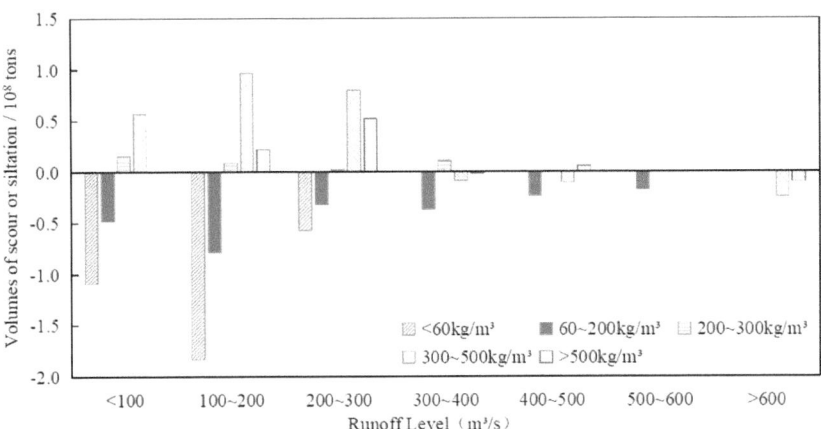

Fig. 2 Scouring and siltation by floodwaters of different runoff levels and sediment content levels entering the downstream channel of Weihe River at Zhangjiashan station of Jinghe River

Since 1974, a total of 133 non-floodplain floods occurred in the lower reaches of the Weihe River, which lasted 758 days. During this period, a total of 4.79×10^8 tons of sediment was washed out in the lower reaches of the Weihe River. According to the total value of Xianyang station and Zhangjiashan station entering the lower reaches of the Weihe River, the river channel erosion and deposition during the flood period of the lower reaches of the Weihe River are analyzed according to different flow levels and different sediment concentration levels, as shown in Fig. 3. It can be seen that when the sediment concentration of the flood in the lower reaches of the Weihe River is below 100 kg/m^3, the lower reaches of the Weihe River basically shows scouring. When the flow rate is above 600 m^3/s, the lower reaches of the Weihe River are scoured. When the flood sediment concentration is more than 100 kg/m^3, the lower reaches of the Weihe River are seriously silted up.

Therefore, in order to realize the erosion of the lower reaches of the Weihe River, it is necessary for the Dongzhuang Reservoir of Jinghe River to impound the water and sediment process that is prone to cause the deposition of the lower reaches of the Weihe River during the sediment retention period, and release the water and sediment process that is conducive to the erosion of the lower reaches of the Weihe River, i.e., when the inflow of the Jinghe River is less than 300 m^3/s and the sediment concentration is greater than 300 kg/m^3, the sediment deposition in the lower reaches of the Weihe River is serious, and the Dongzhuang Reservoir should retain such floods. When the flow rate of the Jinghe River is greater than 600 m^3/s and the sediment concentration is greater than 300 kg/m^3, the erosion of the lower reaches of the Weihe River is more obvious. Dongzhuang Reservoir should empty this kind of flood appropriately, so as to minimize the sediment deposition of the reservoir and flushing the downstream channel of the Weihe River.

(2) Normal operation period of the reservoir sediment discharge to maintain the reservoir capacity adjustment method

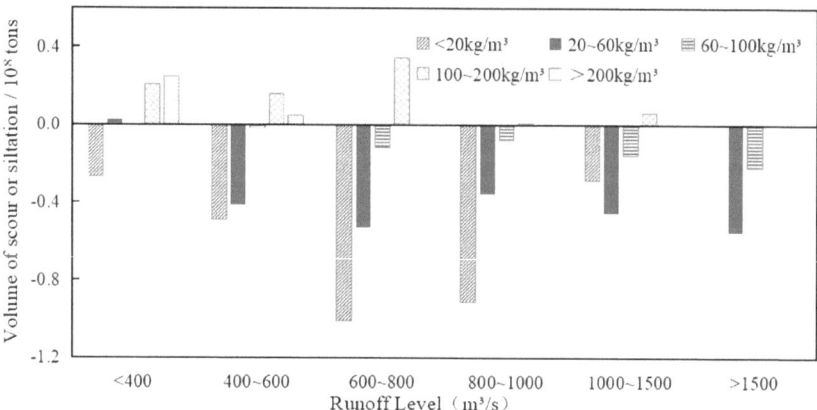

Fig. 3 Scouring and siltation of the lower Weihe River in floods of different runoff levels and sediment content levels

Fig. 4 The scour and
siltation process of
Dongzhuang Reservoir under
different sediment discharge

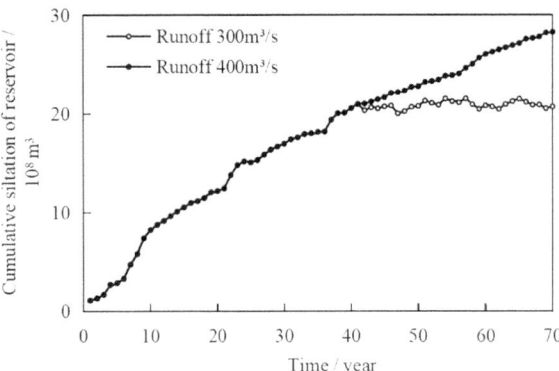

During the normal operation period of the reservoir, Dongzhuang Reservoir should discharge sediment openly when the inflow is large, scour the sediment deposited in the reservoir area, and maintain the effective storage capacity of the reservoir for a long time. During the open discharge period of sediment, the sediment concentration of the outflow water is high. If the sediment discharge flow is below 300 m³/s, the process of high sediment concentration and small flow out of the reservoir can easily cause siltation in the lower reaches of the Weihe River. However, the greater the sediment discharge flow of the reservoir, the less the chance of open discharge of sediment, and the reservoir may not be able to maintain effective storage capacity for a long time. In this paper, the two schemes of sediment discharge flow of 300 m³/s and 400 m³/s are compared and calculated. When the inflow flow is greater than the sediment discharge flow, the reservoir is open to discharge sediment.

Based on the water and sediment series from 2000 to 2021, the sedimentation volume of Dongzhuang Reservoir under different sediment discharge schemes is calculated by mathematical model, as shown in Fig. 4. From the diagram, it can be seen that after the reservoir has been used for 40 years, the sediment storage capacity of 2.053×10^9 m³ has been silted up, and the sediment discharge flow of 300 m³/s scheme, when the flood season inflow is greater than 300m³/s, the open discharge of sediment can maintain the effective storage capacity for a long time. In the scheme of sediment discharge flow of 400 m³/s, the opportunity of open discharge of sediment of the reservoir is reduced, and it is difficult to discharge the sediment deposited in the reservoir area, and the effective storage capacity cannot be maintained for a long time. Therefore, during the normal operation period of the reservoir, when the inflow in the flood season is greater than 300m³/s, the reservoir discharges sediment openly.

3.2 Reservoir Runoff and Sediment Joint Regulation Mode

The mathematical model is used to calculate the water supply guarantee rate of the reservoir according to the above method. The industrial water supply guarantee rate is only 56.6%, and the agricultural irrigation guarantee rate is only 32%, which does not meet the requirements of 95% industrial water supply guarantee rate and 50% agricultural irrigation guarantee rate. If the Dongzhuang Reservoir is mainly used for water supply during the flood season, the model calculates the sediment erosion and deposition changes in the reservoir area and the water supply guarantee rate as shown in Figs. 5 and 6. A large amount of sediment is deposited in the reservoir area. Only 24 years later, the design sediment interception capacity of the reservoir is 2.053×10^9 m^3, which is silted up and continuously silted up and uplifted, and the effective storage capacity cannot be guaranteed for a long time.

In order to achieve the dual goals of Dongzhuang Reservoir's effective storage capacity and water supply, it is proposed that Dongzhuang Reservoir should be jointly regulated with the surrounding four reservoirs, such as Shanhe Reservoir, Xijiao Reservoir, Longtan Reservoir, and Helan Reservoir. Dongzhuang Reservoir is connected with each storage reservoir through pipelines, as shown in Fig. 7. During the high sediment-laden flood period, the water level of the Dongzhuang Reservoir is lowered to discharge sediment openly, and the water supply is provided by the four reservoirs, such as the built Ganhe Reservoir, as the storage reservoirs. During normal

Fig. 5 The siltation in the reservoir area of Dongzhuang Reservoir with different utilization modes

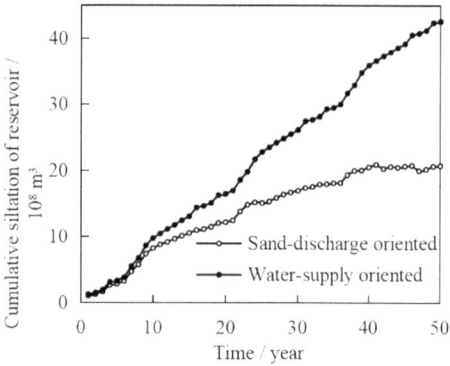

Fig. 6 Guaranteed water supply rate of Dongzhuang Reservoir for different utilization modes

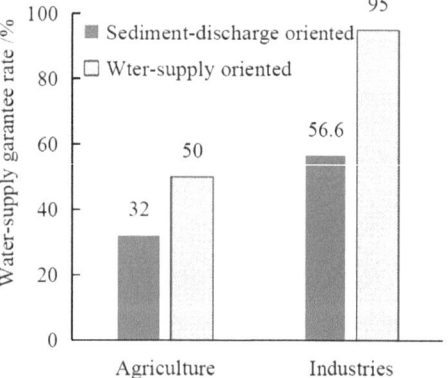

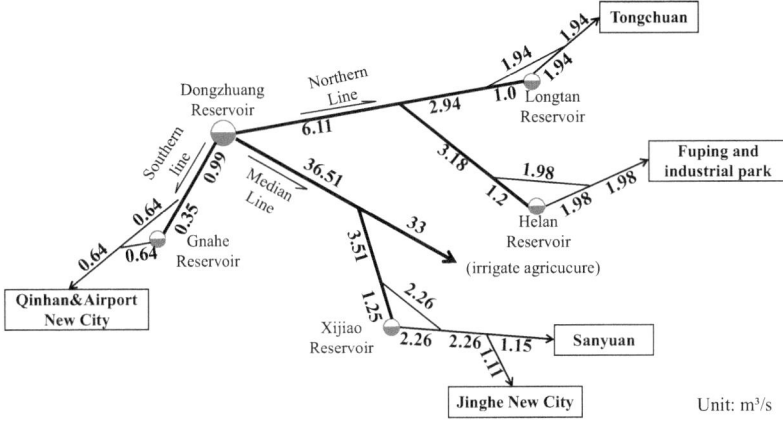

Fig. 7 Dongzhuang water supply district water transmission project design flow diagrams

or non-flood seasons, Dongzhuang Reservoir regulates runoff and introduces fresh water into the storage reservoir. Dongzhuang Reservoir and the storage reservoir supply water at the same time. The specific regulation methods are as follows:

(1) The sediment retaining period

From July to September in the main flood season, when the inflow is greater than $600 m^3/s$ and the sediment concentration is greater than $300 kg/m^3$, the sediment concentration of the inflow is high, and such a water and sediment matching process is beneficial to the deposition reduction of the lower reaches of the Weihe River. The Dongzhuang Reservoir discharges sediment openly and regulates the external water supply of the storage reservoir.

During other periods of the main flood season and from October to June of the next year, the Dongzhuang Reservoir is used to store water. While releasing the downstream ecological flow, it is filled with water to the storage reservoir, and the Dongzhuang Reservoir and the storage reservoir are simultaneously supplied with water.

(2) The normal operation period

From July to September in the main flood season, when the inflow is greater than $300 m^3/s$, in order to maintain the effective storage capacity of the reservoir for a long time, Dongzhuang Reservoir discharges sediment openly and regulates the external water supply of the reservoir.

During other periods of the main flood season and from October to June of the next year, the regulation mode of Dongzhuang Reservoir is used in the same way as the sediment retention period.

3.3   Effect of Joint Regulation

(1)   Effect of sediment reduction in the downstream river channels

Dongzhuang Reservoir changes the water and sediment conditions that are discharged from Jinghe River into Weihe River through reservoir regulation and reduces the deposition of the lower reaches of Weihe River. Using the mathematical model to predictive simulation the sedimentation in the lower Weihe River under Dongzhuang Reservoir or non-Dongzhuang Reservoir.

According to the calculation results of the mathematical predictive model, it can be seen that before the Dongzhuang Reservoir came into effect, the lower reaches of the Weihe River continued to silt up, and the accumulated silting sediment in 50 years was 3.87×10^8 tons, as shown in Fig. 8. After the Dongzhuang Reservoir was put into operation, the cumulative erosion of sediment in the lower reaches of the Weihe River was 1.76×10^8 tons, which completely reversed the situation of continuous siltation in the lower reaches of the Weihe River. Compared with no Dongzhuang reservoir, the cumulative sediment reduction of Dongzhuang reservoir is 5.63×10^8 tons, the average annual sediment reduction is 1.1×10^7 tons, and the cumulative sediment deposition thickness is reduced by 1.19 m, which is equivalent to 90 years without sedimentation in the lower reaches of the Weihe River under the condition of Dongzhuang reservoir.

(2)   Reservoir water supply

The Dongzhuang Reservoir is an important water source for industrial and agricultural production and urban and rural life in the Guanzhong area of Shaanxi Province. According to the calculation, the average annual water supply of Dongzhuang Reservoir is 4.35×10^8 m^3, of which the agricultural irrigation water supply of Jinghuiqu Irrigation District is 2.73×10^8 m^3, which increases the irrigation guarantee rate from 30 to 50%. Supply Tongchuan New Area, Fuping County and Industrial Zone,

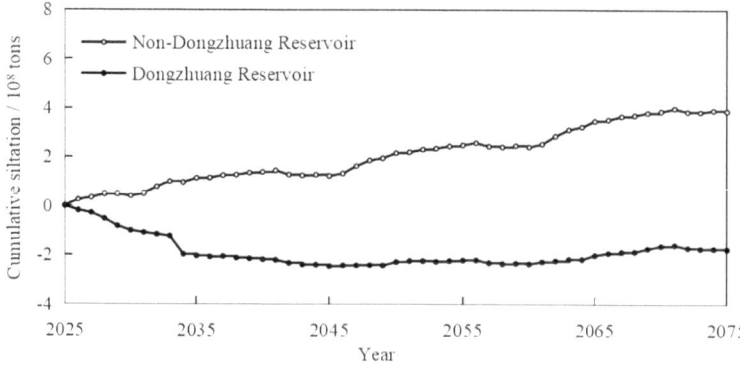

Fig. 8 Changes in channel scouring and siltation in the lower Weihe River

Xixian New Area and Sanyuan County urban life and industrial water 1.62×10^8 m^3, water supply guarantee rate to meet the requirements of 95%.

(3) Ecological base flow guarantee effect

Before the Dongzhuang Reservoir took effect, by analyzing the measured runoff data of Zhangjiashan Station of Jinghe River from 1960 to 2021, it was found that there were 124 days in which the river flow was less than 5.33 m^3/s. From April 15 to June 15, the sensitive period of fish reproduction, the average annual number of days in which the river flow was lower than 10.66 m^3/s of the ecological base flow was 25 days. From June 16 to August 31, the average annual number of days in which the river flow was lower than 15.99 m^3/s of the ecological base flow was 29 days. According to the mathematical model, after the operation of Dongzhuang Reservoir, the annual ecological flow is greater than 5.33 m^3/s, the flow in the river during the sensitive period of fish reproduction is greater than 10.66 m^3/s, and the flow in the river during the flood season from June 16 to August 31 is greater than 15.99 m^3/s, which can effectively improve the ecological impact of insufficient ecological flow and disconnection.

4 Conclusions

In this paper, Dongzhuang Reservoir of Jinghe River is taken as the object to study the joint regulation mode of runoff and sediment in sediment-laden river reservoirs.

(1) The sediment regulation of Dongzhuang Reservoir during the sediment retaining period is mainly to reduce the deposition of the lower reaches of the Weihe River. The reservoir should retain small floods with high sediment concentration when the inflow is less than 300 m^3/s and the sediment concentration is greater than 300 kg/m^3, and discharge floods when the inflow is greater than 600 m^3/s and the sediment concentration is greater than 300 kg/m^3. During the normal operation period of the reservoir, the sediment regulation is mainly to maintain the effective storage capacity of the reservoir for a long time. When the inflow is greater than 300 m^3/s, the reservoir is open to discharge sediment.

(2) During the period of high sediment-laden floods and the Dongzhuang Reservoir lowers its water level to openly discharge sediment, the water is supplied by four surrounding storage reservoirs that have been built, such as Hunhe Reservoir, the Xijiao Reservoir, the Longtan Reservoir, and the Helan Reservoir. During the normal or non-flood season, the Dongzhuang Reservoir regulates and stores runoff, releases the ecological flow of the lower reaches of the Jinghe River, introduces clean water into the storage reservoir, and supplies water to the outside world with the storage reservoirs simultaneously. According to the calculation of the mathematical model, the average annual sedimentation of the lower reaches of the Weihe River can be reduced by 1.1×10^7 tons after the operation of the reservoir, and the average water supply for many years is 4.35 ×

10^8 m^3. The discharge of the reservoir meets the requirements of the ecological base flow of the river.

Acknowledgements This study was funded by the Young Elite Scientists Sponsorship Program of the Henan Association for Science and Technology (No. 2022HYTP022), and the National Key Research and Development Programme of China (Grant No. 2022YFC3080300).

References

1. J. Zhang, C. Hu, J. Liu, Operation mode and design technology of storing clean water and regulating muddy flow of reservoir in sediment-laden river [J]. J. Hydralic Eng. **53**(1), 1–10 (2022). ((in Chinese))
2. M.L. Abdelhadi, Environment and socio-economic impacts of erosion and sedimentation in North African countries. 6th Int. Symp. River Sedimentation [J]. 1141–1153 (1995). Central Board of Irrigation and Power, New Delhi
3. G.R. Basson, A. Rooseboom, *Dealing with Reservoir Sedimentation* (Water Research Commission, South Africa, 1997)
4. G.R. Basson, A. Rooseboom, *Dealing with Reservoir Sedimentation-Dredging* (Water Research Commission, South Africa, 1999)
5. M.C. Farmer, A. Randall, Policies for sustainability: lessons from an overlapping generations model. Land Econ. **73**(4), 608–622 (1997)
6. G.B. Asheim, W. Buchholz, B. Tungodden, Justifying sustainability. J. Environ. Econ. Manag.Manag. **00**, 1–17 (2000)
7. R. Harboe, Multi-objective decision making techniques for reservoir operation. Water Resour. Bull.Resour. Bull. **28**(1), 103–110 (1992)
8. L.A. Rossman, A nonlinear programming algorithm for real-time hourly reservoir operations. J. Am. Water Resour. Assoc.Resour. Assoc. **15**(4), 1178–1180 (1978)
9. J. Kelman, J.R. Stedinger, L.A. Cooper et al., Sampling stochastic dynamic programming applied to reservoir operation. Water Resour. Res.Resour. Res. **36**(3), 447–454 (1990)
10. Y. Wang, D. Zhong, B. Wu, Multi-agent group decision-making for the objectives of flood control, power generation and navigation of cascaded reservoirs [J]. 36th IAHR World Congr. (2015)
11. X. Jinming, *Studies on the Evaluation of Reservoir Sedimentation Management [D]* (Tsinghua University, Beijing, 2012). ((in Chinese))
12. W. Li, *Study on Multi-objective Optimization of Sanmenxia Reservoir Dispatching [D]* (HoHai University, Nanjing, 2006). ((in Chinese))
13. W. Xinhong, G. Liyao, W. Wei, T. Yongpeng, Reservoir operation patterns of water-supply reservior in sedimen-landen river—a case study on the Wangyao Reservoir [J]. J. Sedim. Res. **43**(2), 33–39 (2018). ((in Chinese))
14. C. Chen, F. Jian, W. Moxi et al., High-efficiency sediment-transport requirements for operation of the Xiaolangdi Reservoir in the Lower Yellow River. Water Supply **22**(12), 8572 (2022). https://doi.org/10.2166/ws.2022.397

Mechanism and Stability Analysis of Karst Collapse in Jingquan Water Source Area of Tengzhou City

Tangwu Feng

Abstract To study the development and collapse mechanism of karst collapse in Jingquan Water Source Area of Tengzhou City, a two-dimensional fluid solid coupling model was established using the finite element software ABAQUS. A comprehensive analysis was conducted on the development process of karst collapse, and the mechanism of soil cave collapse under the effects of rainfall and excessive groundwater exploitation in the study area was summarized. The results indicate that the periodic fluctuation of groundwater level caused by rainfall and excessive groundwater exploitation is manifested as a decrease in soil strength, ultimately leading to karst collapse. The use of strength reduction method can effectively conduct quantitative analysis of karst cave collapse stability.

Keywords Karst collapse · Finite element analysis · Mechanization · Strength reduction · Stability analysis

1 Introduction

China has a wide distribution of karst landforms and is one of the most typical countries in the world [1], with a land karst area of nearly 3.4 million square kilometers. Karst collapse has the characteristics of concealment, suddenness, and uneven distribution in time and space. Studying the conditions and mechanisms of karst collapse development can carry out time and space prediction and disaster prevention and control for karst collapse. It has important theoretical and practical significance in preventing and controlling karst collapse disasters [2].

The fluctuation of groundwater level caused by groundwater exploitation and heavy rainfall is prone to geological disasters such as collapse in karst developed areas [3], which has a serious impact on human production, life, and engineering safety operation. Su et al. [4] established a three-dimensional model experiment to study the development and failure process of soil caves. Jia et al. [5] elaborated the

T. Feng (✉)
The First Exploration Team of Shandong Coalfield Geologic Bureau, Qingdao 266500, China
e-mail: fengtangwu@163.com

W. Wang et al. (eds.), *Hydraulic Structure and Hydrodynamics*, Lecture Notes in Civil Engineering 608, https://doi.org/10.1007/978-981-97-7251-3_20

formation and development process of karst collapse and its influencing factors, and concluded that long-term excessive extraction of karst water is the main factor of karst collapse in the area, and reasonable reduction of groundwater mining output is the most effective method to prevent karst collapse.

2   Karst Collapse Mechanism

The formation of karst collapse is often influenced and influenced by various factors [6]. The occurrence of collapse occurs when the collapse force (such as soil self-weight, underground lateral permeability, positive and negative pressure in the gaps of rock and soil) exceeds the anti-collapse force (such as the buoyancy force of groundwater in the rock and soil). Therefore, analyzing the formation conditions and main inducing factors of karst collapse is an important basic work for its prevention and control.

The change in groundwater level has become a limiting factor for soil cave collapse. If the permeability coefficient of the cover layer is relatively small, the stability of the soil cave is greatly affected by the change in groundwater level and the presence of a free surface. When there is a free surface at the base cover inter-face, with the passage of geological time, the soil collapses during continuous stress adjustment. The impact of groundwater on the stability of the cave mainly includes the following aspects:

(1)   When the groundwater level drops, the buoyancy decreases, which relatively increases the weight of the upper soil layer;
(2)   When the groundwater level rises, the pore water pressure increases, the strength of the soil decreases, and the tunnel wall is prone to deformation;
(3)   In a sealed cavity, the rise and fall of groundwater level causes changes in the cavity pressure, resulting in compressive or tensile forces on the wall and weakening the strength of the soil on the wall;
(4)   The changes in groundwater level and the switching between saturation and unsaturation of soil make the soil continuously undergo stress adjustment, and the distribution of stress will change accordingly, manifested as a decrease in the strength of the soil.

As for the Jingquan water source area, the existence of shallow karst cave gaps, a certain thickness of loose cover, and a significant decrease in groundwater level are the basic conditions for the occurrence of karst collapse. The groundwater level in the Jingquan water source area is relatively shallow and has a strong fluctua-tion [7]. The formation and development of karst collapse are closely related to the activity status of groundwater and changes in groundwater level, and are the main factors determining the formation and development speed of soil caves [8]. Enhanced rainfall or surface water supply, resulting in an increase in groundwater levels; Overextraction of groundwater, long-term drought, increased industrial and agricultural water consumption, and a decrease in groundwater levels. As the groundwater level increases and decreases, the overlying rock and soil on the rock and soil cave are in a frequent alternating state of dry and wet conditions [9], causing lateral erosion, erosion, and hollowing out of the loose filling material and covering layer in karst and fissure channels. In addition, the decrease in groundwater level will cause a vacuum negative pressure effect on the soil [10], and the filling material will be transported away, creating voids at the opening of the hole at the bottom of the covering layer, becoming the prototype of the soil cave. The buoyancy of the overlying soil decreases or even disappears [11], enhancing the influence of groundwater on soil seepage, making suspended particles easy to be washed away by water flow, leading to the formation and expansion of soil cavities. Under the combined action of the subsurface erosion and erosion of fissured karst water, as well as the vertical infiltration and subsurface erosion caused by the seepage pressure of water flow, accompanied by the repeated rise and fall of water level, the soil cave continues to expand upwards, ultimately leading to collapse and affecting the property of surface personnel and public facilities.

Therefore, overexploitation of groundwater resources in the study area is the main factor causing karst collapse in the area, with inducing factors including farmland irrigation and atmospheric precipitation. Based on the understanding of the formation mechanism of collapse in the Jingquan water source area, relevant measures can be taken to eliminate or weaken the impact of karst collapse and minimize its impact on people.

(1) In terms of surface anti-seepage, before the rainy season, drainage work should be done in advance near buildings in areas prone to karst collapse where the terrain is low and prone to water accumulation. For karst collapse pits that have already formed and there are no important buildings nearby, backfilling treatment should be carried out.

(2) In terms of groundwater level, excessive exploitation of groundwater resources should be controlled. The unreasonable exploitation and utilization of groundwater resources is the most fundamental cause of karst collapse, so rational planning of groundwater exploitation is the most effective method to prevent karst collapse [12].

(3) In terms of underground reinforcement, if engineering construction cannot avoid areas prone to collapse, grouting treatment should be carried out on the construction site and its surrounding underground karst development based on the investigation. The grouting method can fill the lower karst cave, and after the slurry solidifies, it can enhance the bearing capacity of the foundation.

3   Simulation of Cave Stability

ABAQUS is an excellent numerical simulation software widely used in geotechnical engineering [13], and it also has many built-in material models, making it very suitable for numerical simulation calculations in geotechnical engineering. This article uses ABAQUS software and combines the engineering geological conditions of the site to conduct modeling analysis.

To explore the quantitative analysis method for soil cave stability, this study selected a soil cave exposed by drilling F11-3 in a certain foundation pit project in Tengzhou City for instability mechanism analysis. Both two-dimensional simulation and three-dimensional modeling can effectively use the strength reduction method to determine the development process of plastic zones [14]. A two-dimensional model was established using strength reduction method for numerical simulation analysis [15–18]. The established model is shown in Fig. 1. The karst is located in the middle of the ③ layer of limestone rock layer, with a thickness of 3.7 m for the top plate of the cave and a height of 0.9 m. This article selects the Mohr–Coulomb ideal elastic–plastic constitutive model built into ABAQUS to describe the stress–strain relationship, assuming that the entire soil layer is an isotropic elastic–plastic body [19–22]. The soil layer is composed of topsoil, hard topsoil, medium-weathered claystone, limestone, and strongly-weathered claystone from top to bottom.

The parameters of each layer are shown in Table 1:

The reduction range of soil layer strength is 0.5–4 times, and the vertical deformation results of the soil layer are shown in Fig. 2.

As the strength of the soil layer decreases, the karst cave undergoes significant deformation and the surface subsides, indicating that the karst cave is destroyed and unstable when the strength of the soil layer is reduced by four times.

Figure 3 shows the process of cumulative plastic strain in karst caves as the soil strength weakens. In theory, the deformation process of karst caves is always accompanied by the emergence and development of some physical quantities, such as plastic zone, plastic strain, generalized shear strain, and stress level. When these

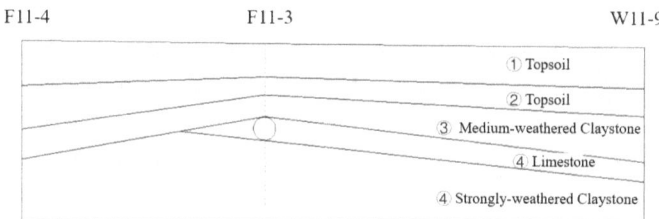

Fig. 1 Model of F11-3 stratum profile in boreholes

Table 1 Parameter table of each layer

Layer number	Characteristic values of bearing capacity (kPa)	Deformation modulus
① ②	200	12.99
③	1500	–
④	3000	–
⑤	260	20

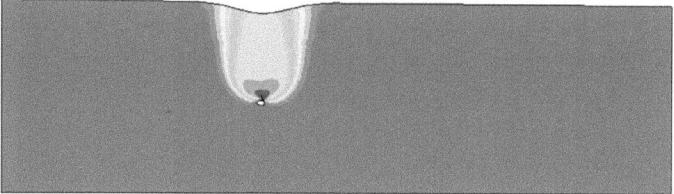

Fig. 2 Vertical deformation diagram of soil layer

physical quantities reach a certain value, karst caves become unstable. Therefore, the criterion for determining slope instability in this study is whether a continuous plastic connection zone is formed. The instability process of the karst cave can be divided into four stages:

(1) Plastic zones appear on both sides of the karst cave;
(2) The plastic zones on both sides of the cave gradually expand;
(3) The plastic zones on both sides of the cave extend upwards, forming a "U" shape;
(4) The plastic zone continues to extend upwards to the ground, and the plastic zone is connected.

When the strength reduction coefficient of the soil layer in the karst cave is 3.465, the plastic zone is connected, and the karst cave is unstable in this state. Therefore, the safety factor of F11-3 karst cave is 3.465, and the karst cave is stable. The reason is that the karst cave is located in a limestone layer, with a moderately weathered mudstone layer above it. The soil yield strength is high, and plastic strains on both sides of the karst cave are difficult to appear. It is also difficult to extend to the ground, and the plastic zone is difficult to penetrate. The karst cave exhibits stability. This method can quantitatively analyze the stability of karst caves and provide a basis for survey, design, and foundation treatment during engineering construction.

4   Conclusions

Unreasonable exploitation of groundwater will lead to a decrease in groundwater level, while short-term heavy precipitation and other factors will cause an increase in groundwater level. Under the influence of both factors, periodic fluctuations in groundwater level are important reasons for karst collapse.

The impact of unreasonable groundwater extraction and rainfall on soil is manifested as a decrease in soil strength. The strength reduction method can be used in the finite element software ABAQUS to quantitatively analyze the stability of karst caves.

When the strength of the soil decreases, the changes in the soil during karst collapse process are as follows: plastic zones first appear on both sides of the karst

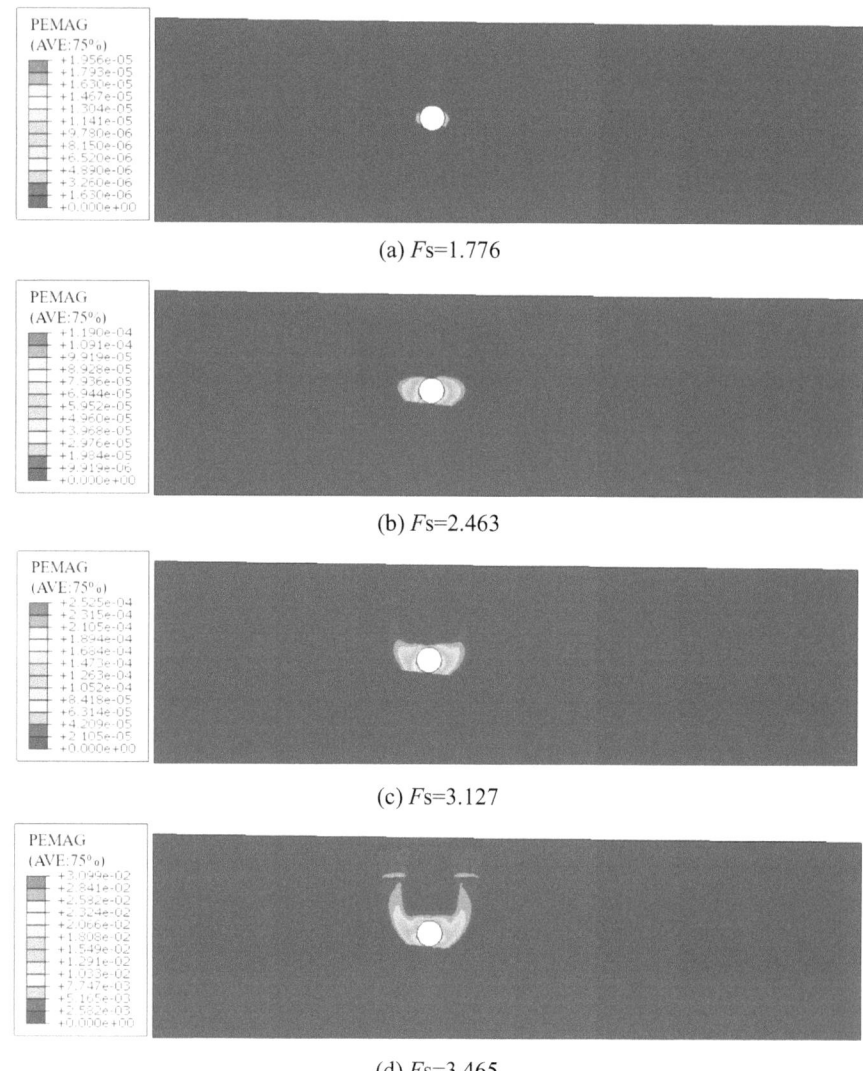

(a) $F_s=1.776$

(b) $F_s=2.463$

(c) $F_s=3.127$

(d) $F_s=3.465$

Fig. 3 Plastic strain diagram of soil

cave. As the strength of the soil further decreases, the plastic zone gradually expands and extends upwards, and finally the plastic zone connects, leading to karst collapse.

Acknowledgements This study is jointly supported by 2022–2023 Shandong Coalfield Geological Bureau Research Special Project (Lu Mei Di Ke Zi (2022) No. 23). Its support is gratefully acknowledged. Author would like to thank anonymous reviewers for their valuable and constructive suggestions.

References

1. S. Zeng, Z. Liu, Karst-related carbon sink and the carbon neutral potential by carbonate liming in non-karst areas in China, Kexue Tongbao/Chinese. Sci. Bull. **67**, 4116–4129 (2022)
2. N. Ravbar, J. Mulec, C. Mayaud, M. Blatnik, B. Kogovšek, M. Petrič, A comprehensive early warning system for karst water sources contamination risk, case study of the Unica springs, SW Slovenia. Sci. Total. Environ. **885**, 163958 (2023)
3. X. Chen, M. Xue, Y. Song, Analysis of the influence of groundwater level fluctuation on Karst soil cave collapse. Geofluids **2023**, 1468–8123 (2023)
4. Y. Su, J. Liao, T. Huang, Study on development and failure characteristics of soil caves in urban surface subsidence, Hunan Daxue Xuebao. J. Hunan Univ. Nat. Sci. **48**, 177–184 (2021)
5. J. Jia, X. Meng, P. Bai, Analysis on characteristics and origins of karst collapse in Jing Quan water source in Tengzhou City. Shandong L. Resour. **34**, 41–46 (2018)
6. K. Papadopoulou-Vrynioti, G.D. Bathrellos, H.D. Skilodimou, G. Kaviris, K. Makropoulos, Karst collapse susceptibility mapping considering peak ground acceleration in a rapidly growing urban area. Eng. Geol. **158**, 77–88 (2013)
7. J. Pan, X. Rong, J. Chang, B. Lin, Study on karst collapse in Jingquan water source area in Shandong Province. Shandong L. Resour. **38**, 56–62 (2022)
8. H. Keqiang, J. Yuyue, Z. Min, C. Weigong, Comprehensive analysis and quantitative evaluation of the influencing factors of karst collapse in groundwater exploitation area of Shiliquan of Zaozhuang, China. Environ. Earth Sci. **66**, 2531–2541 (2012)
9. T. Liu, Y. Liu, Z. Deng, X. Ji, Z. Zhang, Study on the safety of tunnel in filling karst area under dry-wet cycle **18**, 983–992 (2022)
10. H. Chen, R. Guo, X. Chen, Collapse model and influence factor analysis of covered karst soil cave induced by vacuum erosion. J. Eng. Geol. **30**, 1284–1291 (2022)
11. C. Tianwen, W. Jin, H. Dawei, Analysis of the settlement behaviors for pile group with low cap under precipitation. Chin. J. Undergr. Sp. Eng. **12**, 1659–1666 (2016)
12. B. Lan, Mechanism and prevention measures of karst ground collapse in a certain area. Eng. Des. Gr. **14**, 77–78 (2020)
13. C.E. Augarde, S.J. Lee, D. Loukidis, Numerical modelling of large deformation problems in geotechnical engineering: a state-of-the-art review. Soils Found. **61**, 1718–1735 (2021)
14. Y. Li, *A Contrastive Study on Numerical Simulation of Landslide Deformation and Failure in Different Dimensions: Illustrated by the Case of Leijiaping Landslide in Three Gorges Reservoir Area*. Master Degree thesis, China University of Geosciences (Beijing, 2017)
15. X. Zhang, *Degree Study on Collapse Test and Numerical Simulation of Soil Cave in Karst Area under Heavy Rainfall*. Master Degree thesis, Guilin University of Technology, 2022
16. W. Shi, *Research on the Development Law and Stability of Soil Caves at Wulong Airport*. Master Degree thesis, Guizhou University, 2021
17. F. Lu, *Research on Safety Degree of Existing Tunnel Based On Strength Reduction Method* (Southwest Jiaotong University, 2014)
18. Y. Zhang, *Study on the Mechanism of Karst Cave Instability Induced by Shallow Buried Coal Seam Mining in a Mine in Guizhou*. Master Degree thesis, Guizhou University, 2022
19. G. Liu, *The Stability Analysis of Cavern Under the Pile Foundation of Bridge in Karst Area*. Master Degree thesis, University of Science and Technology Beijing, 2005
20. S. Ren, *Simulation of Subsidence Deformation Law of Highway Subgrade over Shallow Coal Seam Subsidence Area*. Master Degree thesis, AnHui University of Science and Technology, 2021
21. Q. Ye, *Study on the Collapse Mechanism of Karst Soil Cave under the Action of Precipitation Funnel*. Master Degree thesis, Anhui University of Science and Technology, 2021
22. Y. Xue, *The Airport Karst in Bijie of Guizhou Collapses on the Cause Analysis and Appraisal of Stability*. Master Degree thesis, Chengdu University of Technology, 2009

Study of Discharge Capacity of Curved Weir

Peisheng Qiu, Yu Zhou, Linjie Wang, Yuan Guo, Tinghui Ge, Dongxue Wang, Minrong Xu, Xin Mao, and Shuxian Zhu

Abstract The curved weir, as an innovative weir design, offers an extended upstream edge compared to orthogonal weirs, enhancing discharge capacity across its width and ensuring river flow safety. However, for rivers of fixed width, the effects of the curved configurations, partically the central angles, on hydraulic performance, warrant further investigation. This research uses FLOW-3D software to numerically simulate curved weir and orthogonal weir with different central angles (120°, 150° and 180°), and comparatively analyses the hydraulic characteristics of the two types of weirs under different upstream head conditions in terms of flow pattern, flow velocity, discharge coefficient and so on. The results show that under the condition of low upstream head, the discharge capacity of curved weir is obviously stronger than that of orthogonal weir, and when the degree of central angle $\alpha = 120°$, which provides the necessary theoretical basis for the practical popularization of curved weir.

Keywords Curved weir · Discharge capacity · Flow pattern · Flow velocity · Numerical simulation

1 Introduction

Curved weir, as a common water retention and drainage structure, has been widely used in irrigation, flood control and water ecological landscape [1]. According to the geometric distribution characteristics of the weir top axis and the river channel, the curved weir can be divided into: orthogonal weir, diagonal crossing weir, labyrinth weir and curved weir etc. [2]. Among them, the curved weir is widely used in the

P. Qiu · Y. Zhou (✉) · Y. Guo · T. Ge · D. Wang · M. Xu · X. Mao · S. Zhu
College of Water Conservancy and Environmental Engineering, Zhejiang University of Water Resources and Electric Power, Hangzhou, Zhejiang, China
e-mail: zhouyu@zjweu.edu.cn

L. Wang
School of Civil Engineering and Architecture, Zhejiang Sci-Tech University, Hangzhou, Zhejiang, China

© The Author(s) 2025
W. Wang et al. (eds.), *Hydraulic Structure and Hydrodynamics*, Lecture Notes in Civil Engineering 608, https://doi.org/10.1007/978-981-97-7251-3_21

river improvement project in the plain area because its unfolding length is longer than that of the orthogonal weir, and the discharge capacity of the curved weir is obviously larger than that of the orthogonal weir under the same upstream water head. However, the theoretical level of curved weir is much lower than the engineering design level. In addition, there is a lack of comprehensive understanding of the hydraulic characteristics of curved weir. Consequently, the inability to effectively utilise and promote the use of curved weirs has led to limitations in their widespread and meaningful application. In the theoretical study of curved weirs in terms of flow pattern, water velocity, flow coefficient and other aspects of hydraulic characteristics, it is essential to elucidate their inherent principles. This clarification serves several important purposes: to assist in the selection of appropriate curved weirs, to inform the hydraulic and structural design processes, and to provide the necessary technical support and theoretical basis for the practical implementation of curved weirs.

In recent years, many researchers have focused on different types of weir designs, seeking to understand their hydraulic characteristics under different inflow conditions. Hu [3] studied the discharge capacity of the inclined weir by combining physical model test and numerical simulation analysis, and made a comparative analysis with the orthogonal weir, and the results showed that the discharge capacity of the inclined weir was more effective than the orthogonal weir under the condition of equal width of the river unit, and the flow coefficient formula of the inclined weir was derived by theoretical analysis, which verified the accuracy and reliability of the numerical simulation software. Sorensen [4] found through model test that in the case of low upstream water level, the falling water will occur the phenomenon of overstep falling, that is, when the water falls from the step on the weir, it will occur jumping and falling to the step a few steps apart. Kumar et al. [5] carried out a comparative study of the flow coefficients of the trapezoidal weir and the Qin key weir through model test and numerical simulation, the results show that under the same upstream head condition, the flow coefficient of the Qin key weir is higher than that of the trapezoidal weir, and this difference is especially obvious in the case of high water level. Peng [6] found that the discharge capacity of the labyrinth weir is several times larger than that of the traditional linear weir through the test, which has good value for popularization and application, and adjusted the flow coefficient formula of the labyrinth weir based on the test results and theoretical analysis. Huang [7] simulated the effect of arc radius, number of arcs and other parameters on the flow coefficient of curved weir under different upstream head conditions through physical modelling test, and the test results show that curved weir can greatly improve the discharge capacity compared with orthogonal weir, and under the condition of low head, the discharge capacity of curved weir is more good. Guo [8] found that the discharge capacity of curved weir with different arc angles is analysed through physical modelling test and numerical simulation analysis, which shows that the discharge capacity of curved weir is inversely proportional to the number of arc angles under the same upstream head condition.

2   Numerical Simulation

FLOW-3D is a powerful three-dimensional simulation software, users can create a variety of different models and mesh creation to achieve fluid flow simulation calculations. The basic control equations are mainly the conservation of mass and energy equations. This research focuses on the discharge capacity of the curved weir, so the energy equation of the fluid may not be considered.

2.1   Governing Equations

Continuity equations.

$$V_F \frac{\partial \rho}{\partial t} + \frac{\partial}{\partial x}(\rho u A_x) + R\frac{\partial}{\partial y}(\rho v A_y) + R\frac{\partial}{\partial z}(\rho w A_z) + \xi \frac{\rho u A_x}{x}$$

$$= R_{DIF} + R_{SOR} \tag{1}$$

In Eq. (1): v_F is the volume micrometabolite; ρ is the density; u, v, and w are the velocity components in each direction; A_x, A_y, and A_z are the fluid areas in each direction; R_{DIF} is the turbulent diffusion term; and R_{SOR} is the mass source.

Momentum equation.

The Navier–Stokes equations (N-S equations for short) are a momentum conservation equation used to describe the motion in an incompressible viscous fluid, and the specific expression of this equation is shown in Eq. (2):

$$\frac{\partial u}{\partial t} + \frac{1}{V_F}\left\{ uA_x\frac{\partial u}{\partial x} + vA_yR\frac{\partial u}{\partial y} + wA_z\frac{\partial u}{\partial z} \right\} - \xi\frac{A_y v^2}{xV_F}$$

$$= -\frac{1}{\rho}\frac{\partial p}{\partial x} + G_x + f_x - b_x - \frac{R_{SOR}}{\rho V_F}(u - u_w - \delta u_s)$$

$$\frac{\partial v}{\partial t} + \frac{1}{V_F}\left\{ uA_x\frac{\partial v}{\partial x} + vA_yR\frac{\partial v}{\partial y} + wA_z\frac{\partial v}{\partial z} \right\} + \xi\frac{A_y uv}{xV_F}$$

$$= -\frac{1}{\rho}\frac{\partial p}{\partial x} + G_y + f_y - b_y - \frac{R_{SOR}}{\rho V_F}(v - v_w - \delta v_s)$$

$$\frac{\partial w}{\partial t} + \frac{1}{V_F}\left\{ uA_x\frac{\partial w}{\partial x} + vA_yR\frac{\partial w}{\partial y} + wA_z\frac{\partial w}{\partial z} \right\}$$

$$= -\frac{1}{\rho}\frac{\partial p}{\partial x} + G_z + f_z - b_z - \frac{R_{SOR}}{\rho V_F}(w - w_w - \delta w_s) \tag{2}$$

In Eq. (2): A_x, A_y, A_z are the fluid acceleration in the x, y, z direction; f_x, f_y, f_z are the viscous acceleration in the x, y, z direction; b_x, b_y, b_z are the fluid losses blocked by the baffle in the x, y, z direction.

2.2 Geometric Modeling

The curved weir model shown in Fig. 1 is a thin-walled weir with a weir thickness of $W = 0.025$ m. The weir width is $B = 1.2$ m, the weir height is $H = 0.1$ m, and the number of curves is $n = 3$. In order to compare the effects of different central angles α on the hydraulic characteristics of curved weirs, three different central angles α were designed on the basic model in this research, namely $\alpha = 120°$, $\alpha = 150°$ and $\alpha = 180°$. In order to compare and verify the hydraulic characteristics of the curved weir and the conventional orthotropic weir, a three-dimensional model diagram of orthotropic weir was designed with the weir thickness $W = 0.025$ m, the weir width $B = 1.2$ m, and the weir height $H = 0.1$ m (Fig. 2).

Fig. 1 3D model view of curved weir with different central angles of each related arc

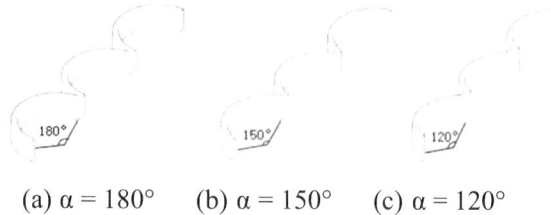

(a) $\alpha = 180°$ (b) $\alpha = 150°$ (c) $\alpha = 120°$

Fig. 2 3D model drawing of orthogonal weir

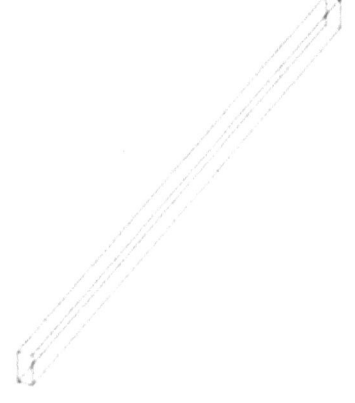

Fig. 3 Grid division map

2.3 Division of the Grid

The number of grids has a greater impact on the accuracy of the calculation, the more the number of grids, the more accurate the calculation is, the more the calculation results are in line with the actual situation, but the time spent on the calculation is also longer. This research in the numerical simulation using FLOW-3D software rectangular grid division processing technology, grid block total length 2.4 m, width 1.2 m, height 0.1 m, grid size 0.015, so the total number of curved weir grids for 166,400. The specific grating diagram takes the angle $\alpha = 120°$ of the curved weir as an example, as shown in Fig. 3.

2.4 Grid Independence Analysis

When carrying out numerical simulation analysis, the water flow characteristics of continuous media often produce certain errors, mainly due to the use of finite-length discretization to approximate the continuous system, but when the error size of numerical simulation tends to zero, the finer the mesh precision, the error is negligible [9, 10], so the model establishment should be considered for grid independent problem.

Roache [11] in his study proposed a unified metric applicable to the calculation of discrete errors called The Grid Convergence Index, or GCI for short. The size of the value of the Grid Convergence Factor is related to the accuracy of the calculation, i.e. the smaller the value of the GCI, the more accurate the result is, and is calculated as in Eq. (3).

$$GCI_{i+1,i} = F_s \frac{|\sigma_{i+1,i}|}{f_i(r^\beta - 1)} \tag{3}$$

In Eq. (3): F_s is the safety factor, $F_s = 1.25$ for the three grid schemes; β is the convergence accuracy, $\beta = 1$ for the first command and $\beta = 2$ for the second command; f is the representation of the different variables flow rates and flow velocity; D is the grid size, $r = \frac{D_i}{D_{i+1}}$; $|\sigma|$ is the absolute error, $|\sigma| = \frac{(f_i - f_{i+1})}{f_{i+1}}$.

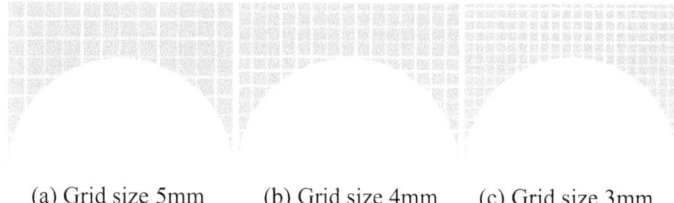

(a) Grid size 5mm   (b) Grid size 4mm   (c) Grid size 3mm

Fig. 4 Model meshing diagram

Table 1 Calculated values of grid convergence (GCI) values

Grid size (mm)	$r(D_i/D_{i+1})$	Q(L/s)	σ	GCI (%)
5	–	4.561	–	–
4	1.25	4.597	0.00974	2.16
3	1.33	4.639	0.00914	1.49

In this research, the Grid Convergence Index (GCI) method is used to perform mesh independence analysis, which is calculated for three different mesh size division schemes, with mesh sizes of 5 mm, 4 mm and 3 mm, and the corresponding total number of meshes are 176,440, 262,375 and 482,470, respectively, as shown in Fig. 4.

According to Eq. (3), the value of GCI at upstream head H = 0.1750 m was calculated, as shown in Table 1. The results show that the value of the GCI decreases as the mesh size is reduced. When the mesh size is 3 mm, the mesh convergence factor decreases to 1.49%, indicating that the accuracy of the mesh calculation is more accurate at this time and the influence of mesh size on accuracy decreases [12, 13]. As the grid size decreases, the error in flow rate between experimental and simulated values decreases. With this grid size, the measured flow error for the curved weir α = 120° is only 5.11%, indicating that this grid size can well characterise the flow of the curved weir and can be used in the next step of the discharge capacity analysis.

2.5 Model Validation

To verify the accuracy and feasibility of the numerical simulation, the simulated value is compared with Guo's [8] experimental value in this research. The experimental flow velocity, simulated flow velocity and relative errors of the curved weirs for α = 120°, 150 and 180° at the cross section along the x-direction (X = 0.44 m) are shown in Table 2.

In Table 2: Experimental and simulated value are in m/s; errors are in %.

According to Table 2, the experimental flow velocity and simulated flow velocity have errors in five different upstream head with different central angles of the curved weir, in which the experimental flow velocity is slightly smaller than the simulated

Table 2 Comparison of experimental and simulated values of flow velocity

Upstream head (m)	$\alpha = 120°$			$\alpha = 150°$			$\alpha = 180°$		
	Experimental value	Simulated value	Errors	Experimental value	Simulated value	Inaccuracies	Experimental value	Simulated value	Errors
0.1250	0.7951	0.8083	1.66	0.8450	0.8758	3.65	0.8716	0.8927	2.42
0.1375	1.0985	1.1287	2.75	1.0949	1.1130	1.66	1.0352	1.0452	0.97
0.1500	1.1603	1.2111	4.38	1.1758	1.1878	1.02	1.1505	1.1722	1.89
0.1625	1.2896	1.3254	2.78	1.2559	1.2672	0.90	1.2093	1.2349	2.12
0.1750	1.3513	1.3760	1.83	1.3294	1.3317	0.17	1.3002	1.3140	1.06

flow velocity. The maximum error between the experimental flow velocity and the simulated flow velocity is 4.38%, and the minimum error is 0.17%, which indicates that the numerical simulation and grid division adopted in this research have certain accuracy and feasibility, and the calculation value is reliable.

3 Results and Analysis

3.1 Flow Patterns

The following upstream head H = 0.1750 m as an example, through the FLOW-3D were simulated in the case of the same upstream head, three different central angles α of the curved weir of the overall flow situation, the flow pattern shown in Fig. 5.

According to Fig. 5, it can be found that the central angles α of the curved weir α = 120° when the underflow will form a apparent cavity near the top of the curved weir, and when the central angles (α) of the curved weir is increased to α = 150° or α = 180°, the cavity of the curved weir disappears when the discharge occurs. It can be seen that as the central angles α of the curved weir increase, the unfolding length (discharge length) of the curved weir also increases, resulting in less water flowing through both sides of the arc of the weir, and its discharge capacity is gradually reduced. Therefore, the discharge capacity of the curved weir is higher compared to the other two types of curved weir when the degree of central angle is α = 120°.

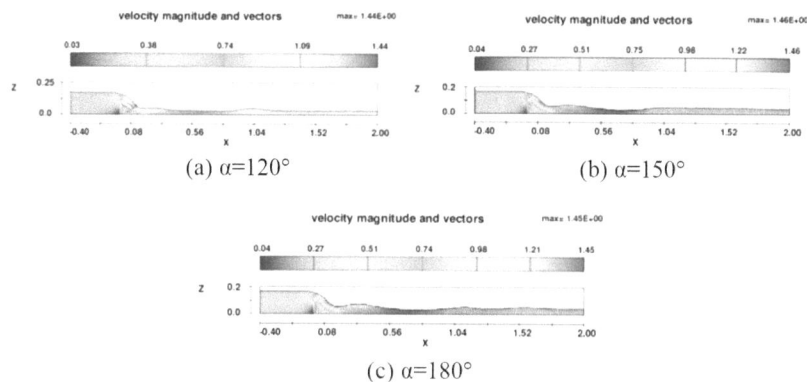

Fig. 5 Schematic flow pattern of curvilinear curved weir with different center-of-circle angles for upstream head H = 0.1750 m

3.2 Flow Velocity

Comparative analysis of flow velocity of different weir types for the same upstream head case.

According to Fig. 6, it can be clearly seen that: under the same upstream head, the flow velocity of the underflow through the curved weir and the orthogonal weir along the x-axis direction have different changes, and the flow velocity of curved weirs with different central angles α change only slightly. For the curved weir with different central angles α, when the upstream head $H = 0.1250$ m and $H = 0.1375$ m, the flow velocity of the water over the weir vary between 0.4–1.2 m/s and 0.85–1.3 m/s, respectively. And when the upstream head is $H = 0.1625$ m and $H = 0.1750$ m, the variation range of the water flow velocity of water over the weir is between 1.0–1.35 m/s and 1.1–1.4 m/s respectively. It shows that under low flow conditions, the variation range of the flow velocity of water over the weir with different central angles α is greater, as the upstream head increases, the variation of variation in its flow velocity decreases.

By comparing the flow velocity with that of orthogonal weir, it can be found that the curved weir is affected by the weir type, which leads to a very sophisticated flow pattern of falling water, which is a ternary flow, and the flow velocity of curved weir increase and then decrease after the interaction of water flow. And orthogonal weir due to the weir type is very simple, the water in the fall did not have obvious flow changes, the interaction between the water is negligible, so the water flow velocity will be mainly affected by the impact of friction after the fall and slowly decrease, and gradually stabilised after the water flow, in the case of medium and high water level, the flow velocity of curved weir is greater than the orthogonal weir. It is worth noting that after the water flow has stabilised, the deceleration rate of the flow velocity of the curved weir is basically the same as that of the orthogonal weir.

Comparative analysis of flow velocity at different upstream heads with the same weir type.

According to Fig. 7, it can be clearly seen that for the same type of weir, the flow velocity along the river are basically positively correlated with the values of the upstream head. The flow velocity along the river increase with the increase of the upstream head. In addition, it can be obviously found that due to the influence of the curved weir type, the curve expansion length of the α = 120° curved weir is the smallest, so its discharge capacity front is the lowest, and the water flow through both sides of the arc is more. Therefore, under the condition of the same upstream water head, the flow velocity of α = 120° curved weir is larger than that of α = 150° curved weir and α = 180° curved weir, and α = 180° curved weir has the largest curve expansion length, so its discharge capacity front is the highest, more water flows through the central of the arc, and the flow in the central of the arc has a violent collision with the lateral flow, eliminating part of the kinetic energy of the flow. The result is a relatively low flow velocity.

Fig. 6 Comparison of flow velocity curves of different weir types with the same upstream head

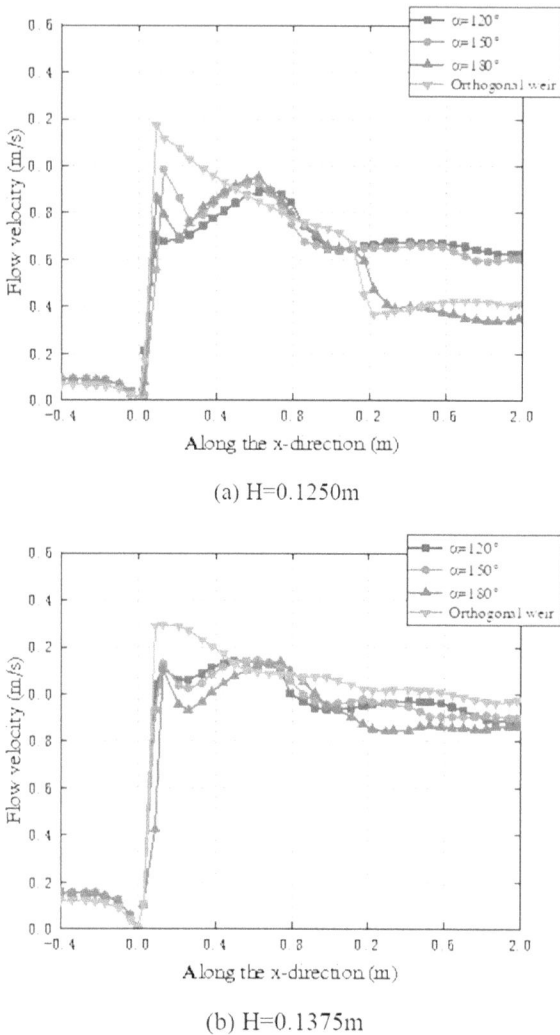

(a) H=0.1250m

(b) H=0.1375m

As the weir type of orthogonal weir is common, the flow velocity of the discharged water are almost unaffected by the weir type, the flow velocity are increased immediately after crossing the weir and then the flow velocity are slowed down relatively uniformly due to the influence of friction between the water flow and the bottom of the river channel, the flow velocity of the orthogonal weir are almost linear along the x-axis under the condition of large upstream head. Overall, the flow velocity fluctuated more along the curved weir, while the flow velocity fluctuated less along the orthogonal weir. At low head, the flow velocity of the curved weir and the orthogonal weir are almost the same, and at high head, the curved weir still maintains high flow velocity while the flow velocity of the orthogonal weir decrease significantly.

Fig. 6 (continued)

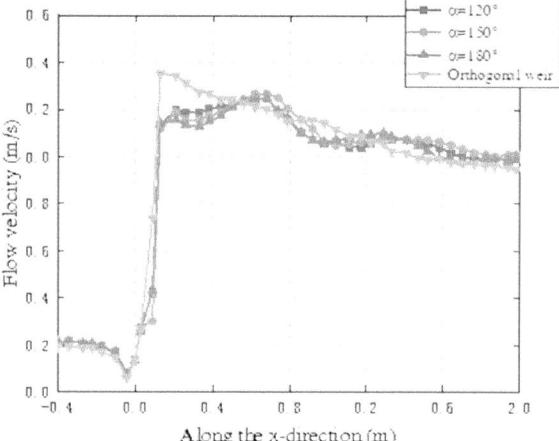

(c) H=0.1500m

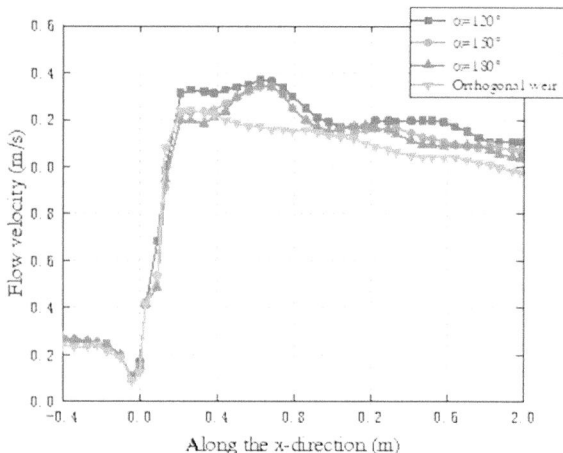

(d) H=0.1625m

Fig. 6 (continued)

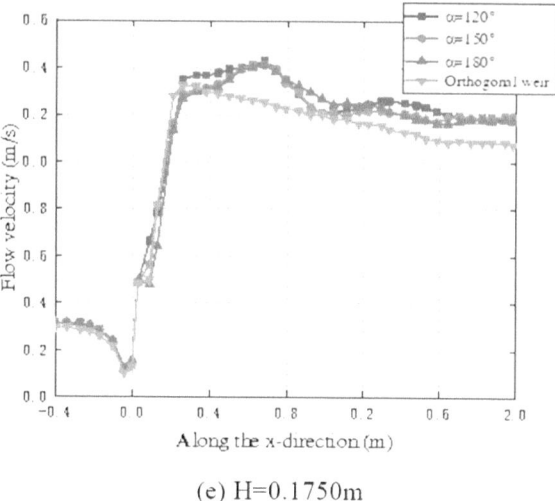

(e) H=0.1750m

3.3 Flow Coefficients

The flow coefficient m can comprehensively respond to the influence of curved weir geometry and hydraulic parameters on flow [14]. Due to the many factors affecting the flow of the curved weir, there is no unified standard expression formula at present. At present, for the curved weir research status can be found in the literature based on the weir flow formula of the conventional weir (Eq. 4) and the formula calculated according to the actual weir length of the curved weir (Eq. 5) can be obtained by the weir flow formula of the curved weir, this research will follow the above two formulas on the curved weir flow coefficient calculation, and will be the two schemes compared and analysed.

$$Q = m_1 B \sqrt{2g} H^{2/3} \tag{4}$$

$$Q = m_2 L \sqrt{2g} H^{2/3} \tag{5}$$

In Eq. (4): m_1 for the flow coefficient required based on the weir flow formula of conventional weir, B for the weir width, L for the actual weir length, H for the weir height.

In Eq. (5): m_2 for the flow coefficient calculated by the actual weir length formula of the curved weir, B for the weir width, L for the actual weir length, H for the weir height.

The variation curves of the two flow coefficients with increasing upstream head H under different upstream head conditions have been measured experimentally and are shown in Figs. 8 and 9. As can be seen from Figs. 8 and 9:

Fig. 7 Comparison of flow
velocity curves of different
same upstream head with the
weir types

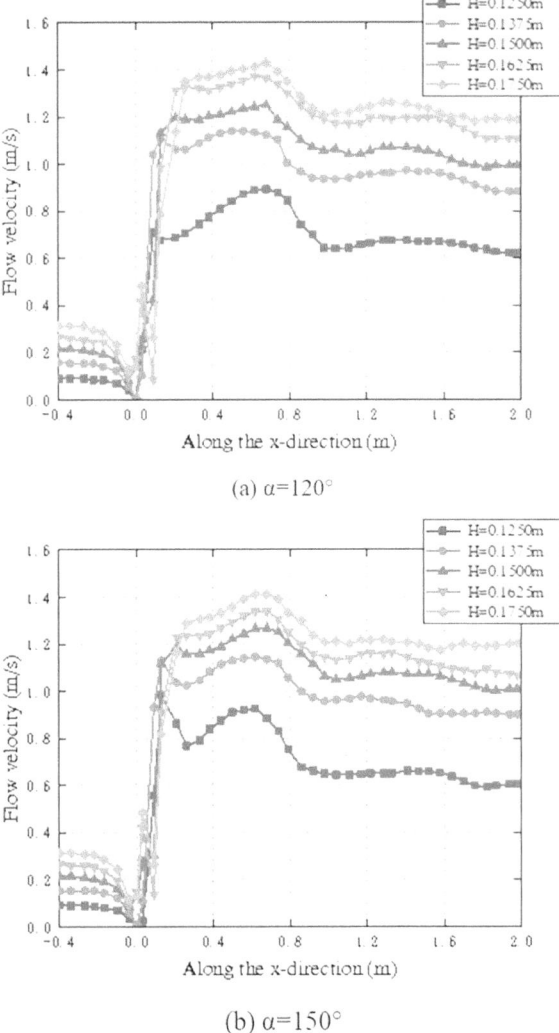

(a) α=120°

(b) α=150°

(1) When calculating the required flow coefficient based on the weir flow formula
    of conventional weir, it is determined by the weir type, upstream head, and
    the degree of central angle. In general, the flow coefficient of the curved weir
    is significantly larger than that of the orthogonal weir, especially at low water
    levels. As the upstream water level increases, the flow coefficient of the curved
    weir with three angles decreases overall, while the flow coefficient of the orthog-
    onal weir first increases and then decreases. When the central angle α = 180°,
    the flow coefficient of the discharge capacity weir is significantly lower than
    that of the curved weir with a central angle α = 120° and that of the curved

Fig. 7 (continued)

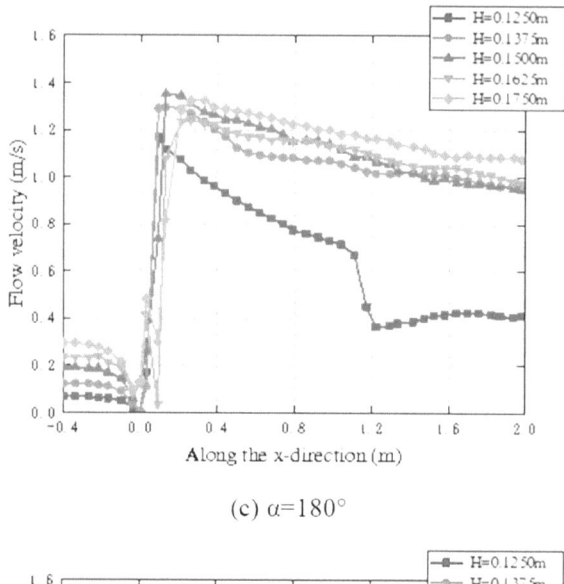

(c) α=180°

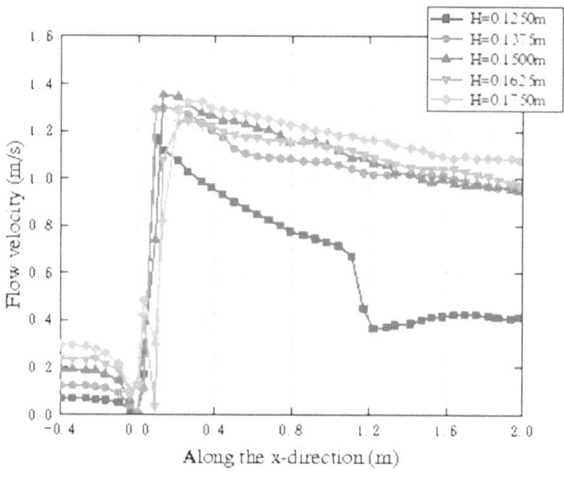

(d) orthogonal weir

weir with a central angles α = 150°. The flow coefficient is the highest when the central angle is α = 120°.

(2) When calculating the flow coefficient calculated by the actual weir length formula of the curved weir, it is also influenced by the type of weir, the upstream water head, and the central angles of the curve. However, this relationship is subject to change. As the upstream water head increases, the flow coefficient of the curved weir with a central angle α = 120° is greater than that of an orthogonal weir. Conversely, the flow coefficient of the curved weir with a central angle α = 180° is smaller than that of an orthogonal weir. For the curved weir with a

Fig. 8 Flow coefficient required based on the weir flow formula of conventional weir plotted against head over weir

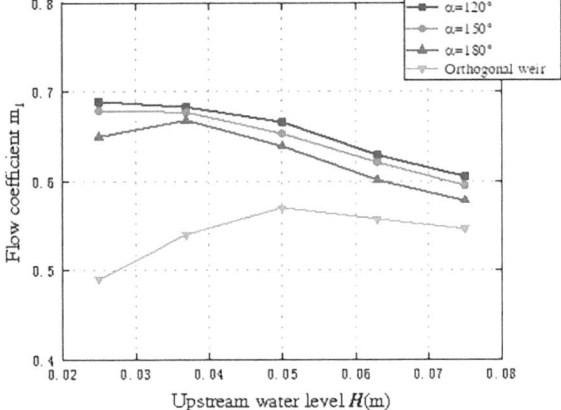

Fig. 9 Flow coefficient calculated by the actual weir length formula of the curved weir plotted against head over weir

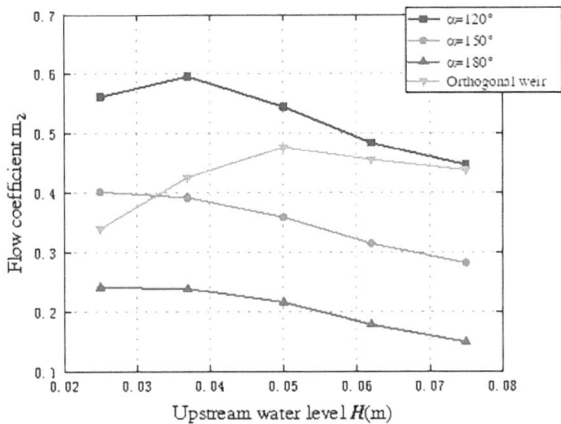

central angle $\alpha = 150°$, the flow coefficient at low water levels is higher than for an orthogonal weir. However, at high water levels, the flow coefficient is lower than that of an orthogonal weir. It should be noted that curved weirs have a higher flow coefficient than orthogonal weirs. Under identical upstream conditions, the flow coefficient of the curved weir decreases steadily with increasing central angles α, indicating that larger central angles α correspond to poorer discharge capacity performance.

4 Conclusion and Prospect

4.1 Conclusion

In this research, we use FLOW-3D software to numerically simulate and analyse the discharge capacity of the curved weir, and make a comparative study with an orthogonal weir, and come to the following conclusions:

(1) The discharge capacity of the curved weir is related to the upstream head and the degree of the central angle. For the same type of weir, the discharge capacity of the curved weir is directly proportional to the upstream head. If the upstream head is the same, the increase in the degree of the central angle will cause the expansion length of the curved weir (discharge capacity length) to increase, resulting in less water flowing through both sides of the arc of the weir, and its discharge capacity will be weakened.

(2) From the analysis of the longitudinal flow velocity, in the case of low water level, the longitudinal flow velocity of the curved weir is almost the same as that of the orthogonal weir, and due to the influence of the weir type, which makes to the flow pattern of its water fall is very complicated, the longitudinal flow velocity fluctuates a lot, and in the case of high water level, the curved weir can still maintain the high longitudinal flow velocity, and the longitudinal flow velocity of the orthogonal weir obviously decreases.

(3) When the flow coefficient is calculated according to the width of the river channel, the curved weir has a larger flow coefficient than the orthogonal weir, and the curved weir adopts the curve type under the condition of low head to obtain a larger discharge capacity.

(4) The flow coefficient of the curved weir shows an overall trend of first increasing and then decreasing, when the flow coefficient is calculated according to the actual weir length, the flow coefficient of the curved weir will be smaller than the flow coefficient of the orthogonal weir under the condition of high head, due to the curved weir has the effect of landscape when the head is low, so it is more widely used at present.

4.2 Prospect

In this research, the discharge capacity of the curved weir is studied by numerical simulation, and two parameters, the upstream head H and the central angles α, are defined. Combined with the simulation value, the relationship between the upstream head H and the central angle α and the discharge capacity of the curved weir is derived, and the curved weir and orthogonal weir are compared with each other for the purpose of research, and certain research results are obtained, which are of great theoretical significance and engineering value for the promotion of the curved weir in the development and application of the project in the whole world. There are many aspects of the curved weir that need to be explored further in this research due to time constraints:

(1) In this research, only three kinds of the curved weirs with the central angles are numerically simulated in FLOW-3D software, so further numerical simulations for other curved weirs with the central angle are needed, and it is recommended to carry out relevant experimental researches to explore the flow patterns and discharge capacity in depth.

(2) In this research, the discharge mode is free discharge, but there is no comparative research on the influence of different degrees of flooding on the discharge capacity of the curved weir in the case of submerged discharge, so the influence of different degrees of flooding on the overflow capacity of the curved weir needs to be followed up by in-depth study.

(3) The curved weir due to its special curved structure, the degasification efficiency tends to be higher than the traditional weir. With the progress of of river regulation work, the quality of water body has become the key object of of river regulation work. Maintaining the oxygen content in the water body at a reasonable level is of great importance for improving the water body quality, so further research is needed to study the degasification efficiency of the curved weir.

Acknowledgements The authors would like to express the appreciation for the National Undergraduate Training Program for Innovation and Entrepreneurship under Grant No. 202311481001, No. 202211481034, No. 202311481006, and No. S202311481002, and the Zhejiang Province New Seedling Plan of China under Grant No. 2023R423002.

References

1. H. Yang, *Study on Energy Dissipation Characteristics of Curved S-type Step Weir*. D. https://kns.cnki.net/kcms2/article/abstract?v=Mw9W0jY1lXC_4HtOWaB9F-TSRT9BCgBHMTbx4p-l_gOFbcuFEFaOZJl7fyI_a32Jk4lXPP301WW51uZ-AM2Fxviw6k0hY68WxKIEhFU7mVioygHtveGddygc5WQ_sgEzowq4QMX4udA=&uniplatform=NZKPT&language=CHS (2022)

2. Z. Yang, *Experimental Study on Hydraulic Characteristics of Diagonally Intersecting V-Type Wide Top Weir*. D. https://kns.cnki.net/kcms2/article/abstract?v=Mw9W0jY1lXDY-qV7QhkPm62c0MP9V8vvmyK9tIu2MeoeugmIKMtAznD2A_xSP-y077t1yqaB11VAv33tZwigZ6o4KN8AxQPzXvSOLuPNmWfTSnqZrpc5LcCTrsDQylmiQGysfG_HKNY=&uniplatform=NZKPT&language=CHS (2021)

3. L. Hu, *Experimental Study and Numerical Simulation of Overflow Capacity and Hydraulic Characteristics of Inclined Utility Weir*. D. https://kns.cnki.net/kcms2/article/abstract?v=Mw9W0jY1lXBKNKM7FTNc1-6hCkE0SPeaX_eq-mL-CZ9WuiouaivpxIatyxw6yV_OdWmss8lvw8SMi6NjPQt8brD90HZcgDOppQ7z7aaXNGkfquc4DZ03HcZahouckoZhc-Li9sB2sH4=&uniplatform=NZKPT&language=CHS (2019)

4. R.M. Sorensen, Stepped overflow capacity hydraulic model investigation. J. Hydraul. Eng. **111**(12), 1461–1472 (1985)

5. M. Kumar, P. Sihag, N.K. Tiwari, S. Ranjan, Experimental study and modelling discharge coefficient of trapezoidal and rectangular piano key weirs. J. Appl. Water Sci. **10**(1), 43 (2020)

6. X. Peng, G. Cui, S. Jia, Experimental study on discharge capacity and hydraulic characteristics of labyrinth weir. J. Tianjin Univ. **06**, 727–730 (2003)
7. X. Huang, B. Zhi, W. Liu, Physical modeling experimental study on the hydraulic properties of curved weir. J. China Water Transp. **16**(06), 189–190, 195 (2016)
8. Y. Guo, *Study on the Overflow Characteristics of Multi-stage Curved Weir*. D. https://kns.cnki. net/kcms2/article/abstract?v=Mw9W0jY1lXCQRJjvuMsvihVzkIj8LCpVjKWCtPYhqSi7F afJrZ3kTYZjYmSjtfBhIRXGiO2bPYfsY2PDiEUn5R0JINjL9v5ivz7RDccoUkoS9B8phTttI fAzIhA4r8G30iEPPwY03ak=&uniplatform=NZKPT&language=CHS (2023)
9. J.F. Christopher, The issue of numerical uncertainty. J. Appl. Math. Model. **26**(2), 237–248 (2002)
10. L. Kwaśniewski, Application of grid convergence index in FE computation. J. Bull. Pol. Acad. Sci: Techn. Sci. **61**(1), 123–128 (2013)
11. P.J. Roache, *Verification and Validation in Computational Science and Engineering* (Hermosa, Albuquerque, 1998)
12. K. Elsayed, C. Lacor, Numerical modeling of the flow field and performance in cyclones of different cone-tip diameters. J. Comput. Fluids. **51**(1), 48–59 (2011)
13. B.M. Savage, B.M. Crookston, G.S. Paxson, Physical and numerical modeling of large headwater ratios for a 15 labyrinth spillway. J. Hydraul. Eng. **142**(11), 04016046 (2016)
14. J. Mohammadzadeh-Habili, M. Heidarpour, H. Afzalimehr, Hydraulic characteristics of a new weir entitled of quarter-circular crested weir. J. Flow Meas. Instrum. **33**, 168–178 (2013)

Study on Evolution and Prediction of the Yellow River Estuary Wetland

Linjing Wu and Zhenmin Zhou

Abstract Estuary wetland, as an independent "living body", has the process of geographical evolution, i.e. juvenile stages (juvenile), maturation stage (adult) and aging stage (later). Taking the Yellow River estuary wetland as a case study, this paper calculates the evolution stages of the estuary wetland by using the area-elevation integral curve and fractal theory. The following conclusions are obtained. (1) In the juvenile stage, the basin erosion are strong, estuary wetland develops quickly due to high sediment transport into the estuary area. When the basin geomorphology develops to the adult stage, the river sediment transport is stable, the estuary wetland area continues to expand at a stable speed. In the later stage, erosion from the river basin weakens, the river sediment transport decreases, the sediment transport to the estuary wetland decreases, and the wetland begins to shrink gradually. (2) Human activities directly affect the ecological environment of estuary wetlands and shorten the evolution process of estuary wetlands. (3) According to the temporal and spatial variation trend calculation on Yellow River estuary wetland, the total area of wetland in 2025 will be 2518.4 km^2, of which natural wetland accounts for 1252.8 km^2, the area of constructed wetland accounts for 1265.6 km^2. Natural wetland begins to shrink. Therefore, it is necessary to take measures to protect the natural wetland in the Yellow River Delta, to control the development of constructed wetland reasonably, and to realize the sustainable development of wetland ecological environment in the Yellow River Delta.

Keywords Estuary wetland · Eco-environment · Wetland evolution · Trend prediction · Yellow River

L. Wu (✉)
College of Water Conservancy and Civil Engineering, Shandong Agriculture University, Taian, China
e-mail: 464353994@qq.com

Z. Zhou
North China University of Water Resources and Electric Power, Zhengzhou, China

© The Author(s) 2025 255
W. Wang et al. (eds.), *Hydraulic Structure and Hydrodynamics*, Lecture Notes in Civil Engineering 608, https://doi.org/10.1007/978-981-97-7251-3_22

1 Introduction

Estuary wetland is an very important natural resource in nature. It is a unique ecosystem formed by the interaction of river flow with sea water. It is also one of the most abundant ecological landscapes in nature. Estuary wetland has the ecological functions of stabilizing environment, protecting species genes and utilizing resources [1]. Estuary wetland variation is one of the important factors affecting human living space. In 2016, the State Council of China issued a plan for the wetland conservation and restoration system [2]. According to the programme, in 2025, the national wetland area will be not less than 533,000 km^2, of which the natural wetland area is not less than 466,000 km^2, the new developed wetland area is 2000 km^2, and the wetland protection rate will be increased to more than 50%, which shows a new breakthrough in the reform of ecological civilization construction in China, and also reflects the high requirement for wetland resources protection.

Estuaries have advantages such as water conservation, climate regulation, drought resistance and flood storage, as well as strong purification functions and important socio-economic value [3]. However, due to human activities during recent years, wetland is suffered from not only area shrinkage, but also species diversity reduction. Therefore, estuary wetland has become research focus of scholars at home and abroad during recent years [4]. There are four large scale wetlands in China, i.e. Liaohe Estuary Delta, Yellow River Estuary Delta, Pearl River Estuary Delta and Yangtze River Estuary Delta, which have formulated the corresponding wetland protection plan [5, 6].

In recent years, a lot of researchers studied the wetland evolution of the Yellow River Delta. In 2022, Jiao Ruifeng and others used 15 years of satellite remote sensing images from 1986 to 2018 as data sources to analyze the wetland area, distribution pattern, and change characteristics in the Yellow River Delta using ERDAS 9.0 and ArcGIS 10.3 platforms [7]. In 2023, Niu Xinqing et al. conducted intensive observations of the Yellow River Delta wetlands over 15 periods from 1986 to 2021 based on Landsat images, studying the spatiotemporal changes of wetlands and quantitatively analyzing driving factors [8]. In 2023, Fan Yanguo et al. used Landsat optical images and Sentinel-1 orbit reduction VV polarization SAR images to extract information from the wetland area of the Yellow River Delta, and combined it with socio-economic and natural environment data for driving factor analysis [9]. In 2016, Liu J. F. et al. used the TM image data (1989–2014) to classify and calculate the wetland index by using artificial visual identification, which reflects the intensity of human activities [10]. The impact of human activities on the Yellow River delta was analyzed from a quantitative point.

The natural evolution of estuary wetland experiences five stages: i.e. development, growth, maturity, stability, and decline [11], each stage is closely related to the sediment transport from the river to the estuary, and the sediment transport capacity of a river is directly related to the evolution process of the watershed geomorphology [12].

In the past few decades, many scholars studied the evolution and driving mechanism of estuary wetland [13, 14]. In 2015, Olmedo M. T. et al. compared three runs of models that simulate transitions among land categories. Pattern validation compared a reference map of transition to maps from pairs of runs. Quantity and allocation are helpful concepts to describe models and to compare maps [15], In 2021, Wang Yini et al. analyzed the dynamic changes in the landscape pattern of estuarine tidal flats using Landsat5 TM images and Landsat8 OLI images as the main data sources. Based on eCognition and ArcGIS software, they used object-oriented methods to interpret remote sensing images and conducted long-term (1986–2018) analysis of the landscape pattern at both the type and overall levels [16]. In 2022, Xu Zhentian et al. utilized Landsat remote sensing images as data sources and comprehensively utilized methods such as MNDWI, NDVI, visual interpretation, PCA, etc. to construct a hierarchical classification and discrimination method, extract land cover and wetland classification information from the study area, and analyze the dynamic changes of land cover and wetland in the Yellow River Delta over the past 30 years [17].

In above studies, manual vector and image-based classification methods are mostly applied. Unfortunately, the manual vector classification method has the disadvantages of complex artificial interference factors, high cost and heavy workload. The traditional image-based classification methods result in large degree of fragmentation. The minimum unit of object-oriented image analysis is a single object rather than an entire images [18].

To overcome the disadvantages of manual vector methods and traditional image-based classification methods, and according to the Yellow River basin sediment transportation, this paper established the correlation between the estuary wetland development process and the basin geomorphologic evolution to reveal the estuary wetland evolution, which could provide scientific and reasonable support to judge the evolution and variation of estuary wetland. Taking the evolution process of the Yellow River estuary wetland as an example, the wetland evolution evaluation methods in different stages were proposed and the relationship between wetland ecological environment and development was explored.

2 Materials and Methods

2.1 Data

The data used in this paper mainly include the high-precision DEM data of the Yellow River Delta and the water system map of 1:250,000 Yellow River networks, as Table 1. The vector files which include boundary data of modern Yellow River Delta, road traffic data, statistical yearbook of Dongying City from 2000 to 2019 [19], Dongying City history records and other field survey data were used for verification.

Table 1 Main data parameters

Data types	Data scope	Data precision	Unit	Data pattern	Spatial parameters
DEM	Basin scale	1/10000	m	GRID	WGS-2010-transverse
River systems	Basin scale	1:250,000	km	Map	Network

2.2 Calculation Methods

The estuary wetland evolution stages were calculated and determined by the evolution process of river basin geology and geomorphology. In this paper, the area-altitude integral curve method and the river fractal dimension algorithm in the fractal theory are used to determine the geological evolution stages of river estuary wetland.

(1) Integral curve method

Integral curve method is derived from the theory of topographic evolution stage proposed by American geographer Davis [20]. Here, we have improved the forms through elevation analysis, called area-height analysis method, as Eq. (1).

$$\frac{V}{HA} = \int_0^1 \frac{a}{A} d\frac{h}{H} \text{ or } \frac{V}{HA} = \int_0^1 x dy \tag{1}$$

where V represents the basin topography volume, m^3; H represents the height difference of the basin, m; A represents the total area of the basin, km^2; a represents the area of the horizontal section, m^2; and h represents the relative height of the contour line, m.

This method considers that the evolution stages of a watershed topography can be represented by the elevation integral curve which can reflects the landform of the basin evolution. Hence, the stages of the basin topographic evolution can be judged based on the shape of the basin landform elevation curve. When the elevation integral value is greater than 60%, the geomorphologic features change rapidly, the river system continuous to expand and strong erosion happens, which is regarded as an unbalanced juvenile topography (juvenile). When the elevation integral value is greater than 35% and less than 60%, the erosion process becomes slower, the geomorphologic form of the basin tends to be stable, and the topographic characteristics are no longer obviously changed, which is regarded as a balanced adult terrain (adult). When the integral value of elevation is less than 35%, the erosion process is very weak, the geomorphology is stable, and the terrain is no longer changed, which enters into the later stage (later). The elevation curve are shown in Fig. 1.

In Fig. 1, the vertical ordinate represents the ratio of the relative height (h) of the contour to the height difference (H) of the watershed, i.e. y = h/H. The horizontal coordinates represent the ratio of the horizontal section area (a) cut by the contour line to the total basin area (A), i.e. x = a/A.

(2) Fractal theory

Fig. 1 Elevation curve characteristic diagram

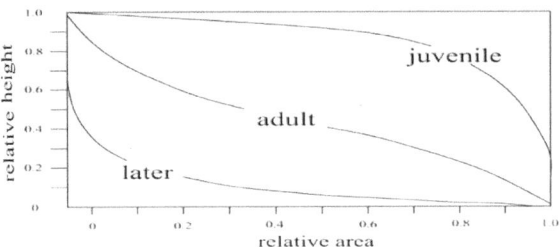

Fractal theory is a discipline, which aims to explore the phenomena and processes of disorder, instability, nonlinear, non-balance and non-certainty in nature, and to study the internal laws of order–disorder, macro–micro, entirety-local in nature [21]. The fractal dimension can be regarded as a complex quantitative parameter in watershed geomorphology to carry out study on watershed geomorphologic characteristics, which can comprehensively reflect the geomorphologic characteristics of a basin. At the same time, the fractal theory is helpful to study the relationship between river length and catchment area, to simulate the form of basin network, and to reflect the difference between the development degree of basin erosion and the geomorphology. The calculation formula is as follows,

$$D = \min\left[\max\left(\frac{\ln R_b}{\ln R_L}, 1\right), 2\right] \tag{2}$$

where D is the river channel fractal dimension; R_L is the ratio of river system length; R_b is the ratio of branch river to the main river.

2.3 Estuary Wetland Evolution Stage Classification and Analysis

According to WANG Lin's definition [22], when the fractal dimension $D \leq 1.60$, it can be classified as juvenile stage with the features of strong erosion and development. In this stage, the river system is not fully developed, the density of river network is small, the landform is relatively completed, the river is deeply eroded, the sediment transportation into the estuary increases gradually, and the estuary wetland is in the development stage. When $1.60 < D \leq 1.89$, the watershed is in the adult stage of geomorphologic evolution, and the topography of the watershed fluctuates greatly. In this stage, under the river channel lateral, gravity and slope erosion, the sharp watershed ridge is continuously eroded and becomes lower, the valley slope becomes gentle and flat, the estuary wetland sediment supply is sufficient and stable, and the estuary wetland area expands rapidly, which is easy to form large scale estuary alluvial plain and wetland delta. When $1.89 < D \leq 2.00$, the basin geomorphologic evolution is in the later stage. In this stage, the basin erosion is weak, the sediment transport to the estuary wetland is reduced, the estuary wetland evolution comes into the later stage, the area of estuary wetland is no longer expanded, whereas under the strong tidal action, the degradation and seawater invasion into the estuary wetland begin to occur.

3 Case Study

3.1 Background

The Yellow River Delta lies in the Bohai Sea to the north and Laizhou Bay to the east. The geographical coordinates spans from $117°31'E \sim 119°18'31'E$ to $36°55'31'N \sim 38°16'31'N$. The climate of the area belongs to the continental semi-humid monsoon climate, with the annual average temperature $11.7 \sim 12.6$ °C, and annual average precipitation $530 \sim 630$ mm, of which 70% concentrates in summer. The annual evaporation capacity is $1900 \sim 2400$ mm.

The Delta formation comes from the large amount of sediment from the Yellow River. The topography of the Yellow River Delta is slightly fluctuated, relatively higher in the west and south, and lower in the east and north. In 1855, The Yellow River broken its left dike flowing into Daqing river and emptied into the Huanghai sea at the Lijin county of Shandong Province. Since it entered the sea in northern Shandong province in 1855, it happened more than 70 times for the Yellow River to divert its main course, in which more than 10 major course diversions happened. The wobble of the Yellow River's course has adversely affected the ecological protection and environment of the Yellow River Delta and severely restricted economic and social development of the Yellow River Delta area. On the other hand, the Yellow River Delta is rich in land resources, with increasing new areas and great potential for economic development [1].

In 1994, The Yellow River Delta Wetland Nature Reserve was listed as one of the 16 important natural reserves in the world. In 2013, the Yellow River Delta was selected in the list of International Importance Wetlands by the Secretariat of the Wetlands Convention. The research area of this paper focus on the present fan-shaped area of the Yellow River Delta, which takes Yuwa of Kenli county as an axis point, from Tiaohekou to Song Chunrong, the total area is 5400 km², Fig. 2.

3.2 Calculation and Analysis

DEM data for the Yellow River Basin was generated using GIS software, as shown in Fig. 3.

To study the topographic features of the watershed, 13 topographic sections (20 ~ 1400 m) of sub-watersheds were taken at different elevations. The section area,

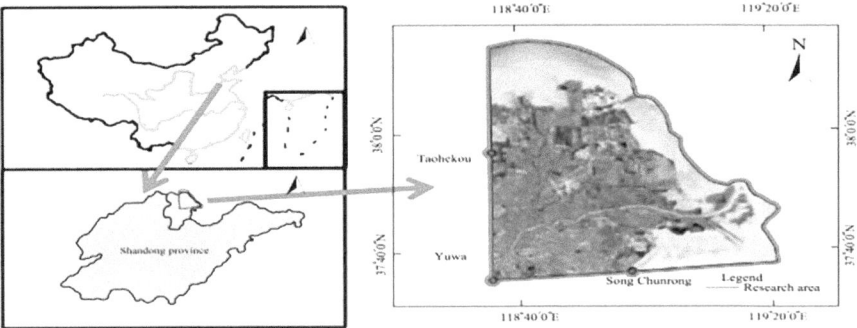

Fig. 2 Schematic map of study area

Fig. 3 The DEM of the
Yellow river

relative height and relative area for different height are calculated, and the results are shown in Table 2.

The relative height h/H and relative area a/A are calculated, and the relationship curve is generated by statistical software, as shown in Fig. 4.

The river network map is digitized, as shown in Fig. 5.

Geometric correction is made using the statistical software SPSS.21, then, the vector classification extraction is carried out. According to the classical water system classification standard, the main courses and tributaries of the Yellow River basin are divided into 5 grades, and the number of watercourses, branch ratio and length ratio of watercourses are calculated. The results are shown in Tables 3 and 4.

According to Table 3, the basin branch ratio $R_b = 1693901.92/356485 = 4.75$.

Substitute the basin branch ratio (R_b) and basin average length ratio (R_L) into formula (2), through calculation, the Yellow River network dimensional index can be obtained as follows,

$$D = \min\left\{\max\left[\frac{4.75}{\ln(1.16 + 8.06 + 1.15 + 0.087)}, 1\right], 2\right\} = 2.1$$

Comparing the watershed area integral curve in Fig. 4 with the curve in Fig. 2, it can be concluded that the curve shape belongs to the later stage of the watershed

Table 2 The different altitude cross-sectional area

Altitudes/m	Cross-sectional area/m^2	Relative height (h/H)	Relative area (a/A)
20	6,882,853,186	0.018	0.90
50	6,446,675,934	0.045	0.84
100	5,719,659,231	0.089	0.72
200	4,041,107,839	0.178	0.47
300	2,729,105,350	0.266	0.27
400	980,296,408	0.355	0.14
500	502,791,716	0.443	0.06
600	260,213,795	0.532	0.02
700	63,771,753	0.621	0.008
800	24,255,490	0.709	0.002
1000	99,089	0.886	0.0001
1200	87,564	0.916	0.00001
1400	56,321	0.946	0.000001

Fig. 4 Basin area-elevation curve

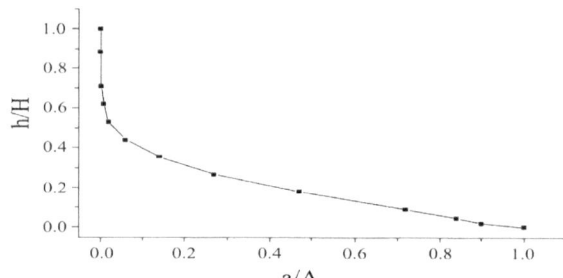

Fig. 5 The Yellow River digital river system

evolution. Besides, according to the definition and threshold of watershed fractal dimension index of Lin [22], the Yellow River basin is also in the later stage of watershed development. Therefore, from the results of area integral curve and river network fractal dimension index calculation, it can be concluded that the Yellow

Table 3 The river system branching ratio

Watercourse grades	Watercourse numbers	Ratio between two watercourses	Number of adjacent rivers	Item3 × Item4
I	111,694	4.63	222,844	1,031,767.72
II	111,150	4.84	122,280	591,835.2
III	11,130	6.20	11,245	69,719.0
IV	115	5.00	116	580.0
V	1	1		
Total			356,485	1,693,901.92

Table 4 The river system length ratio

Watercourse grades	Watercourse numbers	Total length/km	Average length/m	Length ratio
I	111,694	11,688,970.7	12,433.67	1.16
II	111,150	1,653,289.9	14,355.27	8.06
III	11,130	1,487,717.8	115,732.8	1.15
IV	115	1,164,511.8	132,902.4	0.087
V	1	11,514.31	11,514.31	

River basin should be in the later stage of river evolution. At this stage, the basin erosion is weak, the sediment transport to the estuary is reduced, the estuary wetland is also in the later stage of evolution, the area is no longer expanded, under the strong tidal action, the estuary wetland degradation and sea water invasion begin to occur.

4 Discussion

4.1 Human Activities Shorten the Evolution of Estuary Wetlands

In the middle and upper reaches of the Yellow River, such human activities as damming, sand mining, reclamation and aquaculture directly change the sediment content of the river and the landscape pattern of the estuary wetland, shorten the evolution of the estuary wetland, and accelerate the wetland to be declined. On the other hand, agricultural cultivation and animal husbandry activities, such as deforestation and wasteland reclamation in the basin could accelerate soil erosion, increase sediment content in the river flow, provide sufficient sediment sources for estuary wetlands, and promote the growth of estuary wetland vegetation, speed up wetland deposition rate and expansion of wetland area. But from viewpoint of the whole estuary wetland geological evolution process, the human activities actually shorten the estuary wetland growth period and accelerate wetland extinction process.

4.2 The Variation of Sediment Transport is an Important Factor Affecting the Wetland Morphology

Generally speaking, from juvenile to adult stage of watershed geomorphology evolution and watershed erosion are strong and flow sediment content is high, which ensures the sediment and nutrients transport into estuary wetland. In such stages, estuary wetland ecosystem shows the healthy development. In the late stage of river basin evolution, the river mainly provides sediment through lateral erosion, the sediment transport capacity becomes weak, the nutrients transport to the estuary are reduced, the vegetation inthe estuary wetland begin to degenerate, and the wetland area begins to shrink.

4.3 Prediction of Wetland Variation in the Yellow River Delta

Based on the area-elevation integral curve and river fractal theory proposed in this paper, the wetland evolution in the Yellow River Delta can be predicted. The results show that the total area of the Yellow River Delta wetland in 2025 will be 2518.4 km^2, of which the natural wetland area accounts for 1252.8 km^2, the constructed wetland area accounts for 1265.6 km^2. It can be seen that under the current change rate, the area of constructed wetland in the Yellow River Delta in 2025 is still increasing and spreading into the sea, and the area of natural wetland will be decreased. Therefore, in order to realize the goal of wetland protection in China in 2025, it is necessary to take action as soon as possible to protect the natural wetland in the Yellow River Delta, reasonably control the expansion trend of the constructed wetland, and maintain the stability of the ecological environment in the Yellow River Delta.

5   Conclusions

In this paper, both area-elevation integral curve method and river system fractal dimension algorithm are used to determine the geological evolution stages of estuary wetland. According to the river flow sediment transportation capacity, the relationship between the estuary wetland evolution and basin geomorphology were established. The estuary wetland evolution stages were revealed by studying the geomorphologic form of the Yellow River basin, which provides an evaluation basis for judging the evolution on the estuary wetland. The conclusion is scientific and reasonable.

Based on the different evolution stages of a basin, the geological evolution characteristics must be fully considered in the formulation of estuary wetland protection policies and restoration measures, and the formulation of policies should conform to the wetland evolution laws. Generally speaking, from juvenile to adult stage of river basin evolution, the river sediment transport is abundant, and the estuary wetland is in the growth and development stage. It should be strictly prohibited to carry out large-scale reclamation and water projects in the estuary area to ensure the estuary wetland to have enough space for development. In the later stage of the basin evolution, the sediment transport begins to decrease, and the estuary wetland is at the stage of atrophy and degradation. In this stage, the water resources project construction should pay attention to soil conservation in the upstream and middle stream of the basin. The estuary area should make strict management measures to prohibit the conversion of natural wetland into breeding land, industrial and agricultural land use. The shrinking and degraded estuary wetlands can be protected and restored through artificial activities such as artificial sediment transport and wetland conservation projects.

It can be seen from the analysis on the evolution of the Yellow River estuary wetland that human activities such as farmland reclamation, beach development, artificial water resources development projects and environmental pollution in the Yellow River Delta are main factors of wetland degradation.

It shows from study that the present findings would play an important action in the Yellow River delta high quality development and planning in the future. It is still worth of further research according to upstream and downstream soil conservation and climate change in the Yellow River basin.

Declarations

Competing Interests There is no conflict of interest in this manuscript.

Data Accessibility Statement All the data and materials in the current study are available from the corresponding author on reasonable request.

Consent to Participate (Ethics) The authors consent to participate in the works under the Ethical Approval and Compliance with Ethical Standards.

Consent to Publish (Ethics) All the data in the paper can be published without any competing financial interests or personal relationships that could have appeared to influence the work reported in this paper.

References

1. K.X. Chen, Analysis on the evolving progress and driving force of Estuary wetland landscape types, the Yellow River Delta wetland. J. Liaoning Normal Univ. (D) 8–49 (2019)
2. State Council of People's Republic of China, *Wetland Conservation and Rehabilitation Programme*. No. 89. http://www.gov.cn/zhengce/content/2016-12/12/ (2016)
3. Z. Yang, Practical analysis of ecological restoration technology for estuarine wetlands [J]. Metall. Manage. **15**, 34–36 (2023)
4. Y.X. Shi, Y. Li, Y. Meng et al., The pattern evolution and influencing factors of the Jiuduansha wetland in the Yangtze River Estuary from 1989 to 2020 [J]. J. Appl. Ecol. **33**(08), 2229–2236 (2022)

5. N.C. Davidson, How much wetland has the world lost? Long-term and recent trends in global wetland area. Mar. Freshw. Res. **65**(10), 934–941 (2014)
6. D.H. Mao, L. Luo, Z.M. Wang, Conversions between natural wetlands and farmland in China: amultiscale geospatial analysis. Sci. Total. Environ. **634**, 550–560 (2018)
7. R.F. Jiao, L. Ge, C. Huang, Analysis of wetland evolution in the Yellow River Delta based on remote sensing monitoring [J]. Water Resour. Inf. (04), 24–27+34 (2022)
8. X.Q. Niu, X. Zhang, C.M. Zhu, et al., Study on spatiotemporal changes and driving factors of wetlands in the Yellow River Delta based on remote sensing [J]. J. Hebei Eng. Univ. (Nat. Sci. Ed.) **40**(03), 91–98+112 (2023)
9. Y.G. Fan, J. Wang, B.W. Fan et al., Dynamic monitoring of wetlands in the Yellow River Delta based on multi-source remote sensing [J]. Surveying Mapp. Bull. **06**, 27–35 (2023)
10. J.F. Liu, Q.J. Gong, Land-cover classification of the Yellow River Delta wetland based on multiple end-member spectral mixture analysis and a Random Forest classifier. Int. J. Remote Sens. **37**(8), 1845–1867 (2016)
11. X.L. Wang, D.N. Xiao, Landscape pattern analysis of wetland in Liaohe Delta. J. Ecol. **17**(3), 317–323 (1997)
12. H.F. Cheng, P. Xin, J. Liu, F.F. Gu, W. Wang, L. Han, Morphological evolution and dynamic mechanics of the Jiuduansha Shoal (China) during 1959–2018 [J]. Advan. Water Sci. **31**(4), 491–501 (2020)
13. R.C. Frohn, M. Reif, C. Lane, Satellite remote sensing of isolated wetlands using object-oriented classification of Landsat-7 data. Wetlands **29**(3), 931–941 (2009)
14. C.Y. Wang, S. Xing, M.F. Dai, Land object classification method based on multi-scale deep feature fusion of remote sensing images and LiDAR point clouds [J]. J. Surveying Mapp. Sci. Technol. **38**(06), 604–610+617 (2021)
15. M.T.C. Olmedo, R.G. Pontius, M. Paegelow, Comparison of simulation models in terms of quantity and allocation of land change. Environ. Model. Softw. **69**(C), 214–221 (2015)
16. Y.N. Wang, Y.R. Kang, X. Chen et al., Dynamic analysis of the spatial evolution of the wetland landscape pattern in the Liaohe Estuary tidal flat [J]. J. Dalian Ocean Univ. **36**(06), 1009–1017 (2021)
17. Z.T. Xu, A. Shahzad, S. Zhang et al., Extraction of wetlands in the Yellow River Delta based on Landsat data and dynamic research in the last 30 years [J]. Ocean Limnol. Bull. **03**, 70–79 (2020)
18. I. Dronova, Object-based image analysis in wetland research: a review. Remote Sens. **7**(5), 6380–6413 (2015)
19. Dongying Municipal People's Government, *Dongying Wetland Conservation Regulations* (Feb 2019)
20. D.C. Williams, J.G. Lyon, *Use of a Geographic Information System Database to Measure and Evaluate Wetland Changes in the St. Mary's River*. Wetland and environmental applications of GIS. (CRC Press, Inc., 1995), pp. 125–140
21. M.M. Holland, Wetlands and environment gradients. In: G. Mulamoottil, B.G. Warner, E.A. McBean, *Wetland Environment Gradients, Boundaries and Buffers* (CRC Press Inc., 1996), pp. 112–131
22. L. Wang, X.W. Chen, Study on relationship between extracted river network and fractal dimension based on DEM. Geo-Inf. Sci. **9**(4), 133–137 (2007)

Experimental on Distributed Heating Fiber Optic Sensing for Typical Seepage of Embankment

Tao Zhang and Huaizhi Su

Abstract Seepage is an important factor affecting the safety of embankment engineering. It is of great significance to strengthen seepage monitoring, obtain hidden danger information in time and deal with it scientifically for ensuring the safety of the whole embankment. Guided by the heating optical fiber sensing theory, a large scale embankment model optical fiber testing platform was designed and built. With this platform, model tests of large scale embankment are carried out under three typical conditions, including variable water level, different tangential flow velocity and different rainfall patterns. The sensing efficiency of heating fiber to seepage behavior of embankment model driven by different hydraulic conditions is compared and analyzed, and the key results are summarized. The heating fiber method has little effect on expanding the monitoring range, but it can improve the sensitivity. The optical fiber in the infiltrated area has a stronger sensing ability to frequency and amplitude of fluctuations. Tangential flow velocity was positively correlated with temperature variation only in the infiltration zone. The cumulative effect of rainfall on the seepage field can be monitored.

Keywords Embankment · Seepage · Heating fiber · Monitoring model test

T. Zhang (✉)
Power China Huadong Engineering Corporation Limited, Hangzhou 311122, China
e-mail: zhang_t29@hdec.com

H. Su
College of Water Conservancy and Hydropower Engineering, Hohai University, Nanjing 210024, China

W. Wang et al. (eds.), *Hydraulic Structure and Hydrodynamics*, Lecture Notes in Civil Engineering 608, https://doi.org/10.1007/978-981-97-7251-3_23

1 Introduction

There are many embankment projects in China, but most of them have the phenomenon of too long construction period and aging performance. During its service, due to the joint action of internal and external factors, there are seepage, piping, landslide, collapse, cracks, erosion and cavitation, slope protection damage, etc., seepage disaster is particularly serious. At the same time, the diagnosis, identification and treatment of leakage danger is still based on experience or conventional equipment, and there are many shortcomings in inspection and monitoring, such as large blind area inspection, insufficient precision, difficult to accurately locate.

The longitudinal extension of embankment is very long, the conventional monitoring method is mostly point-type monitoring, which is difficult to achieve integral coverage monitoring. Aiming at this feature, many scholars have carried out a lot of useful research, the current research methods include ground-penetrating radar monitoring technology, elastic wave monitoring technology, tracer monitoring technology, distributed fiber optic temperature sensing and so on. As a natural tracer, temperature has unique advantages in seepage monitoring [1]. In recent years, many projects have proved the importance of temperature data in seepage monitoring of embankment [2]. Embankment seepage monitoring with advanced distributed optical fiber temperature sensing technology has attracted great attention in academic and engineering domains [3, 4]. Distributed optical fiber temperature sensing technology uses optical fiber as the medium and optical signal as the carrier, so it is not subject to electromagnetic interference, and has high precision and sensitivity [5–7]. It can accurately monitor the temperature value at any point along the optical fiber in real time. besides, the optical fiber used for temperature monitor of DTS system is a common communication optical fiber, which serves as both a sensor and a transmission medium. It has simple structure, low price, convenient construction, strong maintainability and high reliability, and the survival rate during construction is much higher than that of other bare optical fibers [8]. In the 1980s, the idea of studying the seepage of dam from temperature was introduced into our country, and the exploratory research of this technology was started in Danjiangkou Water conservancy project [9]. Zhu proposed the concept of distributed optical fiber monitoring system for embankments, realized real-time monitoring of seepage [10]. Leng conducted a simulation test on the monitoring of seepage of embankment by fiber optic sensor, focusing on the changes of soil temperature caused by severe seepage around the fiber optic sensor, and concluded that the fiber optic sensor can identify severe seepage with the help of the characteristics of temperature [11]. The seepage monitoring system of Xilongchi pumped storage power station of Tsinghua University is the project background. The application of DTS technology in seepage monitoring of dam is discussed [12].

In the process of service, embankments are affected by many internal and external factors, especially the rapid change of water level, tangential erosion of water flow, short-term heavy rainfall, which have strong driving effects on seepage. These three typical service conditions are the most representative, reflecting the extreme conditions of the embankment to withstand the impact of hydraulic drive damage [13]. Hence, it is of great significance to clarify the characteristics of seepage, study the sensing efficiency of heating optical fiber under those hydraulic conditions, and comprehensively apply the distributed optical fiber temperature monitoring system to seepage of embankments, so as to reasonably warn seepage and take appropriate measures.

2   Monitoring Model Test

2.1   Purpose of Monitoring Model Test

Condition 1–Variable water level (rapid rise and fall of water level)

In this condition, variable water levels such as the rapid rise of river water caused by short-term rainstorm and flood and the sudden drop of water level after flood recede are simulated. By adjusting different water levels, the characteristics of temperature changes in a period of heating fiber are monitored and analyzed.

Condition 2–Different tangential flow velocity

The test was carried out on the basis of working condition 1. Under the specified water level, the drainage pipe combined with the pump was used to produce approximate tangential flow under different flow velocity, and the conditions of different flow velocity in the model river channel were investigated, so as to explore the sensing efficiency of heating optical fiber on the seepage of the embankment under the influence of different tangential flow velocity.

Condition 3–Different rainfall patterns

The test first adjusted the water level in front of the embankment to the specified value, took the saturation permeability coefficient of the embankment as the reference critical point, simulated the rainfall intensity less than and greater than the permeability coefficient, and set different rainfall patterns to monitor the temperature change characteristics of the heating fiber, and analyzed the sensing efficiency of the heating fiber on the seepage of the embankment under the influence of different rainfall patterns.

2.2   Monitoring Model System

In order to study the effectiveness of distributed optical fiber temperature monitor technology based on heating method in monitoring of embankment seepage, a test platform composed of DTS system, heating system, seepage system and embankment

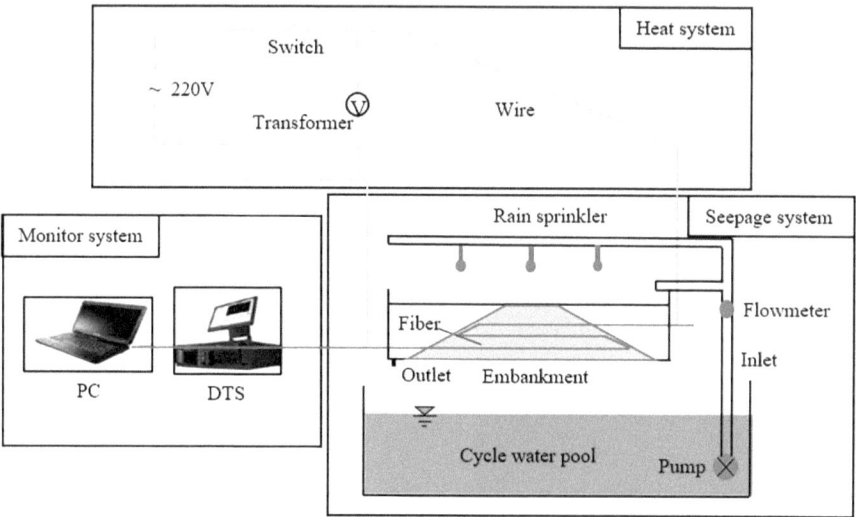

Fig. 1 Embankment seepage monitoring system platform

model was designed and assembled, which can simulate multiple factors in practical projects. The detail of platform is shown in Fig. 1.

2.3 Monitor Point Layout and Monitor Content

In the test of this paper, the optical fiber is laid in three layers. The physical diagram of layered optical fiber is shown in Fig. 2 and the corresponding coordinates are listed in Table 1.

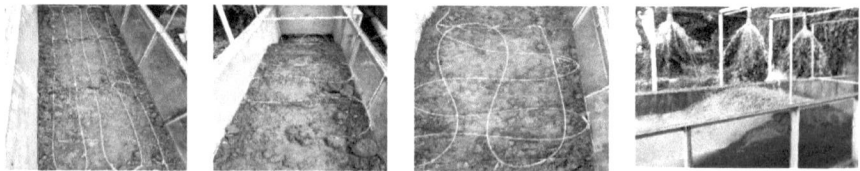

Fig. 2 Optical fiber layout and test model

Table 1 Monitoring point coordinates (bottom , middle , top)

Ponit	coordinate	Ponit	coordinate	Ponit	coordinate	Ponit	coordinate
3#	(0,0,20)	4#	(20,20,20)	5#	(120,20,20)	6#	(220,20,20)
7#	(320,20,20)	8#	(310,40,20)	9#	(210,40,20)	10#	(110,40,20)
11#	(10,40,20)	12#	(60,60,20)	13#	(160,60,20)	14#	(260,60,20)
15#	(355,65,20)	16#	(270,80,20)	17#	(170,80,20)	18#	(70,80,20)
19#	(0,100,20)	20#	(100,100,20)	21#	(200,100,20)	22#	(300,100,20)
23#	(10,0,40)	24#	(50,60,40)	25#	(90,120,40)	26#	(90,20,40)
27#	(130,40,40)	28#	(150,120,40)	29#	(170,40,40)	30#	(210,20,40)
31#	(210,120,40)	32#	(0,0,60)	33#	(10,80,60)	34#	(60,80,60)
35#	(70,0,60)	36#	(80,80,60)	37#	(130,80,60)	38#	(140,0,60)
39#	(75,20,60)	40#	(-10,40,60)	41#	(75,60,60)	42#	(160,80,60)
43#	(75,100,60)	44#	(-10,120,60)				

3   Analysis of Monitoring Model Test Results

3.1   *Condition 1*

In order to explore the sensing effect of distributed optical fiber on seepage under variable water level, the control variable method is adopted. Set three different water levels: H = 40 cm, H = 60 cm, and H = 80 cm. For each type of water level, set 3 groups of heating power, that is, choose the arithmetic sequence of 6 W/m, 12 W/m, 18 W/m. In addition, P = 0 W/m was set as the control group to eliminate the influence of natural environmental factors on the test process, and the influence of natural environment was assumed to be constant in a period. Partial experimental results are presented as Figs. 3, and 4.

(1)   The fluctuation of water level has little effect on the optical fiber in the saturated area of the embankment which has formed stable seepage; The fiber in the infiltrated area has a strong sensing ability to the fluctuation of water level, and its seepage field not only changes greatly but also fluctuates frequently under a certain water level.

(2)   The fiber in the unsaturated area of the embankment is temperature sensitive, and the heating power can be appropriately reduced or the heating fiber laying density can be reduced; The temperature in the convection zone dissipates the

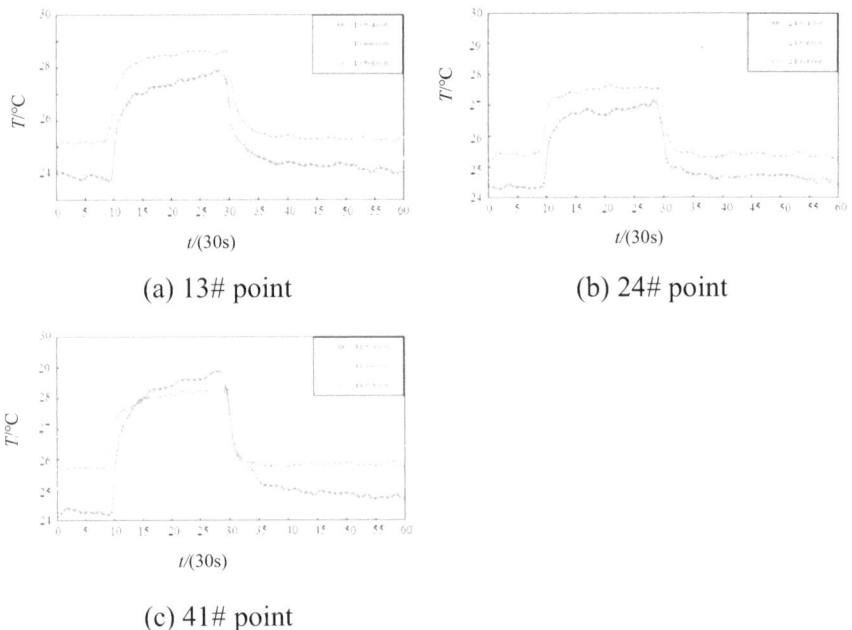

(a) 13# point   (b) 24# point

(c) 41# point

Fig. 3   Temperature of different water levels at the same monitoring point (P = 6 W/m)

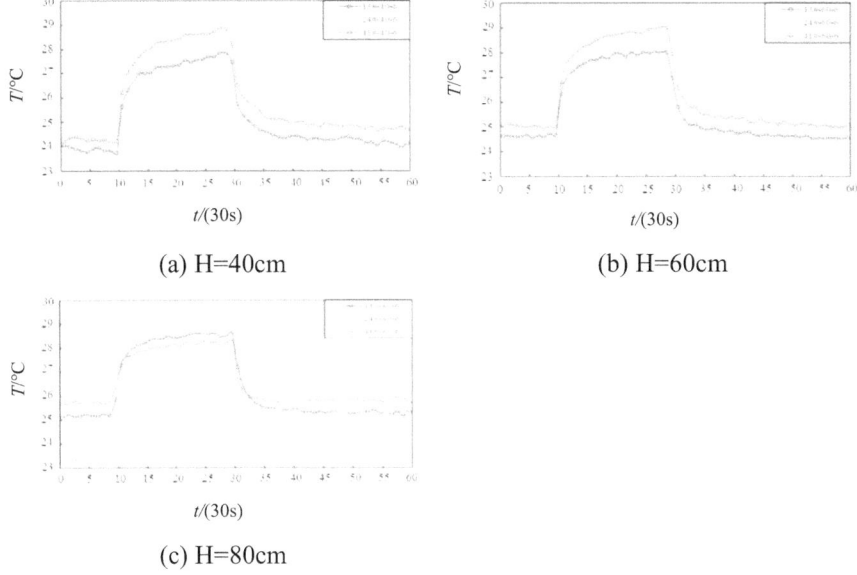

(a) H=40cm   (b) H=60cm

(c) H=80cm

Fig. 4 Temperature of different monitoring points at the same water level (P = 6 W/m)

fastest, and the fluctuation frequency and amplitude are the largest, so the heating power or optical fiber laying density should be appropriately increased. The saturated region is in the middle state.

3.2 Condition 2

In order to understand the effect of DTS sensing the temporal and spatial evolution process of seepage in the embankment under the action of tangential flow, different optical fiber monitoring points were selected reasonably based on the three directions of X, Y and Z. Select 8#, 9# and 10# as a group in the X direction; In the Y direction, 7#, 8# and 22# were selected as a group. In the Z direction, select 14#, 24#, 41# as a group. Base on the DTS monitoring platform, Partial correlation experimental results are presented as Figs. 5, 6, and 7.

(1) H = 40 cm is representative to a certain extent. When H = 60 cm and H = 80 cm, the variation rules of each measuring point along all directions are basically the same.

(2) From the stability of the measured DTS data, tangential flow mainly causes the fluctuation of temperature values near the infiltrating area, and it can be inferred that tangential flow erosion causes the fluctuation of pore water pressure near the infiltrating area. Within a certain range, the larger the tangential flow, the larger the fluctuation amplitude. Compared with the measuring points near the

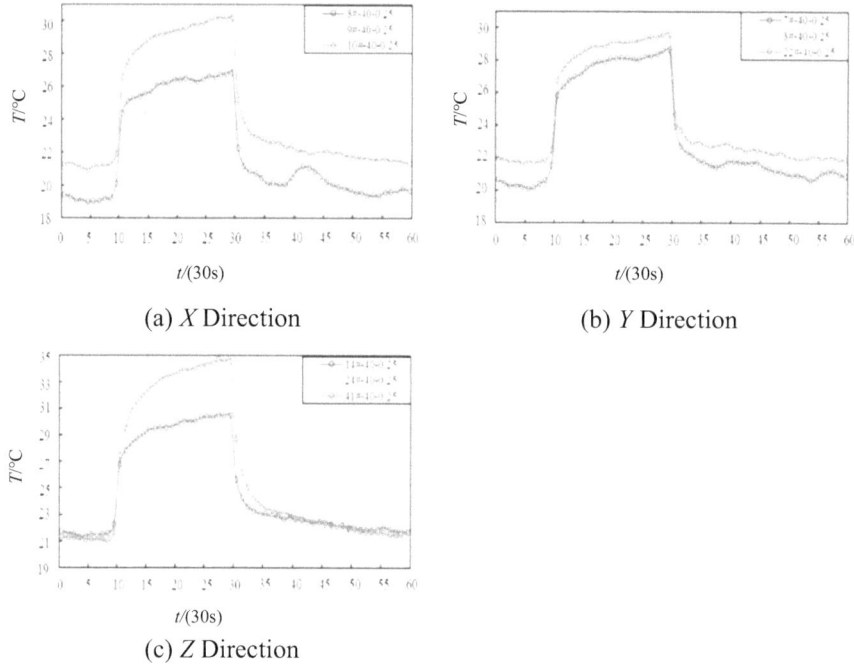

(a) *X* Direction   (b) *Y* Direction

(c) *Z* Direction

Fig. 5 Fiber monitoring value in different directions (H = 40 cm, v = 0.25 m/s)

infiltrating area, the measured temperature time series at the monitoring points in the saturated area has no obvious relationship with the presence or absence of tangential flow and the velocity, indicating that tangential flow has little influence in the saturated area. Of course, there is no seepage in the unsaturated area, and the influence does not exist.

(3) According to this test and relevant results, tangential flow has no direct relationship with the formation of seepage and its range of action, but only generates additional osmotic power at the junction of water level in the infiltrating area, and there is a certain threshold related to soil material properties. Below the threshold, the larger the tangential flow is, the larger the additional osmotic pressure is. After exceeding the threshold value, with the increase of tangential flow velocity, the dissipation of pore water pressure is accelerated, and the additional osmotic pressure decreases gradually. At the same time, the additional osmotic pressure only acts in the vicinity of the infiltrating zone, and has no obvious influence on the measuring point and the water pressure in the saturation zone.

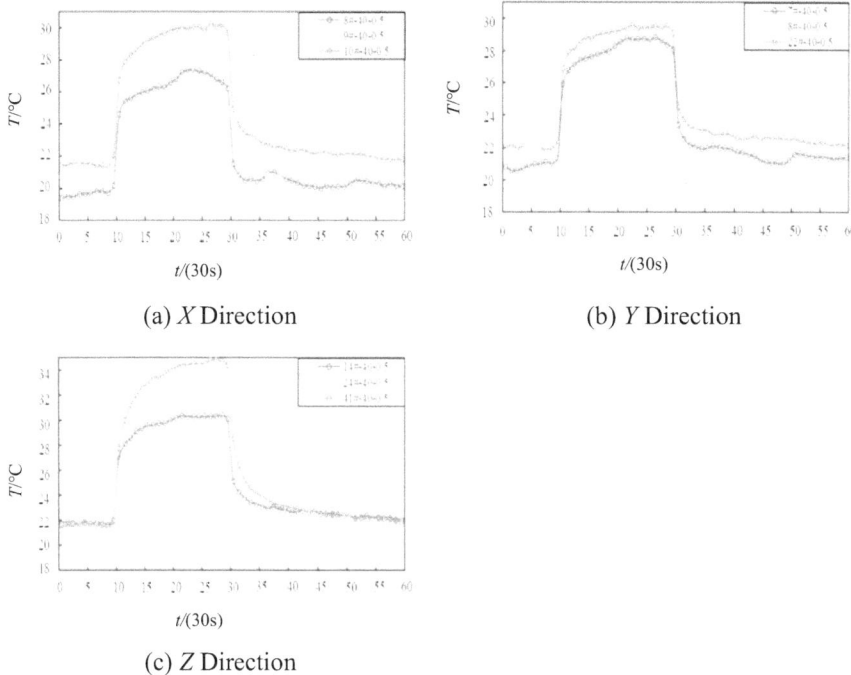

(a) *X* Direction

(b) *Y* Direction

(c) *Z* Direction

Fig. 6 Fiber monitoring value in different directions (H = 40 cm, *v* = 0.5 m/s)

3.3 Condition 3

The basic idea of this test is still the control variable. The influence of rainfall intensity and rainfall duration on seepage inside the embankment was analyzed by setting different groups of test schemes for comparative tests. In order to facilitate the experimental study, it is assumed that the average permeability coefficient of the embankment is = 2.95×10^{-4} cm/s. With the rainfall intensity calculated by the average permeability coefficient = 10.62 mm/h as the boundary, three different rainfall intensity were set, namely 5 mm/h, 10 mm/h and 15 mm/h. In the first model, the rainfall duration was set at 1080 s, 540 s and 360 s respectively, and the total rainfall remained unchanged. The other pattern is that the rainfall duration is 540 s, and the total rainfall is different. In addition, the rainfall intensity was set to 0 mm/h as the control group.

When the total amount of rainfall is constant, different rainfall time and intensity are controlled, and the measured temperature time series is shown in Figs. 8 and 9. Generally, the overall change trendence is consistent with the same rainfall time. Due to the time-lag effect of rainfall infiltration, there is no significant difference under different rainfall time when the total rainfall is controlled uniformly. A new index was defined for this model, that is, the total temperature reduction at each measuring

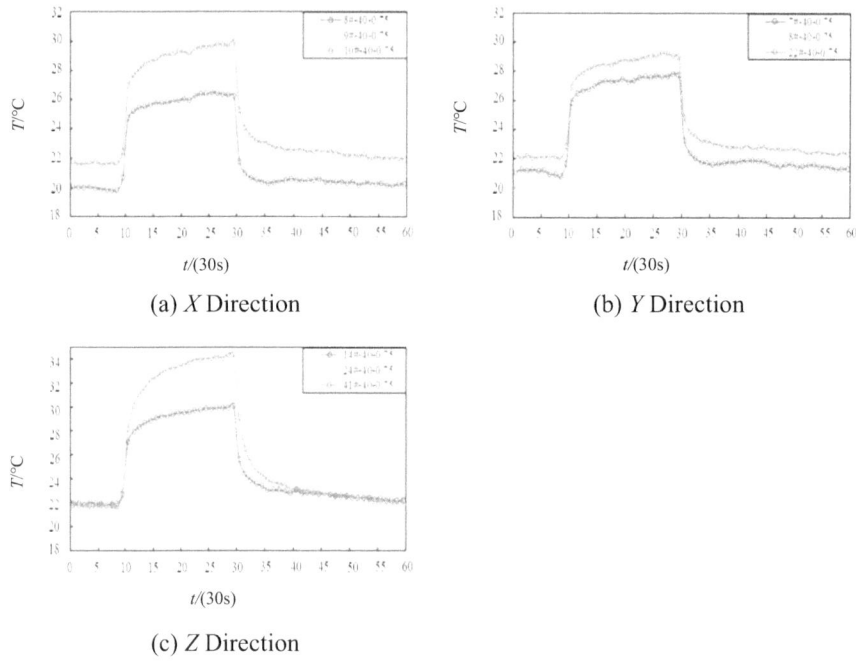

(a) X Direction   (b) Y Direction

(c) Z Direction

Fig. 7 Fiber monitoring value in different directions (H = 40 cm, v = 0.75 m/s)

point during different rainfall times was calculated, as shown in Table 2. When the total amount of rainfall is constant, although the temperature changes at different monitoring points during different rainfall times are different, the total temperature reduction is −85.423 °C, −81.025 °C, and −82.043 °C respectively, which are close to each other, indicating that rainfall is the main factor for temperature change inside the embankment. Due to the different spatial location of different measuring points, the total temperature reduction will be different, but the temperature reduction of the embankment system is stable.

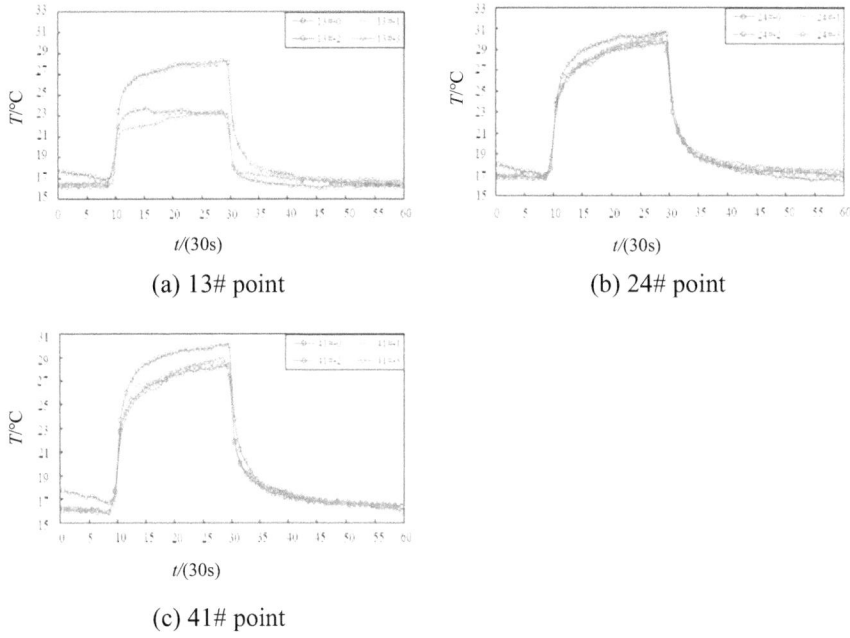

(a) 13# point   (b) 24# point

(c) 41# point

Fig. 8 Temperature time series under different rainfall intensity

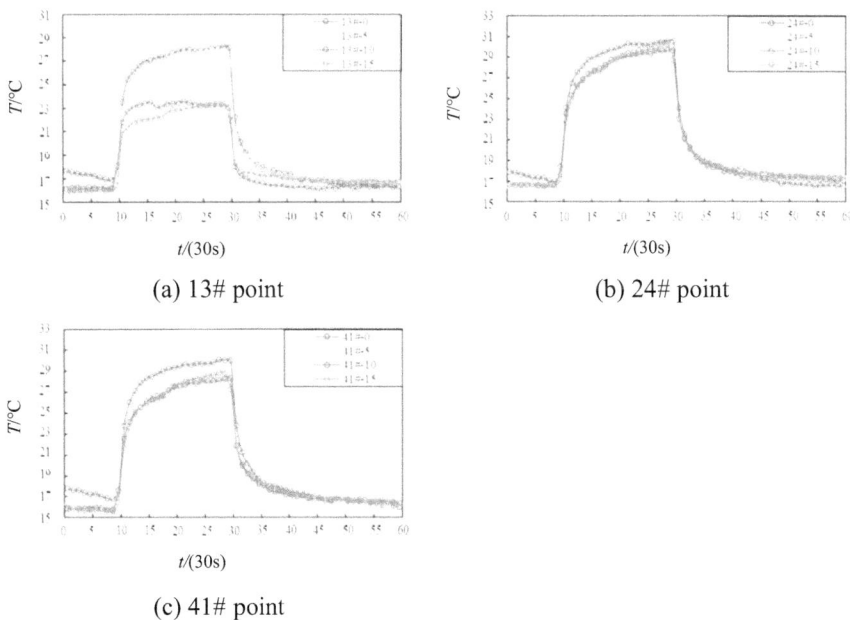

(a) 13# point   (b) 24# point

(c) 41# point

Fig. 9 Temperature time series under different rainfall time

Table 2 Total temperature change in different rainfall models (Unit: °C)

Rainfall pattern	Duration of rainfall (s)	13#	24#	41#	Total
5	1080	− 21.907	− 31.567	− 31.949	− 85.423
10	540	− 35.691	− 16.601	− 28.733	− 81.025
15	360	− 30.272	− 22.177	− 29.594	− 82.043

4 Conclusion

Based on the results and discussions presented above, the conclusions are obtained as below:

(1) The water level change has little influence on the optical fiber monitor in the embankment saturation area where stable seepage has been formed; The optical fiber in the infiltrated area has a strong sensing ability to the fluctuation of water level, and the seepage in this area not only changes greatly but also fluctuates frequently under a certain water level.

(2) There is no obvious relationship between the change of optical fiber monitored value and the water level in front of the embankment under tangential flow condition, and the change trend of value at each monitoring point is basically the same in all directions; Tangential flow velocity mainly causes the fluctuation of temperature near the infiltrating area. The larger the tangential flow velocity is, the larger the fluctuation amplitude of the infiltrating area is.

(3) When the rainfall intensity is greater than the soil permeability, part of the rainfall flows into the river along the embankment, and part of the rainfall penetrates into the embankment along the embankment surface. The former causes temperature fluctuations in the infiltrating area, and the greater the rainfall intensity, the greater the fluctuation amplitude; the latter has a certain limit and is related to the soil permeability coefficient. In the case of the same total rainfall, the embankment system has a relatively constant total temperature reduction, which indicates that the influence of rainfall intensity on the seepage field inside the embankment can be monitored by heating optical fiber, and the cumulative effect of total rainfall on the seepage field can be monitored.

(4) At present, the optical fiber monitoring in the project is mostly uniform layout and lack of key area. In the future, it may be carried out to optimize the optical fiber layout. In the service processes of embankment, it is necessary to distinguish the causes of seepage according to the characteristics of different factors, and formulate necessary measures to treat different seepage modes.

References

1. C. Hao, Quantitative monitoring technology of slope seepage based on DTS [J]. Water Conservancy Sci. Cold Area Eng. **5**(10), 127–130 (2022)
2. Z. Wang, *Study on Seepage Monitoring Method of Embankment Based on Distributed Optical Fiber Temperature Measurement Technology [D]*. (Xi'an University of Technology, 2019)
3. J. Li, X. Lv, W.F. Wang, Leak monitoring and localization in baghouse filtration system using a distributed optical fiber dynamic air pressure sensor[J]. Elsevier Inc. (57), 102–108 (2020)
4. Y. Li, *Research on Multi-parameter Distributed Fiber Optic Monitoring of Embankment [D]*. (Nanjing University, 2021)
5. Y. Fu, Y. Dong, Z. Xu et al., Research progress of distributed fiber-optic temperature tracer identification of fractured groundwater flow [J]. Adv. Water Resour. Hydropower Sci. Technol. **40**(03), 86–94 (2020)
6. A.A. Khan, V. Vrabie, Automatic monitoring system for singularities detection in dikes by DTS data measurement [J]. IEEE Trans. Instrum. Meas. **59**(8), 2167–2175 (2010)
7. J.I. Mars, A.A. Khan, V. Vrabie, et al., *Water Leakage Detection in Dikes by Fiber Optic [C]*. (Barcelona, Spain, June 2010)
8. Y. Jie, W. Zhaohan, C. Lin et al., Experimental study on seepage flow monitoring of sandy soil based on Si-DTS [J]. Water Resour. Hydropower Technol. **50**(07), 168–173 (2019)
9. X. Caizhong, P. Wenchang, Study on seepage field of dam foundation based on temperature field [J]. Yangtze River **30**(5), 21–23 (1999)
10. P. Zhu, Y.B. Leng, Y. Zhou, et al., Safety inspection strategy for earth embankment dams using fully distributed sensing [J]. Procedia Eng. 520–526 (2011)
11. L. Yuanbao, Z. Yuping, Z. Yang, Z. Qingming et al., Study on simulation test of seepage and settlement of soil embankment monitored by fiber optic sensor [J]. Yellow River People **34**(1), 9–10 (2012)
12. W. Shanxun, *Research on Seepage Monitoring System Based on Distributed Optical Fiber Temperature Measurement [D]* (Tsinghua University, Beijing, 2010)
13. H. Su, B. Ou, et al., Dual criterion-based dynamic evaluation approach for dike safety [J]. Struct. Health Monit.—An Int. J. **18**(5–6), 1761–1777 (2019)

The Influence of Rain-Type on the Seepage and Stability of Purple Silty Cohesive Soil Slope

Ou Jiang, Shiyi Yang, Xuancheng Yan, Can Yang, and Tong Liu

Abstract To study the variation law of internal seepage and stability of purple silty cohesive unsaturated soil slope under different rain patterns, we use GeoStudio to calculate and analyze the changes of pore water pressure and safety factor of the slope after undergoing three rain patterns, namely, front peak pattern, back peak pattern and step pattern, under the action of moderate rain, heavy rain and rainstorm. Research has found that under the same conditions, it takes the longest time for the pore water pressure under the back peak type to reach stability, and the slope safety factor decreases the most. Therefore, it is necessary to strengthen the management of slope stability protection after rainfall in the back peak type.

Keywords Purple silty cohesive soil · Rain-type · Seepage · Stability

1 Introduction

Affected by extreme weather conditions, landslides have become one of the main geological hazards affecting the safety of various countries, and rainfall often leads to geological disasters such as landslides [1]. Scholars have conducted extensive research on the impact of rainfall on side slopes. Scholars such as He [2] analyzed the seepage characteristics and the variation law of safety factors of side slopes under heavy rainfall. The study showed that the safety factor decreases with the duration of rainfall, but does not significantly increase after rainfall stops; Li et al. [3] studied the seepage evolution characteristics and stability changes of granite residual soil inside the high side slope of a road cut under rainfall infiltration through numerical simulation. The study found that under the continuous action of rainfall, the pore water pressure and safety factor of the side slope decrease, and after the rainfall stops, the safety factor gradually increase; Cheng et al. [4] studied the effect of rainfall on the stability of loess tunnels based on finite element theory. The analysis showed that tunnel stability is directly related to the thickness of loess cover layer

O. Jiang (✉) · S. Yang · X. Yan · C. Yang · T. Liu
Sichuan Agricultural University, Yaan 625014, China
e-mail: 1752935644@163.com

© The Author(s) 2025   283
W. Wang et al. (eds.), *Hydraulic Structure and Hydrodynamics*, Lecture Notes in Civil Engineering 608, https://doi.org/10.1007/978-981-97-7251-3_24

and rainfall seepage; Wang et al. [5] found that the stability of side slopes decreases with increasing rainfall intensity, and the safety factor of slopes is the lowest for a period of time after rainfall stops.

However, existing research mainly focuses on the impact of different rainfall intensities on the seepage of internal side slope and slope stability under a single rain-type, and there is scarce consideration given to the impact of rainfall patterns on side slopes. Therefore, this article takes purple silty clay as the research object, establishes a side slope seepage model through GeoStudio, analyzes the changes in slope seepage and stability, and provides theoretical reference for landslide instability prevention and control.

2   Research Materials and Objects

2.1   Numerical Analysis Model

Based on previous research [6] and the actual situation of purple silty clay, a homogeneous slope model is established as shown in Fig. 1. The top of the side slope is 15 m long, the height of the slope is 20 m, the bottom of the slope is 40 m long, and the slope ratio is 1:1.5. At both ends of the slope, the water level lines are 6 m and 8 m high respectively; ah, bc, and cd are the seepage ranges; ah, de, and gf are impermeable ranges; gh and ef are the boundary of fixed water level. Considering that rainfall first affects shallow soil, we establish a monitoring point A below 1.5 m from point c.

This study establishes seepage analysis through the SEEP/W module, and takes the distribution of pore water pressure in the initial state of the side slope soil as the initial condition for studying slope seepage and stability. Subsequently, different intensities corresponding to each rainfall pattern were applied to the surface of the slope for analysis, and the safety factor of the slope was calculated using the SLOPE/W module.

Fig. 1 Experimental model

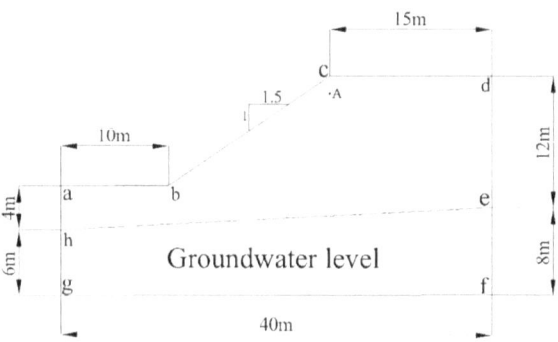

Table 1 Mechanical parameters of purple silty cohesive soil

Mechanical parameters of silty cohesive soil	Cohesive forces/ (kPa)	The angle of internal friction/(°)	Volumetric weight/(kPa)	Saturated cohesive force/(kPa)	Saturated internal friction angle/(°)	Hydraulic conductivity/ $(m \cdot s^{-1})$
Mean value	22	14.7	19.8	13.5	12	7.5×10^{-7}

2.2 Soil Parameters

This soil model adopts a single homogeneous soil, and the soil parameters are shown in Table 1. We established this soil characteristic curves and permeability coefficient curves based on the Van Genuchten model.

2.3 Rain-Type and Calculation Schemes

According to the rainfall grade of the National Meteorological Center: light rain (0 ~ 10 mm/d), moderate rain (10 ~ 25 mm/d), heavy rain (25 ~ 50 mm/d), rainstorm (more than 50 mm/d). Three rainfall levels, namely moderate rain (20 mm/d), heavy rain (40 mm/d) and rainstorm (80 mm/d), are selected for the study. The rain-type is set as front peak, back peak and ladder-type. The rainfall lasts for 24 h, and considering 12 h after the rain, the rain-type is shown in Fig. 2.

3 Analysis of Calculation Results

3.1 Seepage Analysis of Downslope Bodies with Different Rain Patterns

Different rainfall patterns were assigned to each of the three rainfall classes on the developed numerical model. Analytical calculations of pore water pressure over time were carried out for monitoring point A. The curve changes are shown in Fig. 3.

Under the three kinds of rainfall intensities, the changes in pore water pressure are different for different rainfall types. With the front peak rainfall, the pore water pressure first rises rapidly. And the curve changes first show an obvious upward "convex" shape, then the rate of change slowly decreases after a period of time; with the back peak rainfall, the pore water pressure first grows slowly, and then the rate of growth rises rapidly. And the curve shows a downward "concave" shape and upward "convex" shape; with the ladder type rainfall, the rate of change of the pore water pressure is in the middle of the front and back peak rainfall types.

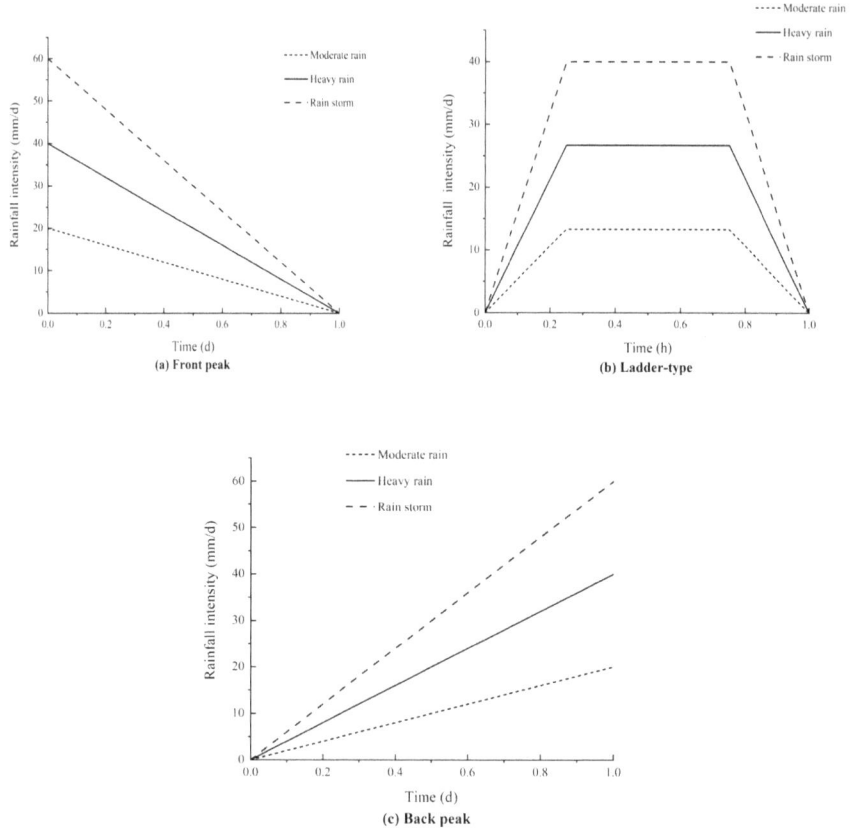

Fig. 2 Rain-type setting

3.2 Stability Analysis Under Different Rainfall Patterns

By using the non-saturated shear strength theory and the limit equilibrium method, the change in safety factor with time which was combined the situations that we have discussed above of the internal seepage changes of the slope body after different rain types under various rainfall levels and used the SLOPE/W module to calculate and analyze the change in safety factor of the slope after 24 h of rainfall and 12 h after the rain is shown in Fig. 4.

During the rainfall process, the coefficient of safety of different rainfall types under each rainfall level shows a decreasing trend with time. And after the end of the rainfall, the coefficient of safety will have a transient rebound, and finally it tends to a stable state. For the front-peak type rainfall, the coefficient of safety decreases firstly fast and then slowly, and the speed change law is shown as a "concave" type; under the back-peak type rainfall, the coefficient of safety changes firstly slow and then fast, and the speed change law is shown as a "convex" type; the change of the

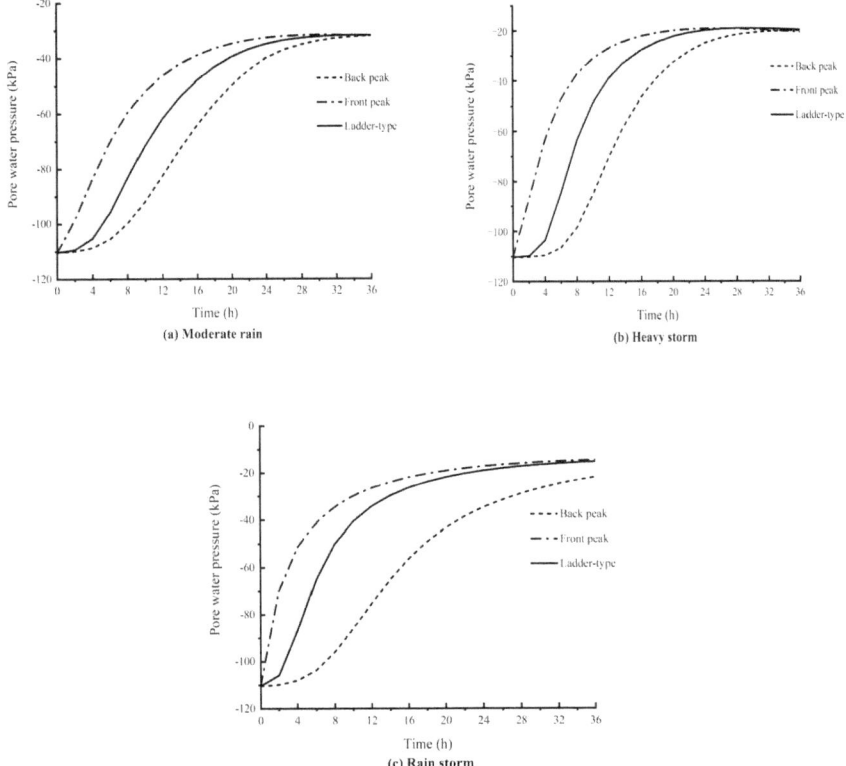

Fig. 3 Pore water pressure change curve with time at monitoring point A

coefficient of safety of the ladder-type rainfall is still between the front-peak type and back-peak type rainfall. After the cessation of rainfall, the coefficients of safety of the three types of rainfall all had a relatively small recovery.

In the initial state, the slope has a coefficient of safety of 1.25556. And the slope has the smallest the coefficient of safety at the end of the rainfall. Taking rainstorm as an example, with the increase of rainfall, the slip surface of different rainfall types changed significantly and the coefficient of safety decreased in all of them. The rate of decrease in the coefficient of safety was 0.406% for the front-peak type rainfall, 0.412% for the ladder type, while the rate of decrease was 0.419% for the back-peak type.

Based on the study of three different rainfall classes, it was found that the three different rainfall types would significantly affect the slope stability when the precipitation and duration of rainfall remain the same. Among them, the front-peak type has the least impact and the back-peak type has the greatest.

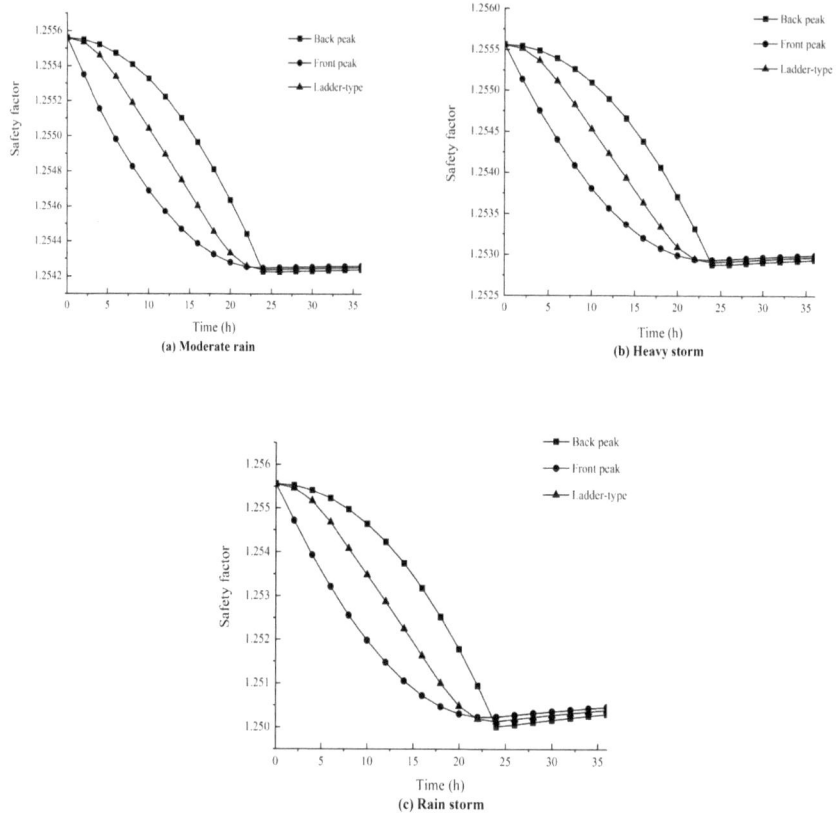

Fig. 4 Curves of different rainfall types' coefficient of safety with time changing for each rainfall level

4   Discussion

Under each rainfall level, there are some differences in the changes of pore water pressure under the function of the three rainfall types. Due to the gradual decrease of rainfall intensity with time in the front peak type, the rainfall intensity is much smaller than the infiltration capacity of the soil body in the late rainfall period, and the growth rate of pore water pressure is firstly fast and then slow, so the change curve of the pore water pressure first shows an upward "convex" type; due to the gradual increase of rainfall intensity in the back peak type, the rainfall intensity is larger than the infiltration capacity of the soil body in the late rainfall period, so the pore water pressure change rate is small in the early stage and gradually increases in the late stage, and the change curve is in the form of a downward concave and upward convex type.

In this paper, the front-peak type has the least effect on the coefficient of safety of the side slope, while the back-peak type has the greatest effect. The possible reason for this is that precipitation infiltrates into the slope body via the slope surface and this increases the pore water pressure of the soil body within the slope, i.e., the matrix suction decreases. And the corresponding shear strength decreases, the infiltration in the slope body increases, and the self-weight of the slope body then increases, leading to an increase in the downward force of the slope body and a decrease in the safety of the slope body [6, 7]. And with the continuation of rainfall duration, the infiltrated rainwater makes the formation of infiltration force in the slope body, and the coefficient of safety decreases greatly. Therefore, in the late stage of rainfall, the intensity of rainfall in the back-peak type is greater than that in the ladder-type, and the ladder-type is greater than that in the front-peak type. At the same period of time, the infiltration volume of the slope body in the back-peak type is the largest, which leads to the largest decrease in the coefficient of safety in the back-peak type, and the smallest one in the front-peak type.

5 Conclusion

Based on the side slope model, this paper investigates the effect of different rainfall types on the change of pore water pressure and coefficient of safety under each rainfall level, and finds that:

(1) Different rainfall types show different laws of change on the growth of pore water pressure. Under the front-peak type rainfall, the growth rate of pore water pressure is firstly fast and then slow, the ladder type is secondary; while under the function of the back-peak type, it is firstly slow and then fast, and it takes a longer time to reach a stable state.

(2) Rainfall causes the slope coefficient of safety to decrease, but there are some differences on the effects of each rainfall type. When the rainfall duration and rainfall precipitation remain the same, the back-peak type rainfall has the greatest influence, and its coefficient of safety decreases the most; the front-peak type has the smallest influence.

References

1. S. Prashant, A.K. Patil, Early prediction framework for a rainfall-induced landslide: validation through a real case study [J]. Sādhanā **48**(03), 187 (2023)
2. Z. He, B. Wang, H. Qingguo, Stability analysis of cohesive soil slopes with weak interlayers under heavy rainfall conditions[J]. J. Chongqing Jianzhu Univ. **40**(05), 109–116 (2018)
3. L. Songtao, N. Yongding, W. Baolin, et al., Influence of rainfall infiltration on stability of granite residual soil high slope [J]. Math. Probl. Eng. (2022)
4. C. Xuansheng, Z. Wanlin, F. Jing, et al., Seismic stability of loess tunnel with rainfall seepage [J]. Adv. Civil Eng. (2020)
5. Y. Wang, J. Chai, J. Cao et al., Effects of seepage on a three-layered slope and its stability analysis under rainfall conditions [J]. Nat. Hazards **102**(03), 01–10 (2020)

6. G. Yang, H. Tao, S. Lei, et al., Seepage and stability analysis of unsaturated soil slopes under different rainfall conditions [J]. Hydroelectric Energy Sci. **40**(06), 166–170 (2022)
7. G. Wang, P. Sun, W. Lizhou et al., Experimental study on the mechanism of rainfall induced shallow loess landslides [J]. J. Eng. Geol. **25**(05), 1252–1263 (2017)

Study on Dynamic Risk Assessment of Urban Flood Based on Improved Genetic Algorithm

Mengqi Zhao, Zhining Wang, Dandong Cen, and Yaoyi Zhang

Abstract Study on risk assessment of urban flood is the key to ensure the operation safety of city. In existing researches, the dynamic impact factors were not taken into account, and the evaluation method based on heuristic algorithm depends on a large number of training samples. To solve the above deficiencies, dynamic risk assessment of urban flood based on improved genetic algorithm was proposed in this paper. In this assessment index system, disaster-inducing factors, hazard-inducing environment, hazard-affected body, disaster prevention and reduction ability are taken into considered. The dynamic assessment of urban flood risk for small sample is realized by using genetic algorithm improved by date augmentation, also the consistency and superiority of the method are verified by case study. The research results of this paper provide a new idea and a technical mean for the dynamic management of urban flood risk assessment, and have very important value of application and promotion.

Keywords Urban flood · Risk assessment index · Dynamic assessment · Improved genetic algorithm

M. Zhao
Hangzhou Binjiang District Comprehensive Administrative Law Enforcement Bureau, Hangzhou 310051, China

Z. Wang (✉)
Huadong Engineering Corporation Limited, Hanhzhou 311100, China
e-mail: iamfromtju@126.com

D. Cen
Hangzhou Water Resources and Hydropower Survey and Design Institute Co., Ltd., Hangzhou 310051, China

Y. Zhang
ZheJiang Keepsoft Information and Technology Corp., Ltd., Hangzhou 310051, China

© The Author(s) 2025
W. Wang et al. (eds.), *Hydraulic Structure and Hydrodynamics*, Lecture Notes in Civil Engineering 608, https://doi.org/10.1007/978-981-97-7251-3_25

1 Introduction

Floods are one of the most frequent and destructive natural disasters. With the acceleration of urbanization, population and economy are highly concentrated in the region, and the dynamic nature of flood influencing factors is enhanced, which exacerbates the uncertainty in risk assessment and poses challenges to urban flood warning. Therefore, a study on urban flood risk based on improved genetic algorithm was proposed in this paper. Taking the dynamic changes of influencing factors into account, a comprehensive urban flood risk assessment is carried out, which can effectively assess flood risks, scientifically support urban flood controlling, and ensure the safe operation of the city.

The risk of urban flood depends on two aspects: the danger of rainfall and the vulnerability of the city. Urban surface catchment area, density of buildings, soil properties, vegetation coverage are important factors that many experts have considered [1–5]. Emily C O'Donnell believed that the economic development status of a region was a component that affected the disaster bearing capacity of urban floods [6]. Bolanle Wahab and Ajayi O believed that if the infrastructure of regional river networks and water systems did not meet the requirements, it would exacerbate the harm of urban floods [7, 8]. However, the impact of dynamic changes in variable factors such as river water level and pumping station drainage capacity on flood risk has not been taken into account, and the timeliness of the evaluation results is relatively insufficient, which affect the guiding effectiveness of evaluation results.

Urban flood risk assessment is a key basis for enhancing the city's flood control and disaster reduction capabilities. Analytic Hierarchy Process [9], Grey Relational Analysis [10], Fuzzy Comprehensive Evaluation [11], Entropy Weight Method [12] and other weighting methods have been proposed to establish the functional relationship between influencing factors and risk index. These methods are simple to calculate and flexible to apply. However, the impact of factor correlation and dimensional differences on the accuracy of results cannot be eliminated using the above methods. Moreover, there is no time to accumulate training data for several years since the urban flood risk index is necessary to launch and operate as early as possible. Therefore, an algorithm suitable for small sample solving is required.

In summary, in this paper, considering disaster-inducing factors, hazard-inducing environment, hazard-affected body, disaster prevention and reduction ability, an urban flood dynamic risk assessment index system is established. The dynamic and timely nature of the assessment results can be ensured by obtaining real-time data through perception devices. Afterwards, an improved genetic algorithm was adopted to compensate for the low accuracy of small sample data training. The dynamic assessment of urban flood risk can be achieved and can effectively supporting the orderly development of the flood controlling work.

2 Dynamic Risk Assessment of Urban Flood Based on Improved Genetic Algorithm

2.1 Evaluation Index System

An integrated dynamic risk assessment index system based on four major aspects: disaster-inducing factors, hazard-inducing environment, hazard-affected body, disaster prevention and reduction ability, was proposed in this paper.

Disaster-inducing factors.

Rainfall intensity, which is the average entropy weight of rainfall and rainfall duration, is taken as the evaluation index of disaster-inducing factor. There is a positive correlation between rainfall intensity and flood risk.

Hazard-inducing environment.

Hazard-inducing environment refers to the natural geographical environment that forms flood disasters. Terrain factors, urban construction conditions, and hydrological factors are the three main aspects of hazard-inducing environment.

Hazard-affected body.

The loss caused by flood is mainly related to population density, GDP density, building density, road density and cultivated land area ratio.

Disaster prevention and reduction ability.

Faced with the same level of waterlogging, cities with better disaster prevention and reduction capabilities are at a lower risk of loss. Engineering defense capability, emergency response and rescue capability, disaster monitoring capability are considered as the evaluation index of disaster prevention and reduction ability.

2.2 Risk Assessment Based on Improved Genetic Algorithm

Establish an urban flood risk assessment model:
Objective function:

$$E(q) = \max \left[\sum_{t=1}^{T} (Q_t - \overline{Q})^2 / (m-1) \right]^{0.5} \\ \times \left[\sum_{t=1}^{T} \sum_{m=1}^{T} (L - l_{tm}) \cdot \partial (L - l_{tm}) \right] \quad (1)$$

Constraints:

$$\sum_{m=1}^{M} q_m^2 = 1 \tag{2}$$

State transition equation:

$$Q_t = \sum_{m=1}^{M} \mathbf{q}_m y_{tm} (t = 1 \sim T) \tag{3}$$

$$y_{tm} = \frac{y_{tm} * - y_{m\ min}}{y_{m\ max} - y_{m\ min}} \tag{4}$$

$$L = 0.1 \left[\sum_{t=1}^{T} (Q_t - \overline{Q})^2 / (m-1) \right]^{0.5} \tag{5}$$

$$l_{tm} = |q_m - q_t| \tag{6}$$

In the formula, y_{mmax} and y_{mmin} are respectively the maximum and minimum values of the indicator sample y_m. $y_m{}^*$ refers to the current indicator value, q refers to the unit vector of projection transformation, Q_t refers to the unit length vector, y_{tm} refers to the normalized result of the evaluation indicator data, L refers to the window radius of local density, and $\partial(L - l_{tm})$ refers to the unit over order function.

Then, the original population, which refers to the projection transformation vector q, can be generated. $E(q)$ is used to indicate fitness. Keep iterating until reaching the best fitness or maximum number of iterations. The final generated q is the optimal solution. The urban flood risk index can be determined according to q:

$$F = \overrightarrow{\mathbf{q}} \cdot \overrightarrow{\mathbf{A}} \tag{7}$$

In the formula, F is the risk assessment index and $\overrightarrow{\mathbf{A}}$ is the vector of influencing factor.

3   Case Study

Taking a certain region in southeastern China as an example, and applying the method proposed in this paper to risk assessment of urban flood in that area.

Using the data of two typical rainfalls from the region in 2022 to train the risk assessment model. Rainfall A has the longest rainfall duration of the year and Rainfall B has the highest rainfall intensity. The specific information for the two rainfalls is shown in Table 1.

Using the data of rainfall from the region in 2023 to verify the accuracy of the dynamic risk assessment model. The rainfall lasted for 85 min. It was concentrated

Table 1 The information for the two rainfalls

Rainfall number		Rainfall A	Rainfall B
Total rainfall of each rainfall station (mm)	K1004[a]	38.9	22.5
	K1121	46.8	30.9
	K1133	44.3	31.8
	K1161	38.4	16.4
	K1171	47.5	43.7
	K1174	37.6	24
	K1175	41.5	26.9
	K1186	39.2	7.7
Rainfall duration		13 h	90 min
Maximum rainfall intensity (mm/h)		11.4	43.7

[a] K*** * refers to the number of rainfall station

in the southwest of the region and the northern part of the region along the Qiantang River. The data of the rainfall was measured by rainfall station K1171 (shown in Fig. 1), with a total rainfall of 83 mm and a maximum hourly rainfall of 82.5 mm. The rainfall caused severe waterlogging.

The comparison between the risk assessment results and the actual locations of waterlogging points is shown in Fig. 2.

The risk threshold is between 25 and 72, and the southern area of the region has a low risk threshold, which is generally consistent with the actual water accumulation situation. Specifically, the highest risk area is located at the northern part of the region along the Qiantang River. This area was the centre of this rainfall, and was also the

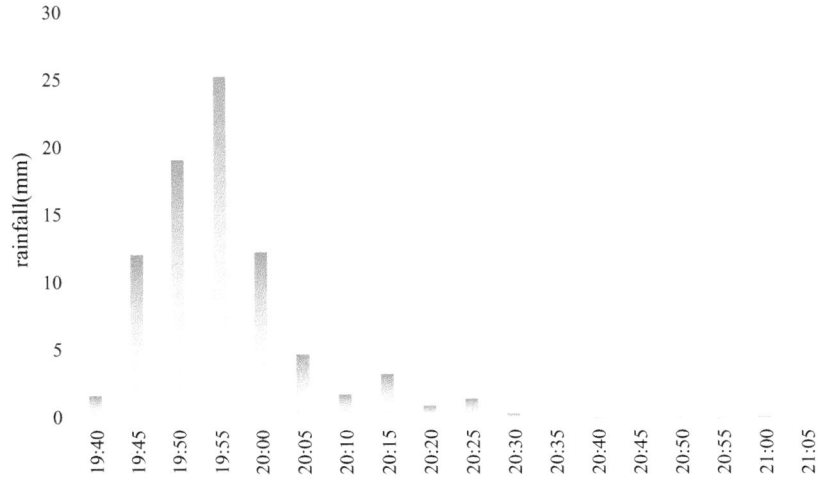

Fig. 1 Actual rainfall data (data from rainfall station K1171)

Fig. 2 The comparison between the risk assessment results and the actual locations of waterlogging points

most severely affected area. Additionally, the southwest area of this region was the second most severely flooded area. but the risk threshold for this area is not high. This is because the population density and economic density here are relatively low, so the losses caused by waterlogging are not significant.

4 Conclusion

The dynamic risk assessment of urban flood is of great significance for the safe operation of cities. In this paper, disaster-inducing factors, hazard-inducing environment, hazard-affected body, disaster prevention and reduction ability, was comprehensively considered. And an urban flood dynamic risk assessment method based on an improved genetic algorithm were proposed. Based on the results and discussions presented in Case Study, the conclusions are obtained as below:

(1) A dynamic Risk Assessment index system for Urban Flood has been established, considering disaster-inducing factors, hazard-inducing environment, hazard-affected body, disaster prevention and reduction ability, was comprehensively considered by obtaining real-time data through sensing devices.

(2) The improved genetic algorithm has been applied for risk assessment. With the support of real-time data and warning data, the accuracy of urban flood risk assessment results under small sample data conditions has been effectively improved.

(3) Applying the proposed method to practice, it has been verified that the overall trend of flood risk obtained using the proposed method in this paper is consistent with the actual situation. Therefore, the method proposed in this paper has good consistency and superiority, and has good practicality and promotional value.

Acknowledgements This work was financially supported by the Science and Technology Plan Project of Department of Water Resources of Zhejiang Province (RC2215).

References

1. H. Zhang, W. Weiming, H. Chunhong et al., A distributed hydrodynamic model for urban storm flood risk assessment [J]. J. Hydrol. **600**(000), 126513–126513 (2021)
2. O.F. Kasin, B. Wahab, M.F. Oweniwe, Urban expansion and enhanced flood risk in Africa: the example of Lagos[J]. J. Hydrol. **21**(2), 137–158 (2022)
3. P. Arun, S. Jason, C. Heejun, Urban flood risk and green infrastructure: who is exposed to risk and who benefits from investment? A case study of three U.S. Cities [J]. Landscape Urban Plann. **223**(000), 104417–104417 (2022)
4. T. Fereshteh, F. Ramin, C. Bahram et al., Urban flood-risk assessment: integration of decision-making and machine learning [J]. Sustainability **14**(8), 4483–4483 (2022)
5. A.M.-M. Miguel, M. Emmanuel, P. Andre et al., Impact of the porosity of an urban block on the flood risk assessment: a laboratory experiment [J]. Sustainability **602**(000), 126715–126715 (2021)
6. E.C. O'Donnell, C.R. Thorne, Drivers of future urban flood risk [J]. Philos. Trans. Ser. A, Math. Phys., Eng. Sci. **2168**(378), 20190216–20190216 (2020)
7. B. Wahab, O. Falola, The consequences and policy implications of urban encroachment into flood-risk areas: the case of Ibadan [J]. Environ. Hazards **1**(16), 1–20 (2017)
8. O. Ajayi, S.B. Agbola, B.F. Olokesusi, et al., Flood management in an urban setting: a case study of Ibadan metropolis [J]. Spec. Publ. Niger. Assoc. Hydrol. Sci. (2012)
9. G. Abhishek, K.S. Kumar, Application of analytical hierarchy process (AHP) for flood risk assessment: a case study in Malda district of West Bengal, India [J]. Nat. Hazards (2018)
10. G. Dong, W. Wei, X. Xia, et al., Safety risk assessment of a Pb-Zn mine based on fuzzy-grey correlation analysis [J]. Multi. Digit. Publishing Inst. (1) (2020)
11. Y. Geng, X. Zheng, Z. Wang et al., Flood risk assessment in Quzhou City (China) using a coupled hydrodynamic model and fuzzy comprehensive evaluation (FCE)[J]. Nat. Hazards **100**(1), 133–149 (2020)
12. X. Lv, N. Shi, J. Wei, et al., *Information System Security Risk Assessment Based on Entropy Weight Method-Bayesian Network [C]. International Conference on Frontiers in Cyber Security.* (Springer, Singapore, 2022)

Hydrodynamic Characterization and Flood Control System Construction

Hydraulic Calculation and Analysis of Head Cover Drainage System of Hydropower Station Based on PIPENET

Zhe Long, Xing Lu, Yingxia Zheng, Lingfeng Shu, and Wenbo Su

Abstract In order to analyze the influence of the siphon effect in the pipeline on the actual operation of the head cover drainage system in different working conditions of a hydropower station, a reasonable and effective siphon failure device is set up. Based on PIPENET, the fluid calculation and analysis software of pipe network, this paper aims at the overall modeling of the head cover drainage system of the hydropower station, and then conducts hydraulic calculation and analysis on it, and studies the design scheme of damage siphon effect, which provides strong support for the optimal design of the head cover drainage system.

Keywords Hydrogenerator · Roof drainage system · PIPENET software · Siphon effect · Air valve

Z. Long (✉) · X. Lu · Y. Zheng · L. Shu · W. Su
Huadong Engineering Corporation Limited, Hangzhou, Zhejiang 311122, China
e-mail: long_z@hdec.com

X. Lu
e-mail: lu_x2@hdec.com

Y. Zheng
e-mail: zheng_yx@hdec.com

L. Shu
e-mail: shu_lf@hdec.com

W. Su
e-mail: su_wb@hdec.com

© The Author(s) 2025
W. Wang et al. (eds.), *Hydraulic Structure and Hydrodynamics*, Lecture Notes in Civil Engineering 608, https://doi.org/10.1007/978-981-97-7251-3_26

1 Project Overview

1.1 Project Background

According to the situation of our country, hydropower is the best way to make full use of the renewable energy and save energy emission reduction [1, 2]. As one of the auxiliary systems of the pumping and storage unit, the top drainage system plays an extremely important role in the safe and stable operation of the pumping and storage power station [3]. During the operation of the hydrogenerator set, the main function of the drainage system of the top cover is to discharge the water leakage at the working seal of the main shaft and the leakage at the sleeve of the movable guide vane [4]. The drainage system of the top cover is not smooth or fails, and there is the risk of flooding [5].

A hydropower station has 6 units in total, and each unit is equipped with two drainage pumps (one main and one standby) in the top cover to pump the water in the top cover along the drainage line to the oil–water separation pool arranged in the auxiliary building at the end. Select the rated discharge water pump 14.4 m³/h, rated lift 8.5 m. Under normal circumstances, only one pump works, the operation time of each pump pumping is 8.4 min (504 s), and the intermittent time is 7.3 min (438 s). When the water level in the top rises abnormally, the standby pump starts and the two pumps run simultaneously. During the actual pipe layout, the discharge pipe of the top drainage pump in each unit is connected to the top drainage main pipe through a 10 m long riser. When the pipe is full of water, the water pressure in the top drainage main pipe is greater than that in the outlet pipe of the drainage pump, which may result in siphon effect. Therefore, the simulation of the top drainage system was carried out to analyze the siphon situation in the pipeline during the operation of the system under different working conditions, and a reasonable and effective siphon destruction device was set up to provide the basis for further optimization design of the system in the future. The drainage line diagram of the unit's top cover is shown in Fig. 1.

1.2 Simulation Software and Governing Equation Introduction

PIPENET is the mainstream fluid calculation and analysis software for pipe network at present [6]. It adopts characteristic line method to solve one-dimensional unsteady flow problem [7] and is widely used in oil, natural gas, shipbuilding, chemical industry, electric power industry and other fields [8]. The basic governing equation of its characteristic type is as follows:

$$C^+ : H_p - H_A + \frac{a}{gA}\left(Q_p - Q_A\right) + \frac{f\,\Delta x}{2gDA^2}Q_A|Q_A| = 0 \tag{1}$$

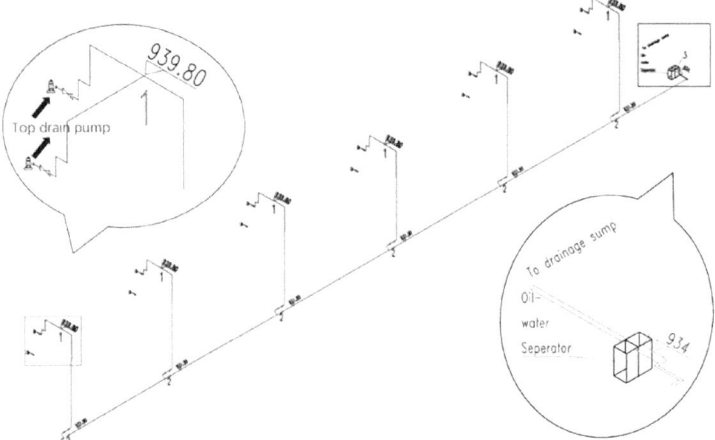

Fig. 1 Pipe diagram of head cover drainage system

$$C^- : H_p - H_B + \frac{a}{gA}(Q_p - Q_B) + \frac{f\,\Delta x}{2gDA^2}Q_B|Q_B| = 0 \qquad (2)$$

where, H and Q are pressure and flow rate; a is the propagation velocity of pressure wave; A and D are the cross section area and pipe diameter; f is the friction coefficient of pipe section; Subscript p is the unknown node in the calculation period; Subscripts A and B are nodes adjacent to P in the previous period. Since the system is initially in A steady state, the pressure and flow in the previous period of nodes (A and B) adjacent to P can be calculated in advance. In this paper, PIPENET software is used to calculate the hydraulic transition process of the top drainage pipe of a hydropower station under various working conditions, so as to verify the existing design scheme. The calculation restores the drainage pipe in strict accordance with the real route parameters and actual pump parameters, as shown in Fig. 2 after the model is completed.

During normal operation, the water in the top cover of each hydrogenerator unit is pumped to the oil–water separation tank of the auxiliary building through two DN100 drainage pipes, and the six units are drained through two DN100 drainage pipes. The simultaneous operating conditions of four, six, seven and twelve top cover drainage pumps were calculated respectively. Among them, the simultaneous operating conditions of twelve pumps only exist in theory, and there is no possibility of maintenance under this working condition. The specific Settings of each working condition are shown in each section.

Relative to the distance of the oil-bearing sump at the end, the six hydroelectric generating Units were named Units 1 ~ Units 6, and each Pump group was named Pump 12 ~ Pump 1 in sequence from near to far.

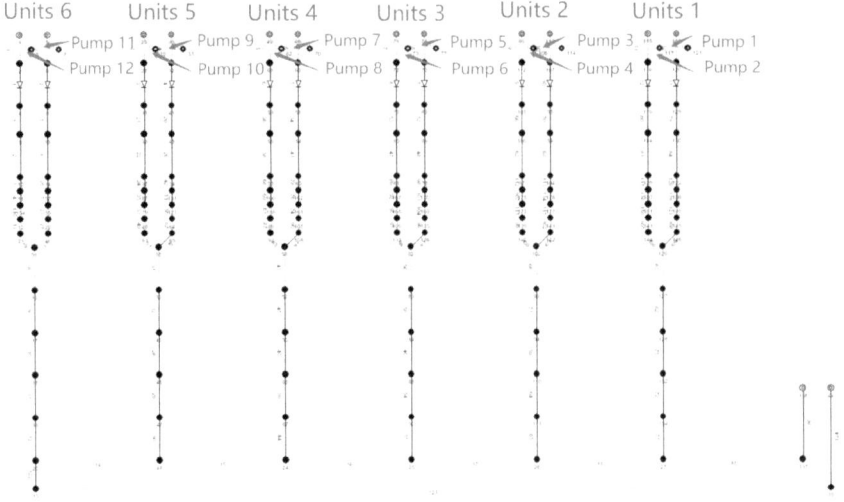

Fig. 2 Schematic diagram of computational modeling

2   Calculation of Constant Flow in Top Drain Pipe Section

The specific calculation conditions of the constant flow of the turbine top drainage are as follows: the water level of the front pool of the top drainage pump is its starting water level of 938.228 m, and the oil-bearing pool at the end of the pipeline is connected with the atmosphere.

Working condition formulation principle:

Case 1.1: Considering the operation of 6 hydrogenerator sets, the pumping time of each pump is 8.4 min (504 s), and the intermittent time is 7.3 min (438 s). Therefore, the probability of the four top drainage pumps running at the same time is the maximum.

Case 2.1: Consider the operation of 6 hydrogenerator units, each unit has 1 head cover drainage pump, and a total of 6 head cover drainage pumps run simultaneously. The probability of this condition is low.

Case 3.1: Considering the operation of 6 hydrogenerator units, each unit has 1 top cover drainage pump to discharge water, and the abnormal rise of the top level of 1 unit leads to the start-up of the standby pump, a total of 7 top cover drainage pumps run at the same time. This working condition is used to simulate the possible accident condition.

Case 4.1: Considering the operation of 6 hydrogenerator units, each unit has 1 head cover drainage pump, while discharging water, the liquid level of the 6 units' top cover rises abnormally, all the standby pumps are started, and a total of 12 head cover drainage pumps run simultaneously. This condition exists in theory, but the probability of actual occurrence is very low.

Table 1 Top drain constant flow operating condition table

Working condition	Operating condition description
Case 1.1	Units 3 ~ Units 6 single pump operation, a total of 4 pumps operation
Case 2.1	Units 1 ~ Units 6 single pump operation, a total of 6 pumps operation
Case 3.1	Units 1 ~ Units 5 single pump operation, Units 6 One main and one standby two pumps running at the same time, a total of 7 pumps running
Case 4.1	Units 1 to Units 6 one active water pump and one standby water pump are running at the same time. The two drainage pipelines are running at the same time. A total of 12 water pumps are running

Table 2 Statistics of pump flow under different working conditions

Working condition	Rate of flow (L/s)					
	Pump 12	Pump 11	Pump 10	Pump 9	Pump 8	Pump 7
Case 1.1	4.12	–	6.38	–	6.37	–
Case 2.1	6..32	–	6.23	–	6.22	–
Case 3.1	5.79	5.79	5.91	—	5.91	–
Case 4.1	4.5	4.5	4.29	4.29	4.27	4.27
Working condition	Rate of flow (L/s)					
	Pump 6	Pump 5	Pump 4	Pump 3	Pump 2	Pump 1
Case 1.1	6.39	–	–	–	–	–
Case 2.1	6.23	–	6.32	–	6.55	–
Case 3.1	5.96	–	6.12	–	6.47	–
Case 4.1	4.31	4.31	4.58	4.58	5.3	5.3

See Table 1 for specific calculation conditions, and Table 2 for calculation results of each working condition.

As can be seen from Table 2, under constant flow conditions, when the top cover drainage pumps of each water turbine unit run in case 1.1 to case 4.1, the actual drainage capacity of each pump is $\geq$ 129 L/min (2.15 L/s).

3 Calculation of Transition Process of Unprotected Pumping Power Outage in Top Drain Pipe Section

The specific calculation conditions of the transition process of unprotected pumping power outage for turbine top drainage are as follows: the water level of the top drainage pump is its starting water level of 938.228 m, and the oil-bearing pool at the end of the pipeline is connected to the atmosphere. After running 8.4 min (the maximum running time of the pump), the water pump will lose power at the same time in all working conditions.

Working condition formulation principle:

Case 1.2: Considering that the probability of the four top cover drainage pumps running at the same time is the maximum, the four top cover drainage pumps drain water under the pumping power failure condition, and all the pumps are powered off at the same time after 8.4 min of operation.

Case 2.2: Consider the operation of 6 hydrogenerator units, each unit has 1 head cover drainage pump, and 6 head cover drainage pumps run at the same time under the pumping power failure condition, and all pumps are powered off at the same time after 8.4 min operation.

Case 3.2: Consider the operation of 6 hydrogenerator units, each unit has 1 top cover drainage pump to discharge water, and the abnormal rise of the top level of 1 unit leads to the start of the standby pump. Under the pumping power failure condition, 7 top cover drainage pumps run at the same time, and all pumps lose power at the same time after 8.4 min operation.

Case 4.2: Considering the operation of 6 hydrogenerator units, each unit has 1 head cover drainage pump, and the liquid level of the 6 units' head cover drainage units abnormally rises at the same time, all the standby pumps start, and 12 head cover drainage pumps run at the same time under the pumping power failure condition. After 8.4 min of operation, all pumps lose power at the same time.

The specific calculation conditions are shown in Table 3. According to the calculation, the maximum negative pressure in the top drain pipe of each unit appears at the highest point of the top drain pipe, and the pressure change at the top drain pipe of the top drain pipe in each working condition is shown in Fig. 4 (as the pressure change trend of the top drain pipe of the top drain pipe in each working condition is similar, only the pressure change in Units 12 in Case 4.2 is shown). The pressure statistics are shown in Table 4.

See Table 4 for the pressure after stable operating conditions:

Table 3 Top drainage without protection pumping power off operation condition table

Working condition	Operating condition description
Case 1.2	Units 1 ~ Units 4 single pump operation, a total of 4 pumps running, 8.4 min after the operation of power failure
Case 2.2	In Units 1 ~ Units 6, a total of 6 pumps run with one pump. After 8.4 min of operation, the water pumps are powered off at the same time
Case 3.2	Units 1 to Units 5 run with a single pump, and Units 6 run with one main pump and one standby pump at the same time. A total of 7 pumps run and power off at the same time after 8.4 min of operation
Case 4.2	Units 1 to Units 6 one water pump and one standby water pump are running at the same time. The two drainage pipes are running at the same time. A total of 12 water pumps are running

Table 4 Pressure statistics of the top drain pipe under different working conditions

Working condition	Pressure (m)					
	Units 12	Units 11	Units 10	Units 9	Units 8	Units 7
Case 1.2	−2.18	–	−1.96	–	−1.92	–
Case 2.2	−2.22	–	−2.02	–	−1.99	–
Case 3.2	−2.60	−2.60	−2.12	–	−2.06	–
Case 4.2	−3.07	−3.07	−2.86	−2.86	−2.83	−2.83
Working condition	Pressure (m)					
	Units 6	Units 5	Units 4	Units 3	Units 2	Units 1
Case 1.2	−1.92	–	–	–	–	–
Case 2.2	−2.00	–	−2.15	–	−2.62	–
Case 3.2	−2.07	–	−2.18	–	−2.62	–
Case 4.2	−2.84	−2.84	−2.99	−2.99	−3.44	−3.44

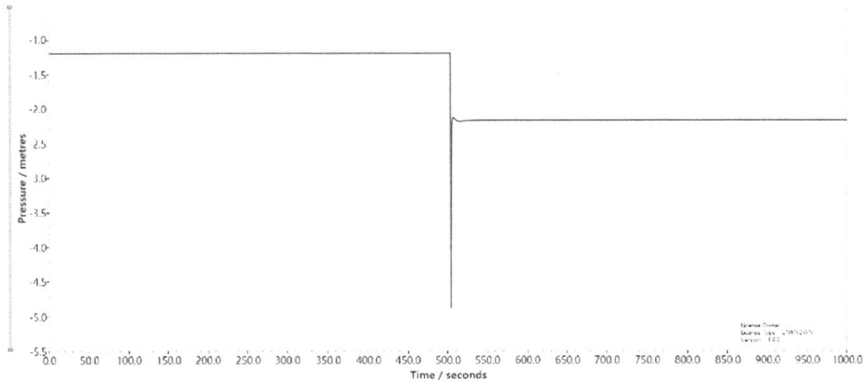

Fig. 3 Pressure variation curve of the top drain pipe

As can be seen from Table 3 and Fig. 3, under all working conditions, the stable value of pressure at the pre-installation point of the air valve is less than −1.92 m, and siphon occurs in the pipeline.

4 Air Valve Protection Scheme Calculation

According to the calculation results of unguarded pumping power outage for the head cover drainage and transmission system of the power station in Sect. 3, after the head cover drainage pump suddenly stops pumping, negative pressure appears at the highest point of the head cover drainage pipe, and siphon occurs in the pipe. Therefore, it is necessary to set up appropriate siphon destruction device on the

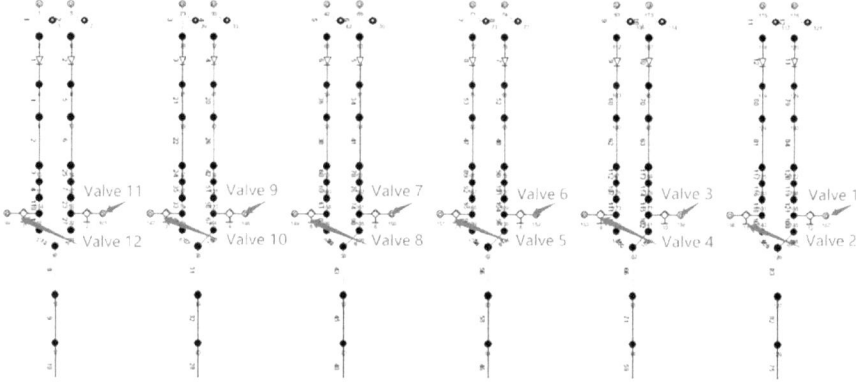

Fig. 4 Air valve installation node diagram

pipeline. According to general protection principles, air valves are arranged at local high points of pipelines [9], It can eliminate the negative pressure of pipeline and form a good protective effect [10]. The calculation model after air valves are arranged is shown in Fig. 4. The protection scheme of system air valves is calculated and analyzed below to check the siphon failure effect.

The specific calculation conditions and the principle of working conditions are the same as the third section. The specific calculation conditions are shown in Table 3, and the pressure changes at the air Valve installation point in all working conditions are shown in Fig. 5 (Because the maximum pressure changes at the air valve installation point are similar in all working conditions, only the pressure changes at the Valve 12 installation point in Case 4.3 are shown). The pressure statistics after the stable operating conditions are shown in Table 5.

See Table 5 for the pressure after the air valve is installed under stable conditions:

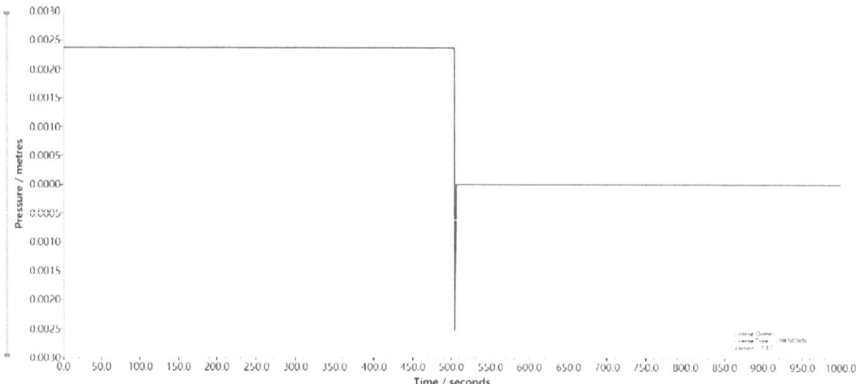

Fig. 5 Valve 12 Installation pressure and flow curve

Table 5 Air valve installation minimum pressure statistics

Working condition	Pressure (m)					
	Valve 12	Valve 11	Valve 10	Valve 9	Valve 8	Valve 7
Case 1.3	0	–	0	–	0	–
Case 2.3	0	–	0	–	0	–
Case 3.3	0	0	0	–	0	–
Case 4.3	0	0	0	0	0	0
Working condition	Pressure (m)					
	Valve 6	Valve 5	Valve 4	Valve 3	Valve 2	Valve 1
Case 1.1	0	–	–	–	–	–
Case 2.1	0	–	0	–	0	–
Case 3.1	0	–	0	–	0	–
Case 4.1	0	0	0	0	0	0

As can be seen from Table 5 and Fig. 5, after the air valve is installed, the negative pressure in the pipeline disappears after the pressure is stabilized, and the siphon is destroyed.

5 Conclusion

(1) Under the constant flow calculation condition, the actual drainage flow of the top drainage pump of a hydropower station in all operating conditions is $\geq$ 2.15 L/s or 7.74 m^3/h, so the selection of the pump is reasonable.

(2) When the water pump pumps at the same time in different working conditions of the top drainage system, the pipeline at the original air valve behind the pump will siphon, and the pressure is all lower than -1.92 m, so it is necessary to install the siphon failure measures.

(3) When the water pump of the top drainage system is pumped and powered off at the same time under different working conditions, and the air valve is installed to protect it, the stable pressure in the top drainage pipe has no negative pressure, and the siphon in the pipe is destroyed. It can be seen that the use of air valves to destroy the siphon can achieve the desired effect.

References

1. C. Liang, Water turbine regulating system summarized research development of the computer simulation [J]. Sci. Technol. Enterp. **276**(3), 89 (2015). https://doi.org/10.13751/j.carolcarroll nkikjyqy.2015.03.083

2. Y. Ni, R. Huren, W. Simin, C. Hongyan, B. Rongrong, W. Yufang. Electric energy structure and trend in hydropower construction [J]. J. Electr. Power **5**(4), 295–301 (2022). https://doi.org/10.13357/j.liuxiaobo.2022.037
3. Z. Ma, Z. Xiao, L. Xu, et al., Turbine roof drainage system health assessment method research [J/OL]. China's Rural Water Conservancy Hydropower 1–12 (17 April 2023). http://kns.cnki.net/kcms/detail/42.1419.TV.20220711.1131.029.html.
4. A. Jia, S. Zhou, L. Li, X. Chen, K. Wang, Optimal operation of Water turbine roof drainage system [J]. Yunnan Hydropower **20**, **36**(02), 162–165
5. J. Hou, T. Tan, D. Tang, Based on siphon principle in the research and application of water turbine roof drainage system [J]. J. Hydropower New Energy **162**(12), 42–43+53 (2017). https://doi.org/10.13622/j.carolcarrollnki/TVcn42-1800.1671-3354.2017.12.010
6. Y. Sun, J. Wu, K. Li, Y. Han, L. Wang, Research on water hammer protection of pressurized water delivery system [J]. Yellow River **201**, **43**(01), 152–155+164
7. W. Liu, Y. Pei, Steam hammer calculation of AP 1000 main steam pipeline based on PIPENET [J]. Nucl. Sci. Eng. **35**(02), 230–235 (2015)
8. W. Zhang, G. Yan, Water shock analysis of pipeline system based on PIPENET [J]. Petrol. Eng. Constr. **37**(S1), 55–57+2 (2011)
9. T. Elias, K. Bryan et al., The crucial importance of air valve characterization to the transient response of pipeline systems [J]. Water **14**(17), 2590–2590 (2022)
10. Ó.E. CoronadoHernández et al., Simplified mathematical model for computing draining operations in pipelines of undulating profiles with vacuum air valves [J]. Water **12**(9), 2544–2544 (2020)

Static Loading Behavior of a Hybrid Pile-Bucket Foundation for Offshore Wind Turbines

Guoliang Dai, Chaoqun Zuo, Weiming Gong, and Jinghui Tao

Abstract Traditional monopile foundations cannot support the greater lateral loads and horizontal displacement needs required by offshore wind turbines due to their increasingly large size. Popularity has increased for a recently created hybrid pile foundation that combines a traditional monopile with a wide-shallow bucket. In this paper, 1-g model tests in soft clay are used to analyze the lateral behavior of hybrid pile foundation under static load. As a standard, the monopile with the same pile diameter and embedded length is employed. According to the experiment's findings, the pile-bucket hybrid foundation's lateral ultimate bearing capacity is more than 70% higher and its initial stiffness is nearly twice as good as that of a monopile foundation under the same lateral load. In addition, this paper compares the pile displacement, pile bending moment, and p-y curve of monopile foundation and hybrid pile-bucket foundation.

Keywords Offshore wind turbine · Hybrid foundation · Static loading · Lateral bearing capacity

G. Dai (✉) · C. Zuo · W. Gong · J. Tao
School of Civil Engineering, Southeast University, Nanjing 211189, China
e-mail: daigl@seu.edu.cn

W. Gong
e-mail: Wmgong@seu.edu.cn

J. Tao
e-mail: dink2000@vip.sina.com

G. Dai · C. Zuo · W. Gong
Key Laboratory of C&PC Structures, Ministry of Education, Southeast University, Nanjing 211189, China

J. Tao
Jiangsu Provincial Architectural Design and Research Institute Co., Ltd., Nanjing 210029, China

W. Wang et al. (eds.), *Hydraulic Structure and Hydrodynamics*, Lecture Notes in Civil Engineering 608, https://doi.org/10.1007/978-981-97-7251-3_27

1 Introduction

With the advantages of steady wind energy supplies, not using up land resources, and no consumption issues, offshore wind power is garnering more and more concerns. Global installed offshore wind power capacity is reported to be 9.4 GW in 2022, with a cumulative global installed offshore wind power capacity of 57.6 GW by the end of 2022 [1]. Offshore wind turbines are mainly subject to horizontal loads such as wind and wave currents, so foundation safety is an important guarantee for the smooth power supply of wind turbines. Monopile foundation is currently the most common form of foundation for offshore wind power applications, with the advantages of simple force forms and mature construction techniques [2]. However, as the size of wind turbines rises, monopile foundations may not be able to meet their load-bearing capacity requirements [3]. Although some projects choose to increase the diameter and length of the monopiles, it certainly increases the difficulty and cost of construction significantly [4–6]. In this regard, scholars have conducted a series of research on the bearing performance of hybrid foundation. Bienen [7] and Peng [8] proposed a new type of foundation combining wings and monopiles, which improved the horizontal ultimate bearing capacity and stiffness. In addition, some scholars believe that installing friction wheels at the foundation surface of the monopile foundation can effectively improve the horizontal response of the monopile foundation [9, 10].

Recently, a new type of hybrid foundation has attracted the attention of scholars for its successful construction in engineering projects, which is composed of a monopile foundation and a wide shallow foundation. Figure 1 shows the construction site of hybrid foundation, the construction process of hybrid pile-bucket foundation is as follows: sink the monopile foundation to the specified position and elevation, then sink the bucket foundation through the reserved hole under the negative pressure, and finally close the gap between the pile foundation and the bucket foundation with high-strength grout to make the two synergistic load-bearing.

Liu [11] applied a numerical analytical approach to analyze the bearing mechanism of the pile and bucket foundation. Additionally, they performed a sensitivity analysis on the pile-bucket size to assess how it affected bearing capacity. Liu [12] obtained the optimum diameter ratio of pile and bucket and the depth of bucket

Fig. 1 Hybrid pile-bucket
site construction

foundation by ABAQUS modal analysis. Li et al. investigated the distribution law of soil displacement field around different size models and the mechanism of horizontal bearing capacity by 1 g model test of pile and bucket composite foundation under water calm load in sandy soil foundation [13]. Lai et al. studied the bearing mechanism of different pile and bucket structures in soft clay by centrifugal model test [14]. There are currently few research on 1 g large scale model testing, and the majority of studies on hybrid pile-bucket foundations concentrate on numerical analysis and small scale model tests. In addition, the hybrid foundation-soil interaction is not thoroughly examined in these research. Therefore, it is critically necessary to conduct 1 g large-scale model studies in order to better understand the mechanism of pile-soil interaction.

Based on this, this article conducted a series of horizontal static tests using a 1 g large scale model for both the large diameter monopile foundation and the hybrid pile-bucket foundation. By analyzing the acquired data, this paper compared the load–displacement curve, pile bending moment, pile displacement and soil resistance distribution between large diameter monopile and hybrid foundation to study the horizontal bearing mechanism of hybrid foundation.

2 Experimental Setup

2.1 Soil Preparation

The soil for this study was a local powder clay in Hebei Province, China, which was filled in layers of 0.2 m and compacted with mechanical devices until it reached an elevation of 2.25 m. Due to the high moisture content of the soil, the soil was left to consolidate for 90 days. Moreover, geotextiles were laid on the surface of the soil in order to prevent water loss and cracking of the soil surface. Furthermore, to prevent water loss and cracking of the soil surface, geotextiles were laid on the surface of the soil layer. After completing the consolidation, CPT tests was conducted on the foundation soil, and a total of 3 measurement points were selected for the tests. The static contact machine was applied to this test, which resulted in the distribution of cone tip resistance and undrained shear strength, as shown in Fig. 2.

2.2 Model Foundation

The model pile is 2.75 m in length, 300 mm in diameter, 6 mm in thickness and the height of the pile above the mud surface is 1 m. The hybrid pile foundation uses piles with the same dimensions as monopile. The bucket skirt height is 0.6 m, diameter is 0.9 m, and thickness is 6 mm. As shown in Fig. 3, ten pairs of FBG fiber optic gratings were installed in the monopile foundation along the depth direction, with one pair at

Fig. 2 CPT results

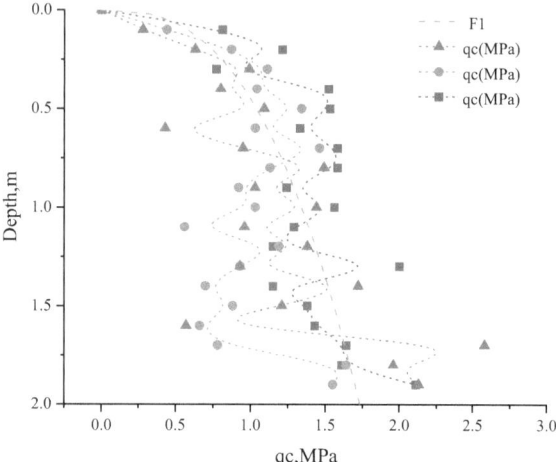

0.1 m and 0.3 m above the mud surface and eight pairs at equal distances from 0.1 m to 1.5 m below the mud surface. The measuring point's position was smoothed down and wiped with alcohol before fiber being adhered with glue. Silicone rubber and modified acrylic adhesive are applied to the surface of the fiber grating to achieve waterproof and protect the fiber. LVDT displacement gauges are positioned on the pile side, and five earth pressure gauges are arranged on the active and passive sides of the bucket, in order to measure the lateral reaction of the foundation under horizontal load.

2.3 Test Program

The test loading adopts graded loading method, taking 1/10 of the estimated ultimate bearing capacity of single pile foundation as the loading amount of each level, using equal gradation step by step, and recording the horizontal displacement of the pile top at the 5th min, 15 min, 30 min, 45 min and 60 min of each level of load application respectively. It indicates that the expansion of pile head displacement is largely constant and the next level of load can be applied when the change in horizontal pile top displacement within a certain time period does not exceed 0.1 mm and happens twice consecutively. It is thought to have reached the maximum bearing capacity of the test loading when the horizontal displacement at the mud surface equals 0.1 times the pile diameter or the pile top displacement curve exhibits a clear inflection point. At this point, loading is stopped.

3 Experimental Result

3.1 Load–displacement Curve

The load displacement curves from the model test are shown in Fig. 4. With increasing load, the hybrid pile-bucket foundation and the monopile foundation go through three stages: elastic, elastoplastic, and plastic phases. The pile-bucket foundation transitions from the elastic phase to the elastoplastic phase later than the monopile foundation, which only has a brief elastic phase. The ultimate bearing capacities of monopile and hybrid foundation are 22.5 KN and 38.5 KN, respectively, as indicated in Fig. 4.The lateral bearing capacity and initial stiffness of the hybrid pile foundation are obviously improved. The hybrid pile foundation's ultimate load bearing

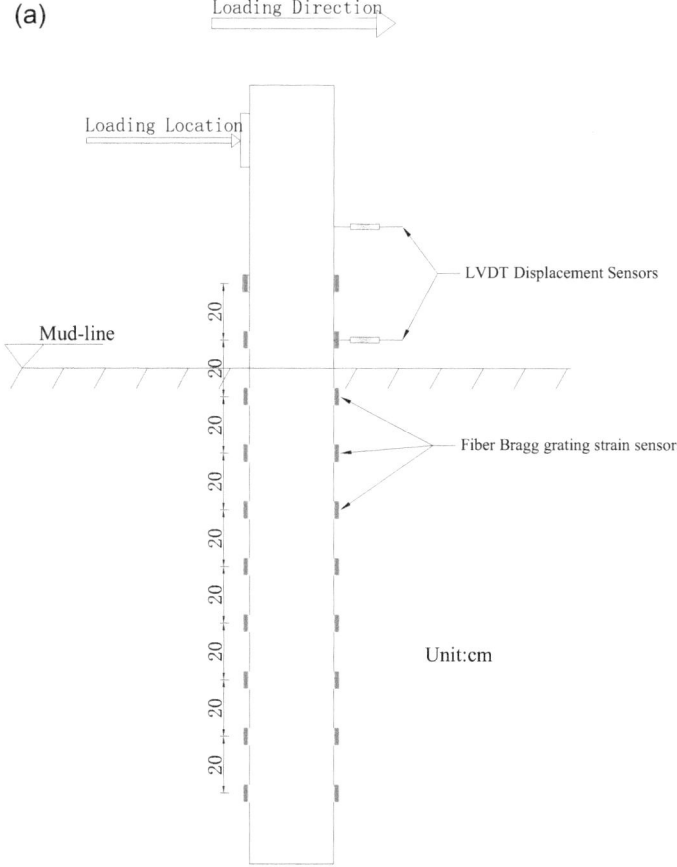

Fig. 3 Sensors layout of model test: **a** monopile; **b** and **c** hybrid pile-bucket

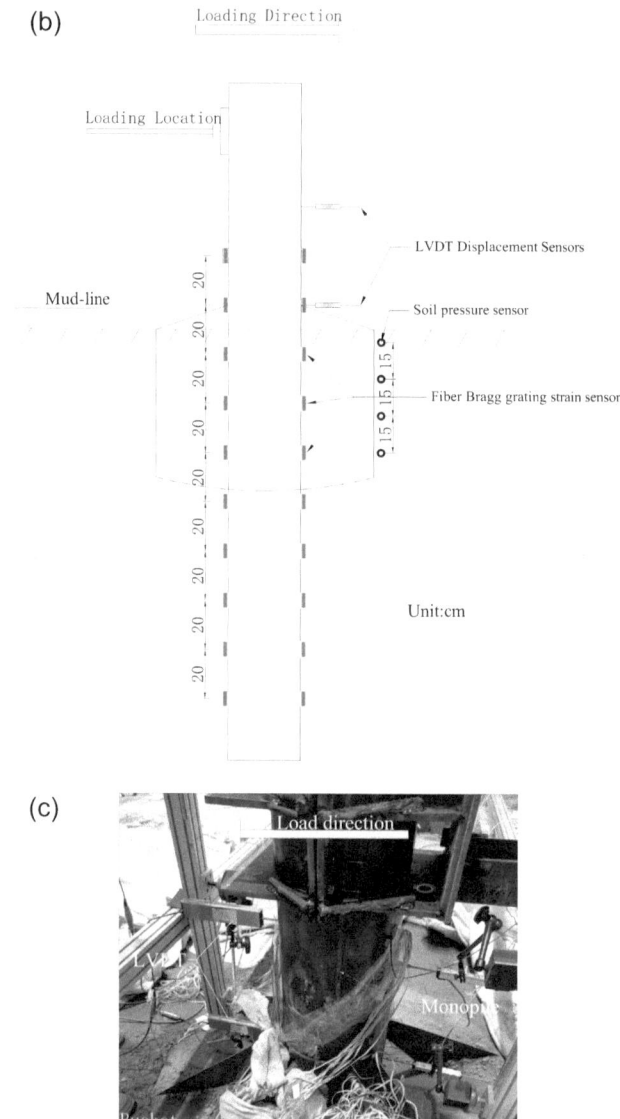

(b)

Loading Direction

Loading Location

LVDT Displacement Sensors

Soil pressure sensor

Mud-line

Fiber Bragg grating strain sensors

Unit:cm

(c)

Load direction

LV

Monopile

Bucket

Fig. 3 (continued)

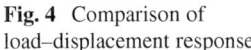

Fig. 4 Comparison of load–displacement response

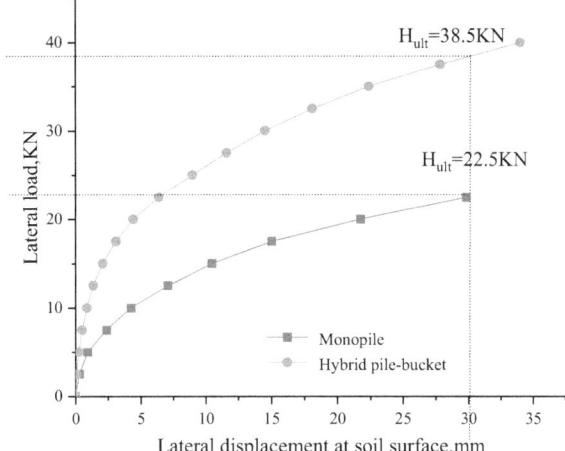

capacity and initial stiffness are both higher than those of the monopile by 71.11% and 216.32%, respectively.

3.2 Pile Bending Moments

In Fig. 5, which depicts the distribution bending moment curves for monopile foundation and pile-bucket foundation along the pile body, the pile bending moment is seen to be paraboliclly distributed throughout the depth and to grow inversely with increasing load levels. The peak bending moment of the monopile under all levels of load is located at 0.5 m below the mud surface, whereas the peak bending moment of the hybrid pile-bucket foundation is located at 0.1 m below the mud surface. This difference shows that the monopile's internal force is transmitted to a deeper soil layer. The hybrid foundation pile bending moment below the mud surface is less than the monopile bending moment, as seen in the comparison of the monopile bending moment and hybrid foundation pile bending moment at the same degree of stress.

3.3 Lateral Displacement

Assuming that the pile behaves as a beam on an elastic base, the relationship between the pile bending moment $M(z)$, the pile lateral displacement $y(z)$, and the soil resistance $p(z)$ can be expressed by the following equation:

$$y(z) = \iint \frac{M(z)}{EI} dzdz \tag{1}$$

$$p(z) = \frac{d^2 M(z)}{dz^2} \tag{2}$$

The pile bending moment is fitted by the sixth order polynomial function $M(z)$. Therefore, the lateral displacement $y(z)$ of the pile is obtained by integrating $M(z)$ twice over the depth z. At the same time, the lateral soil resistance $p(z)$ of the pile is obtained by deriving $M(z)$ twice over the depth z. The lateral displacement $y(z)$ of the pile is obtained by integrating $M(z)$ twice over the depth z.

Fig. 5 Bending moment under different loads: **a** Bending moment of monopile; **b** Bending moment of hybrid pile-bucket **c** Comparison of bending moment with the same load

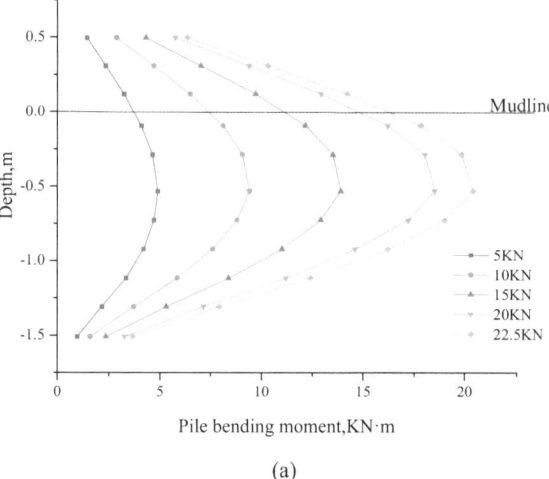

(a)

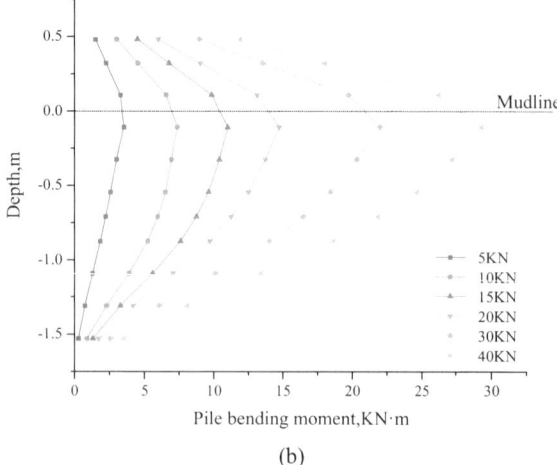

(b)

Fig. 5 (continued)

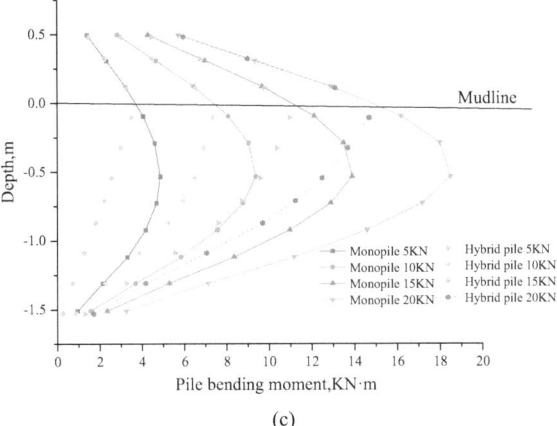

(c)

As shown in Fig. 6, the monopile foundation and the hybrid pile-bucket foundation revolve around their centers of rotation rather than exhibiting deflection deformation as the load increases. The monopile foundation's rotation point is 1.25 m below the soil surface, whereas the hybrid foundation's rotation point is 1 m below the soil surface. After the monopile foundation and bucket foundation were combined, the rotation point's location was raised, indicating that the bucket foundation increased the lateral resistance of the shallow soil on the passive side, leading to a greater active side soil resistance below the rotation point to balance the foundation.

3.4 Soil Resistance and p-y Curves

The lateral soil pressure of the bucket foundation and the soil reaction force of the pile foundation are superimposed to obtain the soil reaction force distribution of the hybrid foundation. Figure 7 shows the lateral soil resistance distribution of single-pile foundation and lateral soil resistance distribution of composite foundation under various levels of loading.

From Fig. 7, we can learn that the passive soil resistance is mainly distributed within 2.5 D below the soil surface, and most of the soil resistance is provided by the bucket foundation.

Figure 8 shows the p-y curves of the hybrid foundation and monopile foundation at various depths. The initial stiffness of the composite foundation is higher than that of the monopile foundation, which can be seen from Fig. 8. At the same depth, the soil resistance of the hybrid foundation with pile and bucket is significantly greater than the monopile foundation, indicating that the bucket foundation significantly increases the soil resistance around the structure.

Fig. 6 Lateral displacement of pile: **a** monopile; **b** hybrid pile-bucket foundation

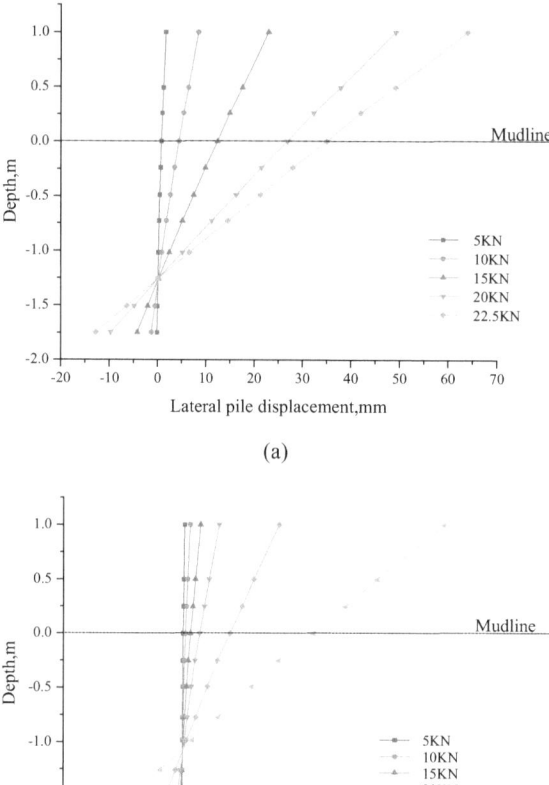

(a)

(b)

4   Conclusion

In this paper, 1 g model tests were carried out for single pile foundations and hybrid pile-bucket foundation in soft clay. The similarities and differences in the structure-soil interaction mechanism and horizontal bearing characteristics of monopile and hybrid pile-bucket foundation under lateral static load were compared and analyzed. The main conclusions are as follows.

- With the combination of monopile foundation and bucket foundation, the initial stiffness and lateral ultimate bearing capacity are greatly improved, the initial stiffness is increased by 216.32% and the horizontal ultimate bearing capacity is increased by 71.11%.

Fig. 7 Distribution of lateral soil resistance: **a** monopile; **b** hybrid pile-bucket foundation

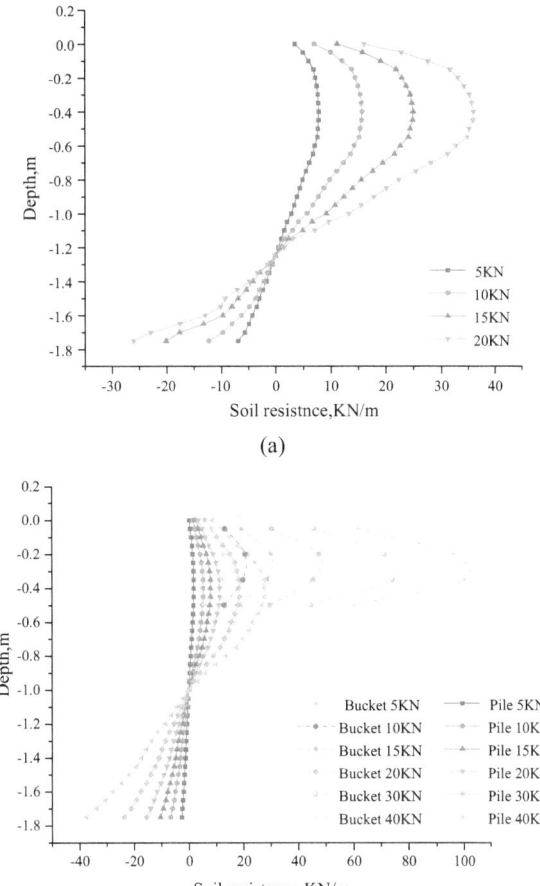

(a)

(b)

Fig. 8 Comparison of monotonic p-y curve response between monopile and hybrid foundation

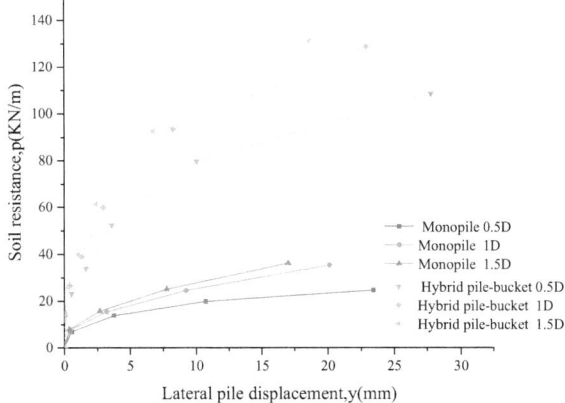

- The pile displacement and pile bending moment of a hybrid foundation are lower than those of monopile foundation under the same load situation, which indicates that the hybrid pile-bucket foundation can control the lateral displacement more effectively than the monopile. Hybrid pile-bucket foundation rotates rigidly around a point below the mud surface under horizontal load, and the depth of the rotation point is smaller than that of monopile foundation.
- The soil resistance is mostly provided by the soil outside the bucket foundation, with the lateral soil resistance of the pile-bucket foundation primarily dispersed within the depth of 2 times the pile diameter. Furthermore, by comparing the p-y curves, it was found that the p-y curve of the hybrid foundation has greater initial stiffness and ultimate soil resistance.

References

1. WFO, *Global Offshore Wind Report*. Technology report. World Forum Offshore Wind (2022)
2. Y. Hu, *Improvement of the Structural Response of Steel Tubular Wind Turbine Towers by Means of Stiffeners* (Doctoral dissertation, University of Birmingham, 2015)
3. G. Murphy, D. Igoe, P. Doherty, K. Gavin, 3D FEM approach for laterally loaded monopile design. Comput. Geotech.. Geotech. **100**, 76–83 (2018)
4. W.C. Tseng, Y.S. Kuo, J.W. Chen, An investigation into the effect of scour on the loading and deformation responses of monopile foundations. Energies **10**(8), 1190 (2017)
5. C. Pérez-Collazo, D. Greaves, G. Iglesias, A review of combined wave and offshore wind energy. Renew. Sustain. Energy Rev. **42**, 141–153 (2015)
6. L.J. Prendergast, K. Gavin, P. Doherty, An investigation into the effect of scour on the natural frequency of an offshore wind turbine. Ocean Eng. **101**, 1–11 (2015)
7. B. Bienen, J. Dührkop, J. Grabe et al., Response of piles with wings to monotonic and cyclic lateral loading in sand. J. Geotech. Geoenviron. **138**(3), 364–375 (2012)
8. J. Peng, B.G. Clarke, M. Rouainia, Increasing the resistance of piles subject to cyclic lateral loading. J. Geotech. Geoenviron. **137**(10), 977–982 (2011)
9. X. Wang, S. Li, J. Li, Load bearing mechanism and simplified design method of hybrid monopile foundation for offshore wind turbines. Appl. Ocean Res. **126**, 103286 (2022)
10. X. Li, X. Zeng, X. Wang, Feasibility study of monopile-friction wheel-bucket hybrid foundation for offshore wind turbine. Ocean Eng. **204**, 107276 (2020)
11. R. Liu, B.R. Li, J.J. Lian, H.Y. Ding, Bearing characteristics of pile-bucket composite foundation for offshore wind turbine. J. Tianjin. Univ. (Sci. Technol.) **48**(5), 429 (2015)
12. H.J. Liu, P. Zhang, Q.D. Wang, Q. Yang, Optimum structural design and loading advantage analysis of pile–bucket foundation. J. Harbin. Eng. Univ. **39**(7), 1165 (2018)
13. L. Li, H. Liu, W. Wu, M. Wen, M.H. El Naggar, Y. Yang, Investigation on the behavior of hybrid pile foundation and its surrounding soil during cyclic lateral loading. Ocean Eng. **240**, 110006 (2021)
14. Y. Lai, W. Li, B. He, L. Wang, G. Xiong, T. Liu, Centrifuge modelling of monotonic and cyclic lateral responses of a hybrid monopile-bucket foundation for offshore wind turbines. Ocean Eng. **260**, 111967 (2022)

Study on Permeable Floor Based on Gray Analysis and AHP

Jianguang Bai, Jianjun Wang, and Tianping Zhou

Abstract The urbanization process leads to many problems with resources and environment, and the problem of urban flooding is particularly prominent. The permeable floor can solve the problem of urban flooding from the perspective of ecology. Based on simulated rainfall, load test, freeze–thaw cycle test and so on, a comprehensive evaluation of permeable floor is carried out from the aspects of functional application, ecological effect and economy by means of gray analysis and AHP. It is showed that the permeable floor can reduce surface runoff and meet the functional requirements of floor. The evaluation of permeable floor mainly needs considering functional application and ecological effect index. The major factor of functional application is bearing capacity. The major factor of ecological effect is capability of conserving water. The major factor of economy is maintenance cost and durability. The gray evaluation result is between excellent and good when 4 level evaluation gray class which includes excellent, good, medium and poor is used. This study provides a reference for the solution of urban flooding and the evaluation of ecological floor.

Keywords Grey analysis and AHP · Permeable floor · The urban flooding · Functional application · Surface runoff

J. Bai (✉)
College of Energy and Transportation Engineering, Inner Mongolia Agricultural University, Hohhot, China
e-mail: b_jg@imau.edu.cn

J. Wang
Dean's Office, Inner Mongolia Green Landscape of Mountain and Water Ecological Engineering Institute, Hohhot, China

T. Zhou
Jining Public Works Section, China Railway Hohhot Bureau Group Co., Ltd., Hohhot, China

© The Author(s) 2025
W. Wang et al. (eds.), *Hydraulic Structure and Hydrodynamics*, Lecture Notes in Civil Engineering 608, https://doi.org/10.1007/978-981-97-7251-3_28

1   Introduction

Urbanization is the inevitable process of human development and national economic development [1]. Rapid urbanization results in many problems of resources and environment, such as environmental pollution [2], frequent water damage [3], ecosystem degradation, urban flooding and slope instability [4]. The large urban impermeable surface can change the original natural conditions, and the area of the bare soil is decreasing [5]. Rainwater can not penetrate into the soil, it will form surface runoff [6], which results in urban flooding [7, 8]. Urban flooding is a systematic and comprehensive problem, which requires a comprehensive and overall solution [9]. In order to eliminate the hidden danger of urban flooding and the phenomenon of "watching the sea" in the city, ecological restoration can be done by the means of vegetation rebuilding to build sponge city [10]. The sponge city can improve the urban ecological environment comprehensively, which can conserve, penetrate and purify water naturally [11, 12]. Prevention and control of urban waterlogging has become a research hotspot in urban ecological environment protection at the present stage [13, 14]. At the same time of ecological slope protection and vegetation restoration with cast-in-place grid technology, it can reduce surface runoff through scale-like grids [15–18], and vegetation rugs can reduce the surface runoff and soil erosion greatly by increasing vegetation coverage [19]. The floor, as an important part of the city, is the focus of the urban flooding management. The urban floor should not only satisfy the basic functional requirements, but also not destroy the environment and slow down the urban flooding. The permeable floor came into being under these conditions.

The stampede body of permeable floor is made up of the concrete grids which is casted by use of the special mould. The steel is laid in the grid and plant can be planted in the middle of the grids. There is grading sand and gravel cushion instead of conventional concrete cushion, which can promote the infiltration of water and growth of roots, as Fig. 1. In this paper, the permeable floor is studied according to the indexes of the functional application, ecological effect and economy.

2   Material and Methods

2.1   Material

According to the function of the floor, it is divided into parking test area and sidewalks test area, each of them is 90 m^2. There are 6 parking spaces in parking test area, the size of each parking space is 2.5 m × 6.0 m. The plant of each parking space respectively is Glade, Nuglade, Rugby rose, Kentucky, Wabash, Rugby rose and Nuglade and Wabash. The concrete mark of stampede body is C30, and its shape is bar and stepping shape. Sidewalks test area is divided into 6 sections, and the size of each section is 2.0 m × 7.5 m. The type of plant is the same as the one in the parking test area. The concrete mark of stampede body is C20 in the Sidewalks test area, and

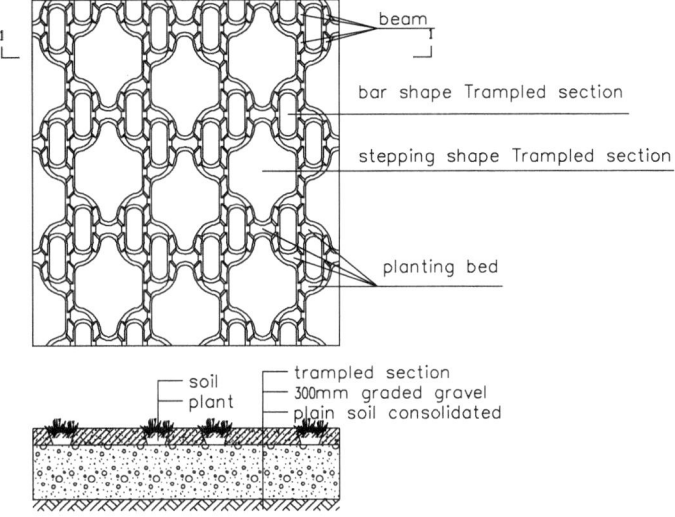

Fig. 1 Permeable floor [20]

its shape is bar which step is 66 cm. Controlled test is set up, concrete floor is for parking space and the bread-brick is for sidewalk area, which is 15 m² respectively.

2.2 Methods

The simulated rainfall device is used to simulate different rainfall intensity, and the rainfall duration before the surface runoff begins is recorded. It sets up eight rainfall intensity from 10 mm/24 h to 80 mm/24 h which gradient is 10 mm/24 h. The load test is used to determine the bearing capacity after 28 days of the floor construction. The survival rate and coverage within 12 months of planting are measured once every month. The tread test is carried out by 30 people and half of them are male, and the no-tread rate which is the proportion of number of no-tread step and number of total step is determined. The freeze–thaw cycle test is carried out to measure its durability.

3 Results

Surface runoff can not take place when the rainfall intensity is no greater than 60 mm/24 h and the duration of rainfall is less than 24 h. When the rainfall intensity reaches 70 mm/24 h or 80 mm/24 h and the duration of rainfall exceeds 24 h, compared with concrete floor and bread-brick floor, the surface runoff of permeable floor is reduced by 51% and 63% at least respectively.

The minimum of the bearing capacity of the permeable floor is 260 kPa. The survival rate of plants is 85–96% and the average is 91%. The coverage of plant after 12 months of planting is 65–73% and the average is 71%. The no-tread rate is $0 \sim 4.1\%$ and the average is 1.8%. The cost is 205 yuan/m^2. Sprinkler maintenance should be carried out only in the first year for permeable floor. The plants can grow naturally through the water conserved by the floor one year later. The strength loss is not more than 6% after the freezing and thawing cycle of 20 times.

4 Discussions

4.1 Weight Calculation and Analysis

Considering the functional applications, ecological effects and economy, the weight of each factor is determined according to AHP. The analysis model based on AHP is built, as Fig. 2.

The judgement matrix is built through the comparison each other and 9 scale method [21, 22]. The coefficient of importance at each level, the maximum value of the judgment matrix and consistency ratio are listed in Tables 1, 2, 3, and 4.

The weight order of the B level is $W_{B1} > W_{B2} > W_{B3}$, as Fig. 3. In the evaluation of the permeable floor, the functional application is the most important factor. The second factor is ecological effect, and the last one is economy.

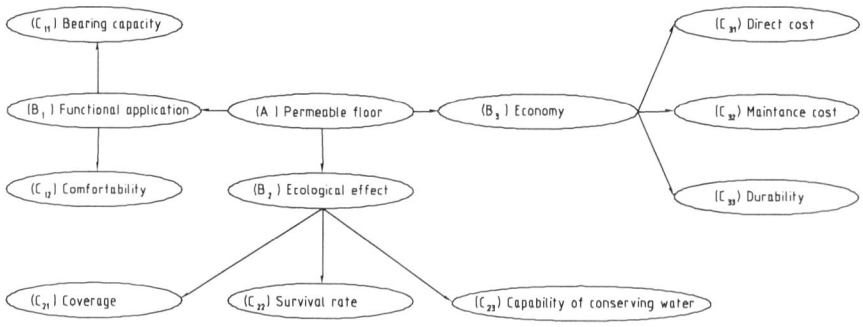

Fig. 2 Analysis model of permeable floor based on AHP

Table 1 The parameter calculation about B1~C layer

B1	C11	C12	W$_{B1}$	
C11	1	5	0.83	λmax $= 2.01$
C12	1/5	1	0.17	C.R$_{B1}$ $= 0$

Table 2 The parameter calculation about B2~C layer

B2	C21	C22	C23	W_{B2}	
C21	1	1/3	1/5	0.11	λmax = 3.03
C22	3	1	1/3	0.26	C.R$_{B1}$ = 0.031
C23	5	3	1	0.63	

Table 3 The parameter calculation about B3~C layer

B3	C31	C32	C33	W_{B3}	
C31	1	1/5	1/5	0.10	λmax = 3.01
C32	5	1	1	0.45	C.R$_{B1}$ = 0.007
C33	5	1	1	0.45	

Table 4 The parameter calculation about A~B layer

A	B1	B2	B3	W_A	
B1	1	3	5	0.65	λmax = 3.02
B2	1/3	1	2	0.23	C.R$_{B1}$ = 0.019
B3	1/5	1/2	1	0.12	

Fig. 3 The weight distribution diagram of B level

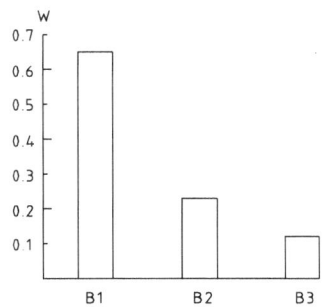

In terms of the B level index, the weight order of the C level is $C_{11} > C_{12}$, $C_{23} > C_{22} > C_{21}$, $C_{32} = C_{33} > C_{31}$, as Fig. 4. The bearing capacity is mainly considered in the functional application. The capability of conserving water is the mainly considered in the ecological effect. The maintenance cost and durability are mainly considered in the economy. The direct cost is only 10%, so it is considered less.

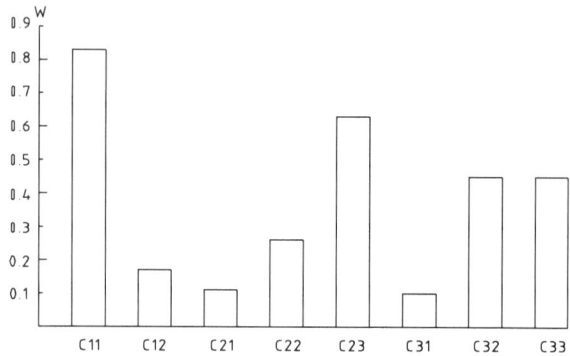

Fig. 4 The weight distribution diagram of C level

4.2 Gray Analysis

The grade standard about gray analysis is set up, as Table 5. According to the test results, the evaluation sample matrix is determined by 6 experts, as formula (1).

$$D = \begin{bmatrix} 3.0 & 2.5 & 2.0 & 4.0 & 4.0 & 3.5 \\ 3.5 & 3.0 & 3.0 & 3.5 & 4.0 & 3.5 \\ 2.0 & 2.0 & 2.0 & 2.5 & 3.0 & 2.5 \\ 3.5 & 3.5 & 3.0 & 3.0 & 3.5 & 3.0 \\ 4.0 & 4.0 & 3.5 & 3.0 & 4.0 & 3.0 \\ 2.0 & 1.5 & 2.5 & 2.5 & 2.0 & 1.0 \\ 3.0 & 3.0 & 2.5 & 2.5 & 3.5 & 1.5 \\ 3.5 & 1.5 & 3.5 & 3.5 & 4.0 & 3.5 \end{bmatrix} \tag{1}$$

Four level evaluation gray class which includes excellent, good, medium and poor is used, that is $e = 1, 2, 3, 4$. The gray evaluation weight vector of eight second-level indexes forms the gray evaluation weight matrix, which is relative to the first-level

Table 5 Grade standard

Index	4	3	2	1
Score				
Bearing capacity	Very high	High	Common	Lower
Comfortability	Very comfort	Comfort	Common	Discomfort
Coverage	Very high	High	Common	Low
Survival rate	Very high	High	Common	Low
Capacity of conserving water	Very strong	Strong	Common	Inferior
Direct cost	Very high	High	Common	Lower
Maintenance cost	Very high	High	Common	Lower
Durability	Very durable	Durable	Common	Inferior

indexes, as formulas (2)–(4).

$$R_1 = \begin{bmatrix} 0.398 \ 0.392 \ 0.210 \ 0 \\ 0.415 \ 0.444 \ 0.141 \ 0 \end{bmatrix} \quad (2)$$

$$R_2 = \begin{bmatrix} 0.266 \ 0.354 \ 0.380 \ 0 \\ 0.386 \ 0.436 \ 0.178 \ 0 \\ 0.469 \ 0.422 \ 0.109 \ 0 \end{bmatrix} \quad (3)$$

$$R_3 = \begin{bmatrix} 0.222 \ 0.296 \ 0.366 \ 0.116 \\ 0.300 \ 0.400 \ 0.263 \ 0.037 \\ 0.420 \ 0.387 \ 0.150 \ 0.043 \end{bmatrix} \quad (4)$$

According to $B_i = W_i R_i$, the evaluation results of the first-level index can be calculated which can be used to build the total gray evaluation weight matrix R, as formula (5).

$$R = \begin{bmatrix} 0.401 \ 0.401 \ 0.198 \ 0 \\ 0.425 \ 0.418 \ 0.157 \ 0 \\ 0.346 \ 0.384 \ 0.222 \ 0.048 \end{bmatrix} \quad (5)$$

$$B = W_A R = \begin{pmatrix} 0.400 \ 0.403 \ 0.191 \ 0.006 \end{pmatrix} \quad (6)$$

$$W = B \cdot C^T = \begin{pmatrix} 0.400 \ 0.403 \ 0.191 \ 0.006 \end{pmatrix} \begin{pmatrix} 4 \ 3 \ 2 \ 1 \end{pmatrix}^T$$
$$= 3.197 \quad (7)$$

Therefore, the total gray evaluation result is between excellent and good.

5 Conclusion

(1) Surface runoff can not take place when the rainfall intensity is no greater than 60 mm/24 h and the duration of rainfall is less than 24 h. When the rainfall intensity reaches 70 mm/24 h or 80 mm/24 h and the duration of rainfall exceeds 24 h, compared with concrete floor and bread-brick floor, the surface runoff of permeable floor is reduced by 51% and 63% at least respectively. So the permeable floor can effectively reduce surface runoff and accumulate rainwater. The minimum of the bearing capacity of the permeable floor is 260 kPa, which can fully meet the requirements of the pedestrian and vehicle. Its functional application is relatively strong. The survival rate of plants is 85–96% and the average is 91%, and the coverage of plant after 12 months of planting is 65–73% and the average is 71%, all of which indicate that the permeable floor can contribute to the growth of plants and play ecological function.

(2) The evaluation of permeable floor mainly needs considering the indexes of functional application and ecological effect, and the economic index is the smallest proportion. The bearing capacity should be considered extremely in the functional application, and the capacity of conversing water should be considered mainly in the ecological effect. Maintenance cost and durability should be considered mainly in economy.

(3) The gray evaluation result is between excellent and good when 4 level evaluation gray class which includes excellent, good, medium and poor is used.

Acknowledgements The authors would like to thank Inner Mongolia Autonomous Region Natural Science Foundation (2023LHMS05048) and Science and Technology Planning Project in Hohhot (2017-Patent-1) for their support for this study.

References

1. C.M. Liu, Y.Y. Zhang, Z.G. Wang et al., The LID pattern for maintaining virtuous water cycle in urbanized area. J. Nat. Resour. **31**(A5), 719–731 (2016)
2. D. Varade, H. Singh, A.P. Singh, et al., Assessment of urban sprawls, amenities, and indifferences of LST and AOD in sub-urban area: a case study of Jammu. Environ. Sci. Pollut. Res. (2023)
3. Q.T. Zuo, Water science issues in sponge city construction. Water Resour. Prot. **32**(A4), 21–26 (2016)
4. S. Paul, V.M. Bondrea, H. Cetean et al., Ameliorative, ecological and landscape roles of faget forest, cluj-napoca, Romania, and possibilities of avoiding risks based on GIS landslide susceptibility map. Notulae Botanicae Horti Agrobotanici Cluj-Napoca **6**(A1), 292–300 (2018)
5. Y.C. Liu, L.C. Liu, J.N. Liu et al., Characteristics and influence factors of summer surface urban heat Island under arid and semi-arid climate. Sci. Technol. Eng. **17**(A28), 160–165 (2017)
6. S. Chen, J.Q. Wang, The theory of the spongy city and its application in the landscape planning. Agric. Technol. **36**(A3), 128–131 (2016)
7. D.J. Wu, S.Z. Zhan, Y.H. Li et al., New trends and practical research on the sponge cities with Chinese characteristics. Chin. Soft Sci. **A1**, 79–97 (2016)
8. L.Q. Chen, Z.S. Wang, L. Shi, Thinking of urban drainage planning and management after rainstorm flooding. Water Supply Drainage **37**(A10), 29–33 (2011)
9. K.J. Yu, D.H. Li, H. Yuan et al., Theory and practice of sponge city. Planning Res. **39**(A6), 26–36 (2015)
10. B.Q. Zhao, Z.Y. Xia, W.N. Xu et al., Review on research of slope eco-restoration technique for engineering disturbed area. Water Resour. Hydropower Eng. **48**(A2), 130–137 (2017)
11. Z.X. Liao, A.G. Gao, E.H. Huang, Enlightenment of rainwater management in foreign countries to sponge city construction in China. Water Resour. Prot. **32**(A1), 42–45, 50 (2016)
12. Y. Yang, G.S. Lin, A review on sponge city. South Architect. **A3**, 59–64 (2015)
13. P. Bai, W.B. Zang, H.P. Zhang, J.S. Zhu, F. Peng, J. Cui, A stratified modeling approach to flooding at urban infrastructures. J. China Inst. Water Resour. Hydropower Res. (2022)
14. J.P. Zhang, Z.Y. Zhang, Q.T. Zuo, Urban waterlogging simulation and emergency response capacity evaluation with extreme rainstorms. J. Zhengzhou Univ. (Eng. Sci.) **44**(A2), 30–37 (2023)

15. K.V. Brix, D.K. DeForest, L. Tear et al., Use of multiple linear regression models for setting water quality criteria for copper: a complementary approach to the biotic ligand model. Environ. Sci. Technol. **51**(A7), 5182–5192 (2017)
16. L. Zhang, Y.F. Liu, X.J. Wei, Forest fragmentation and driving forces in Yingkou Northeastern China. Sustainability **A9**, 51–70 (2017)
17. D. Balazs, V. Orsolya, T. Peter, Factors threatening grassland specialist plants—a multi-proxy study on the vegetation of isolated grasslands. Biol. Cons. **204**(A5), 255–262 (2016)
18. L. Spencer Kate, J. Carr Simon, M. Diggens Lucy, The impact of pre-restoration land-use and disturbance on sediment structure, hydrology and the sediment geochemical environment in restored saltmarshes. Sci. Total Environ. **587**, 47–58 (2017)
19. T.H. Zhao, S.Y. Niu, H.C. Guo et al., Study on effect of new ecological vegetation blanket in soil and water conservation of slope. Yangtze River **48**(A13), 20–22 (2017)
20. J.G. Bai, J.J. Wang, Y.L. Zhang, N. Wen, Decision analysis of slope ecological restoration based on AHP. Sains Malaysiana **46**(A11), 2075–2081 (2017)
21. X.M. Wang, Q. Kang, J.C. Qin et al., Application of AHP-extenics model to safety evaluation of rock slope stability. J. Central South Univ. (Sci. Technol.) **44**, 2455–2462 (2013)
22. Z.H. Xu, S. Li, L.P. Li et al., Risk assessment of water or mud inrush of karst tunnels based on analytic hierarchy process. Rock Soil Mech. **32**, 1757–1766 (2011)

Study on Hydrodynamic Load of Permeable Structure for Offshore Oilfield Based on Model Test

Qing Qin and Jingjing Qi

Abstract In order to identify the hydrodynamic load of a new type of permeable structure for the offshore oilfield, a physical simulation test is performed using a scale model. The variation of hydrodynamic load parameters of the new permeable structure under different working conditions is compared. The results can provide basic data for optimization of structural parameters of the new permeable structure for offshore oilfields. The relevant experimental results can provide basic data for optimizing the structural parameters of permeable structures in tidal flats, and provide a basis for the engineering design and construction of permeable structures.

Keywords Permeable structure · Flume test · Hydrodynamic load

1 Introduction

The offshore oilfield is essential to the development of oil and gas in China's coastal area to increase the energy reserves and production. In China, the developed offshore oilfields, such as Liaohe Oilfield, Jidong Oilfield, Dagang Oilfield and Shengli Oilfield, are distributed in the offshore areas of Liaodong Bay, Bohai Bay and Laizhou Bay [1–3]. Traditionally, offshore drilling mainly depends on physical structures such as beach-shallow platforms and offshore oil drilling platforms (i.e., artificial islands). These structures can change the original tidal current and flow field in a certain range, including the sediment movement. Accordingly, the sediment scouring and silting environment is changed to different degrees. As the

Q. Qin (✉)
Student Work Department of Guilin Normal College, Guilin, China
e-mail: 251053145@qq.com

Q. Qin · J. Qi
Shengli Oilfield Technology Inspection Center, SINOPEC, Dongying, China

J. Qi
Shengli Oilfield Inspection Evaluation Research Co.Ltd., Dongying 257000, China

© The Author(s) 2025
W. Wang et al. (eds.), *Hydraulic Structure and Hydrodynamics*, Lecture Notes in Civil
Engineering 608, https://doi.org/10.1007/978-981-97-7251-3_29

Chinese government has successively issued laws and policies on marine environmental protection in recent years, fewer traditional beach facilities can be used to develop gas and oil. This creates an urgent need to develop a new environmental-friendly permeable offshore structure. This paper proposes a new type of permeable structure form with wave consumption performance and water exchange capacity. Its hydrodynamic load is analyzed by performing the physical model test, providing a scientific basis for legal compliance and the green and safe production of oil and gas in offshore areas [4, 5].

2   Test Design

2.1   Test Equipment and Instruments

The permeable structure piles proposed in this study are shown in Fig. 1. According to the Gravity Similarity Criterion, the model scale was 1:12. The actual model is shown in Fig. 2. This test was performed in three steps. First, before the model was placed, a calibrated wave height meter was used to calibrate all test conditions to determine the parameters of wave-making files (push plate stroke, period, etc.) that meet the test conditions. Secondly, after the calibration, the structure was placed in a flume, and point pressures at the bottom of breakwater, square tubular pile foundation and wave wall were measured under different working conditions. Thirdly, three water depths were set in this test, 0.65 m, 0.80 m and 0.95 m, respectively. The structure model was placed 29 m away from the wavemaker, where two wave height meters were set to calibrate the wave conditions. See Table 1 for the wave group [6–8].

The offshore permeable structure model was set in the middle and back of the two-dimensional flume, 29 m away from the wavemaker. The arrangement of wave height meters in the flume is shown in Fig. 2. To measure the wave load on the model structure, a total of 28-point pressure sensors were set at the bottom of breakwater, square tubular pile foundation and wave wall, as shown in Fig. 3.

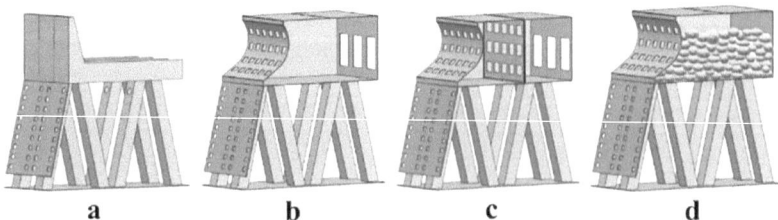

Fig. 1 New offshore permeable structure

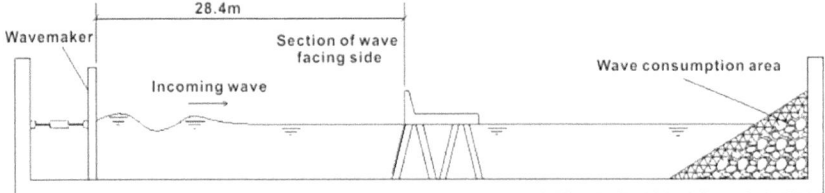

Fig. 2 Two-dimensional flume diagram

Table 1 Wave group in the hydrodynamic load test

Water depth d/m	Incoming wave period T/s	Incoming wave height H/cm
0.65	1.00	10
	1.25	10
0.80	1.50	6, 10, 14, 18, 22
0.95	1.75	10
	2.00	10

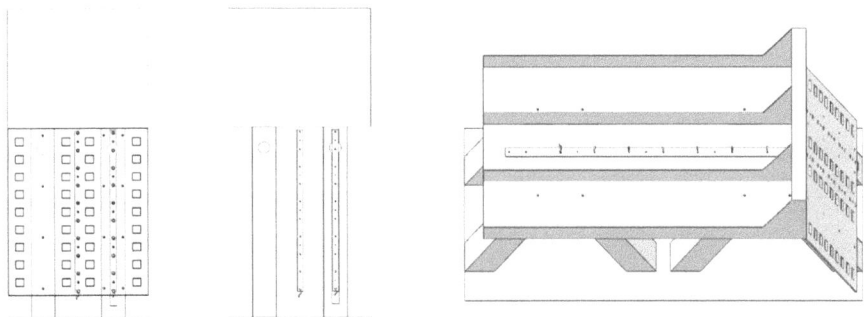

Fig. 3 Schematic diagram of point pressure sensor layout

2.2 Working Condition Setting

The substructure of the studied permeable structure adopted the form of a pile-coupled plate. The pile foundation was a square pile; the breakwater was set with three opening rates of 10%, 20% and 30% in single-layer and double-layer forms [9]. In the test, by changing the typical superstructure form, there were 11 specific structure combinations with three water depths of 0.65 m, 0.80 m and 0.95 m, respectively. The specific working conditions are shown in Table 2.

Table 2 Working condition setting

	Superstructure form	Layer number	Opening rate	Water depth (m)	Wave parameter
1	Vertical wave wall	Single	10%	0.65/0.8/0.95	T = 1.00 s; H = 10 cm
2	Vertical wave wall	Single	20%	0.65/0.95	T = 1.25 s; H = 10 cm
3	Vertical wave wall	Single	30%	0.65/0.95	T = 1.50 s; H = 06 cm
4	Vertical wave wall	Double	10% + 10%	0.65/0.8/0.95	T = 1.50 s; H = 10 cm
5	Vertical wave wall	Double	20% + 20%	0.65/0.8/0.95	T = 1.50 s; H = 14 cm
6	Vertical wave wall	Double	30% + 10%	0.65/0.8/0.95	T = 1.50 s; H = 18 cm
7	Vertical wave wall	Double	10% + 30%	0.65/0.8/0.95	T = 1.50 s; H = 22 cm
8	Reverse arc-shaped wave wall (without partitions)	Single	10%	0.95	T = 1.75 s; H = 10 cm
9	Reverse arc-shaped wave wall (without partitions)	Single	10%	0.95	T = 2.00 s; H = 10 cm
10	Reverse arc-shaped wave wall (Filling to the mean water level).	Single	10%	0.95	T = 2.25 s; H = 10 cm
11	Reverse arc-shaped wave wall (Filling to the mean water level)	Single	10%	0.95	T = 2.50 s; H = 10 cm

3   Test Results and Analysis

3.1   Influence of the Water Depth and Wave Parameters on the Wave Pressure

Test results of Case 1 are studied to analyze the effects of the water depth and wave parameters on the time-domain characteristics of wave pressure. As shown in Fig. 4a–d, at the period T = 1.50 s and the wave height H = 0.10 m, the wave pressure at 01# ~ 10# measuring points of the perforated breakwater decreases with the increase of water depth, and the wave pressure on the breakwater increases with the period. However, the hydrostatic pressure distribution shows an opposite pattern, which increases with the water depth. Besides, in deeper water, the variation of wave pressure at the adjacent measuring point gradually decreases. In addition, the

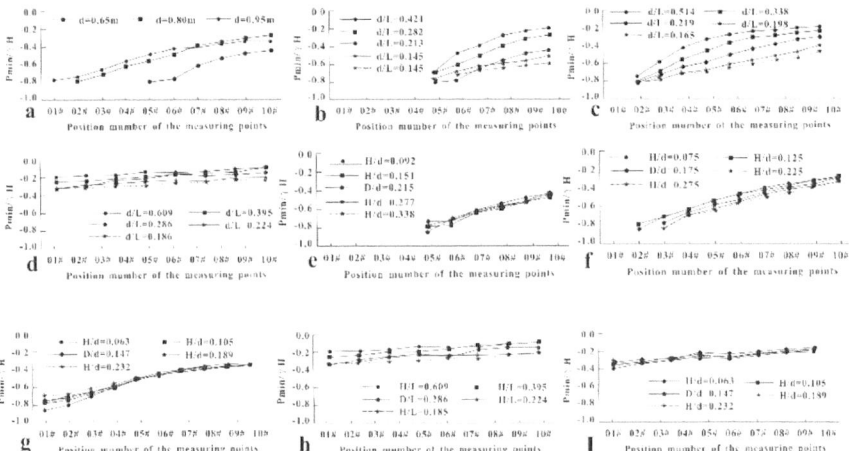

Fig. 4 Influence of the water depth and wave parameters on the wave pressure

variation of wave pressure near the water bottom with the period is more obvious than that near the water surface [10]. According to Fig. 4e–g, when the period is T = 1.50 s, the wave pressure at each measuring point on the breakwater does not change significantly with the wave height. Compared with the period, the wave height has a greater effect on the wave pressure in deeper water. When the wave height changes, the wave pressure decreases linearly along the water depth.

3.2 Influence of the Opening Rate on the Wave Pressure

A single-layer breakwater was arranged on the first row of inclined piles on the wave-facing side, and three opening rates of 10%, 20% and 30% were set to form three combination models of Case 1, 2 and 3. As shown in Fig. 5, the bigger the opening rate, the smaller the wave pressure on the breakwater in shallow water. However, this trend is not obvious in deep water because the wave energy is mainly concentrated in the upper layer of water. Specifically, more than 90% of the wave energy is concentrated in the range of 2–3 times the wave height of the water surface. The perforated breakwater is submerged in deep water, providing a weaker wave consumption effect. The wave energy acting on the breakwater is smaller than that in shallow water. However, compared with the breakwater, the wave pressure variation at the bottom of the wave wall shows an opposite pattern, which increases slightly with the increase of the opening rate.

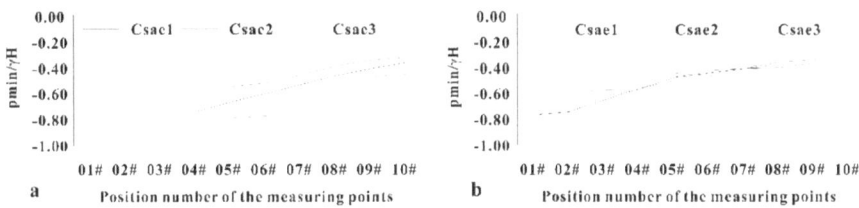

Fig. 5 Influence of the opening rate on the wave pressure

3.3 Influence of Breakwater Layers on the Wave Pressure

Different layers of the breakwater were arranged on the first or the third row of inclined piles on the wave-facing side. After the layout was designed, three combination models of Case 4 to 7 were obtained. According to Fig. 6, at the same opening rate, the wave pressure on the single-layer breakwater in shallow water is lower than that on the front one of the double-layer breakwater, and the difference near the water surface is slightly larger than that at the water bottom. However, in deep water, there is no significant difference. In shallow water, the wave pressure distribution on the rear one of the double-layer breakwater with the water depth and opening rates is consistent with the previous result. Case 4/7 shows that, at the same opening rate of the front breakwater, the wave pressure on the rear one increases with the opening rate, contrary to the changing pattern of the single-layer breakwater. Case 4/6 reveals that at the same opening ratio of the rear one, the larger the opening rate of the front one is, the smaller the wave pressure on the rear one is. However, this difference decreases gradually along with the water depth. In deep water, when the opening rate of the front breakwater is low, the wave pressure on the rear one is less affected by the opening rate of itself and the front one. When the opening rate of the front one is relatively high, the wave pressure on the rear one increases significantly, which is consistent with the previous analysis result.

3.4 Influence of Superstructure on the Wave Pressure

Based on the above ideas, the wave wall on the superstructure was changed accordingly. Research has shown that the influence of changes in the superstructure of the permeable structure on the wave pressure on the bottom of the breakwater and the wave wall is negligible.

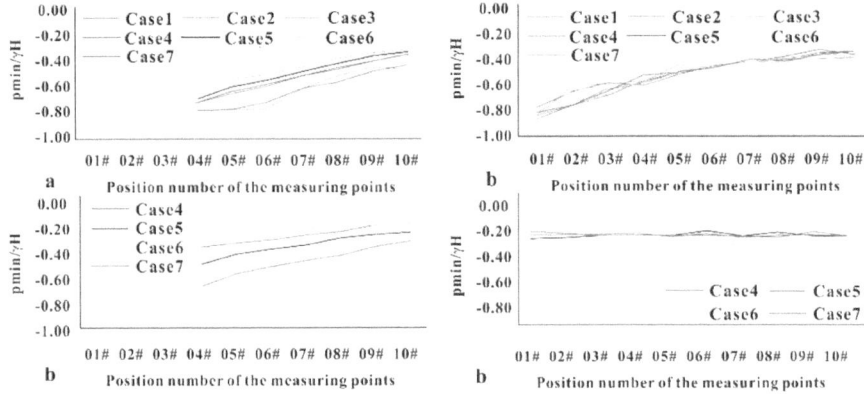

Fig. 6 Influence of breakwater layers on the wave pressure

4 Conclusion

After analyzing the hydrodynamic load of permeable structure for the offshore oilfield based on the model test, conclusions are as follows:

(1) On the simple-layer breakwater, the water pressure increases with the increase of the period, but it has no obvious change with the wave height. The wave pressure on the bottom of the wave wall changes slightly. In shallow water, the bigger the opening rate, the smaller the wave pressure on the breakwater. However, this trend is not obvious in deep water.

(2) For the double-layer breakwater, the wave pressure on the front one is smaller than that on the single-layer breakwater, and the difference near the surface is slightly larger than the water bottom. With the bigger opening rate of the front breakwater, the water pressure on the rear one increases significantly. Overall, the number of the breakwater has less impact on the load distribution on the bottom.

(3) The influence of changing the superstructure of the permeable structure on the wave pressure on the bottom of the breakwater and the wave wall is negligible.

(4) In summary, the hydrodynamic load of permeable structures follows the following pattern. The wave pressure at the single-layer wave baffle increases with the increase of the period, and there is no significant change with the wave height. The change in wave pressure at the bottom of the wave wall is relatively small. As the opening rate increases, the wave pressure at the wave deflector gradually decreases in shallow water conditions, and the difference is not significant in deep water conditions. When there is a double layer wave baffle, the pressure of the front baffle wave is smaller than that of the single layer wave baffle, and the difference near the water surface is slightly larger than that at the bottom; When the opening rate of the front row is high, the wave pressure of

the rear baffle increases significantly; The number of wave deflectors has little effect on the distribution of bottom load.

Acknowledgements Supported by the Natural Science Foundation of Shandong Province. (ZR2022QD074); Sinopec Scientific Research Project (P23164, YKD2303).

References

1. W. Tang, J. Hu, Y.Z. Cheng, Y.C. Hu, Analysis of structure and wave dissipation characteristics of a new type of perforated composite plate breakwater [J]. Hydro-Sci. Eng. **5**, 37–44 (2017)
2. J.B. Zhang, X.Y. Li, Q. Wang, F. Yi, J.Z. Zhang, L.X. Wang, Z.C. Zhang, Q. Li, J.R. Wang, Y.K. Wang, Comparison of hydrodynamic characteristics between single board and single arc board breakwaters based on physical model test [J]. Port Waterway Eng. (06), 9–14+27 (2020)
3. Q. Li, X.Y. Li, Q. Wang, J.B. Zhang, L.X. Wang, Z.C. Zhang, F. Yi, Numerical study on hydrodynamic characteristics of arc-plate open type breakwater based on open FOAM. [J] J. Xiamen Univ. (Nat. Sci.) **59**(03), 412–419 (2020)
4. J.S. Xu, X.Y. Xu, Y.Q. Zhang et al., Experimental study on the influence of pipeline vibration on silty seabed liquefaction [J]. Water **14**, 1782 (2022)
5. X.L. Zhou, J. Zhang, J.J. Guo, et al., Cnoidal wave induced seabed response around a buried pipeline [J]. Ocean Eng. **101**(C), 118–130 (2015)
6. S. Ye, K. Gu, J. Zhang, et al., Design technology of impervious sea-linking structures in riprap foundation [J]. Port Waterway Eng. (05), 41–47 (2021)
7. P. Foray, D. Bonjean, H. Michallet et al., Fluid-soil-structure interaction in liquefaction around a cyclically moving cylinder [J]. J. Waterw. Port Coast. Ocean Eng. **132**(4), 289–299 (2006)
8. ASCE Pipeline Flotation Research Council, ASCE preliminary research on pipeline flotation: report of the pipeline flotation research council [J]. J. Pipeline Div. **92**(1), 27–74 (1966)
9. Y. Tian, B. Youssef, M. Jcassidy, Assessment of pipeline stability in the Gulf of Mexico during hurricanes using dynamic analysis [J]. Theor. Appl. Mech. Lett. **5**(02), 74–79 (2015)
10. S.L. Dunn, P.L. Vun, A.H.C. Chan et al., Numerical modelling of wave-induced liquefaction around pipelines [J]. J. Waterw. Port Coast. Ocean Eng.Waterw. Port Coast. Ocean Eng. **132**(4), 276–328 (2006)

Application of Digital Twin Technology in Assessing the Level of Water Ecological Civilization Construction in Yangtze River Basin

Donghui Hu, Chaomin Zhou, and Feng Xie

Abstract The emergence of digital twin technology has facilitated the assessment of the level of water ecological civilization in the Yangtze River basin. In recent years, the digital twin technology in the Yangtze River basin has led to a gradual improvement in the efficiency of water quality and environment monitoring. In this context, how to flexibly apply digital twin technology to measure the development level of ecological civilization in the Yangtze River basin? In order to solve this problem, this paper firstly builds the evaluation index system of water ecological civilization level in Yangtze River basin based on DPSIRM model, and introduces Entropy weighting technique. Then analyzes the impact of digital twin technology on each index dimension, and finally designs how digital twin construction can better promote the construction of water ecological civilization in Yangtze River basin.

Keywords The Yangtze River basin · Water ecological civilization · Assessment · Digital twin technology

1 Introduction

The Yangtze River basin has always had an important strategic position in the development pattern of China in the new era, and water ecological protection is a basic prerequisite for promoting the development of the Yangtze River Economic Belt by relying on the golden waterway. With the economic and social development and population growth in the Yangtze River basin, the imbalance between water supply and demand and water pollution have gradually become bottlenecks that restrict regional economic and social development [1, 2]. General Secretary Xi Jinping stressed at the

D. Hu · C. Zhou (✉)
Yuyao Water Conservancy Bureau, Ningbo 323000, Zhejiang, China
e-mail: 190425584@qq.com

F. Xie
Zhejiang University of Water Resources and Electric Power, Hangzhou Zhejiang 310018, China
e-mail: xiefeng@zjweu.edu.cn

© The Author(s) 2025
W. Wang et al. (eds.), *Hydraulic Structure and Hydrodynamics*, Lecture Notes in Civil Engineering 608, https://doi.org/10.1007/978-981-97-7251-3_30

symposium on comprehensively promoting the development of the Yangtze River Economic Belt that restoration of the Yangtze River ecological environment should be placed in an overriding position [3]. As can be seen, a correct grasp of the current situation of water ecological civilization is the basis for solving water ecological environment problems. The core of water ecological civilization is based on the concept of human-water harmony, to achieve sustainable use of water resources and support the harmonious development of economy and society. However, since the water ecological environment in the basin is more complex, how to more efficiently assess the construction level of water ecological civilization in the basin and accurately grasp the impact of economic and social production activities on the environment will be an important basis for coordinating economic development and protecting the environment, and achieving green and high-quality development in the basin. The emergence of digital twin technology provides a solution to this problem. In recent years, the development of digital twin technology and its application on the ground in the Yangtze River Basin has led to a gradual improvement in the efficiency of water quality, water quantity and water environment supervision, the continuous improvement of the monitoring and sensing system, and the strengthening of monitoring data aggregation and processing analysis. In this context, how to flexibly apply digital twin technology to measure the development level of ecological civilization in the Yangtze River basin? To solve this problem, this paper firstly builds the evaluation index system of water ecological civilization level in Yangtze River basin based on DPSIRM model, and then analyzes the impact of digital twin technology on each index dimension, and finally designs how digital twin construction can better promote the construction of water ecological civilization in Yangtze River basin.

2 Regional Water Ecological Civilization Level Evaluation Index System Based on DPSIRM Model

2.1 Research Framework

The DPSIRM model is an evolution of the DPSIR model proposed by the European Environment Agency (EEA), and is currently used in studies on ecological security, water environment carrying capacity and lake ecosystem health assessment. The model consists of six systems: "driver-pressure-state-impact-response-management", which can explain the inner connection between economic society and ecosystem in a more scientific way. The construction of regional water ecological civilization involves resources, environment, as well as economic and social factors. These factors have a wide range of multi-level interlinkages, mutual constraints and interactions. The use of DPSIRM (driving force—pressure—state—impact—response—management) model, can better explain the complex structure of the relationship between the two in the process of building water ecological civilization [4, 5]. Specifically: the continuous population concentration and continuous economic

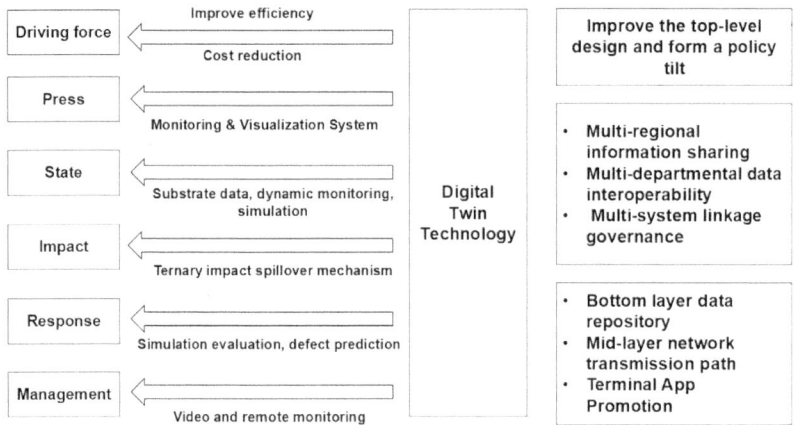

Fig. 1 The use of digital twin technology in assessing the level of water ecological civilization construction in the Yangtze River basin

and social development bring pressure to the regional water ecological environment. When the pressure on the water ecological environment exceeds the carrying capacity of the regional water ecological environment, the state of water resources available and demanded will change. The water ecosystem is an important part of the natural ecological environment and is closely related to the human economy and society, so the deterioration of the water ecological environment will have a significant negative impact on the natural and social environment. In order to alleviate the contradiction between water ecosystem, nature and economy and society, human will take the initiative to respond to water ecological environment management measures, improve water resources utilization efficiency and purify water ecological environment. At the same time, the government will formulate relevant macro-control policies to realize the long-term management of water ecosystems in order to achieve the goal of synergistic economic-social-natural-ecological sustainable development. Therefore, this paper will construct the evaluation index system of water ecological civilization construction in Yangtze River Economic Zone, and analyze the application of digital twin technology in the evaluation process (as shown in Fig. 1).

2.2 Research Method

The methods of determining indicator weights are mainly subjective and objective assignment methods. The objective assignment method determines the weight of an indicator based on the amount of information provided by the original value of the indicator, and the assignment information comes from the objective environment [6]. Among them, the entropy weighting method, as one of the commonly used objective assignment methods, relies only on the discrete nature of the data itself. The higher the discrete degree of the data, the greater the entropy and the more information it contains, the greater the weight it takes [7].

Suppose there are r years, n provinces and cities, and m indicators, then $X_{\theta ij}$ is the value of the jth indicator of province i in the θth year.

Calculating the weight of the sample to the indicator:

$$P_{\theta ij} = \frac{X_{\theta ij}}{\sum_{\theta} \sum_{i} X_{\theta ij}}, i = 1, 2, \ldots, n; j = 1, 2, \ldots, m \tag{1}$$

Combining Eq. (1) yields the entropy value of the jth indicator:

$$e_j = -k \sum_{\theta}^{r} \sum_{i}^{n} P_{\theta ij} \ln(P_{\theta ij}), j = 1, 2, \ldots, m \tag{2}$$

where,

$$k = 1/\ln(n) > 0 \tag{3}$$

$e_j \geq 0$. Based on the relationship between information entropy redundancy (difference) and entropy value, the information entropy redundancy (difference) is derived as,

$$d_j = 1 - e_j, j = 1, 2, \ldots, m \tag{4}$$

Substituting into Eq. (4), the weights of each index are obtained as,

$$w_j = d_j / \sum_{j} d_j \tag{5}$$

Combining Eq. (5) to calculate the comprehensive score of regional water ecological civilization of each province and city in the Yangtze River economic belt:

$$s_j = \sum_{j} w_j X_{\theta ij} \tag{6}$$

3 The Use of Digital Twin Technology in the Assessment of Water Ecological Civilization Construction Level

3.1 Driving Force Indicators (D)

Economic and social development is an important factor driving the construction of water ecological civilization. The digital twin technology can promote the traditional development mode of the Yangtze River basin from modernization, digitalization and intelligence, and promote economic development to high quality. On the one hand, the industrial manufacturing sector is most affected by digital twin technology, which is an important breakthrough to transform the manufacturing industry into a "smart manufacturing industry". Digital twin technology can improve production efficiency, reduce the cost per unit of production, and enhance the competitiveness of products in the market. Specifically, digital twin technology can simulate product design, simulate equipment operating parameters, predict production efficiency, discover potential faults, detect the whole production process, reduce the potential risks of production, save costs and improve input and output efficiency. On the other hand, digital twin technology is an important cornerstone for building smart cities. Through digital twin technology, the management system of cities and watersheds, as well as the allocation of resources, can be planned in an integrated manner to promote the operation of cities and improve the efficiency of resource allocation more efficiently. In summary, digital twin technology can effectively improve the efficiency of economic and social operation, improve the level of economic and social development, and thus drive the construction of water ecological civilization in the Yangtze River basin [8].

3.2 Pressure Indicators (P)

Economic and social development and continuous urbanization will adversely affect the ecological environment of the basin, causing double pressure on the water environment and water resources. During the production process, people will discharge pollutants into the natural water bodies during the production process, which will put great pressure on the water environment purification and restoration. Digital twin technology can alleviate this problem to a certain extent. Digital twin technology has a wide application prospect in the field of water environment monitoring, which can provide more accurate, reliable and efficient water environment monitoring services and help protect and manage the water environment and promote sustainable development [9].

3.3 Status Indicator (S)

Faced with the rapid growth of water demand brought about by economic and social development, mastering the regional water resources volume and supply capacity is a prerequisite and basis for achieving sustainable utilization of regional water resources. Digital twin technology has a great contribution in clarifying the supply capacity of water resources and the carrying capacity of water environment. The prerequisite for the application of digital twin technology is the water resources volume as the base data, and the establishment of the original database for water resources volume monitoring, storing a large amount of data on precipitation, vegetation, topography, groundwater, runoff and other parameters.

3.4 Impact Indicators (I)

Changes in the state of water supply can have an impact on the natural environment and the economy and society. Changes in water conditions will directly affect the formation development and succession of ecosystems such as vegetation and communities; the economic status of the basin cities will also be indirectly affected by changes in water quality and quantity in the basin, mainly in two aspects: on the one hand, the development of traditional agriculture and fisheries has a more important correlation with water quality and supply levels, and on the other hand, the inland navigation capacity determines the transportation capacity of the basin, which in turn affects Regional economic development. Whether through ecological, economic or social networks, digital twin technology can be used to model the linkages between the underlying data and whether changes in the state of water supply will affect the natural environment and the economy and society, the extent and the degree of the impact [10].

3.5 Response Indicators (R)

To avoid further deterioration in the relationship between water ecology, nature and economic and social systems, humans respond to negative feedback from water resources systems mainly through measures such as pollution source management, water resources pressure relief and ecological restoration. Improving sewage treatment capacity and speeding up the construction of drainage networks can help reduce water environment pollution at source and improve water use efficiency; water-saving irrigation in agricultural production activities and increasing the utilization rate of agricultural water-saving irrigation-type machinery can alleviate the pressure of water shortage and ensure the sustainable development of China's agriculture. On the one hand, digital twin technology can test whether human response policies can effectively achieve the corresponding goals. Through augmented reality, mixed reality, engineering simulation technology, etc., the response measures to obtain the presentation of simulation visualization, to carry out fault diagnosis, performance optimization, etc.

3.6 Management Indicators (M)

The government will develop macro-control policies to achieve long-term management of water ecosystems. By increasing the financial investment in environmental protection can strengthen the regional water ecological civilization construction efforts, and improving the level of urban greening can weaken the negative impact of urbanization on the ecological environment. Digital twin technology can provide more accurate video and remote sensing monitoring to obtain accurate and reliable data, which can be applied and measured to improve the government's effectiveness in managing watershed ecosystems and systems [11].

4 Supporting Measures for Using Digital Twin Technology to Assess the Level of Water Ecological Civilization Construction

4.1 Improving Top-Level Design and Forming Policy Inclination

Governments at all levels should formulate relevant macro-control policies for promoting digital twin technology in water ecological civilization construction. The government is supposed to increase its support for digital twin projects, increase the amount of investment in digital monitoring systems and other basic platforms, and encourage the introduction of big data, artificial intelligence and other methods to achieve efficient water resources management.

4.2 Creating Interconnected Systems and Sharing Information Achievements

- Multi-regional information sharing

The construction of water ecological civilization in the Yangtze River basin is a system project, in which provinces and municipalities have to break administrative boundaries and barriers, strengthen joint prevention and control of environmental pollution, promote the establishment of inter-regional and upstream and downstream ecological compensation mechanisms, and realize comprehensive planning of the basin. Therefore, there is a need for inter-regional digital twin platform as an opportunity to build an overall water ecological civilization monitoring system in the Yangtze River basin to guarantee the achievement of water management goals in the basin as a whole rather than in local areas.

- Multi-sectoral data interoperability

External shared data mainly includes socio-economic, population density, emergencies, shipping and other cross-industry data shared by other industries in the Yangtze River basin, which is mainly obtained through dispatching instructions from higher-level departments or accessing related industry shared data by local departments as needed. In accordance with the requirements of intelligent water conservancy and digital twin technology construction, a multi-level database should be established to realize data reporting, dispatching and synchronization in the Ministry of Water Resources and provincial administrative departments, and to realize cross-level data sharing among the management agencies in the basin, provincial water conservancy departments and key water conservancy project management units to realize multi-sectoral linkage of water ecological civilization in the Yangtze River basin.

4.3   Building Sound Infrastructure and a Decision-Making System

- Underlying data repository

Simulation of dynamic changes in the level of water ecological civilization in the Yangtze River basin through simulation technology requires the use of parallel computing acceleration technology and distributed parallel scheduling algorithms to achieve a dynamic simulation of the natural-economic-social ternary water cycle flow field in the basin. The work is based on a large amount of baseboard data. These data not only need to be stored for a long time, but also need to be updated regularly. Therefore, the data repository and update system need to be improved urgently to manage the base data more efficiently and prevent the loss of labor, time and money caused by missing and outdated data.

- Middle-tier network transmission path

Whether it is multi-regional, multi-level or multi-system supervision of water ecological civilization, there is a need to share and interoperate monitoring data, visualization results and related information. Water information network transmission capacity, security, reliability for integrated collaborative governance is particularly critical. Therefore, should vigorously introduce and apply low-power Internet of things, Beidou satellite communication technology, while focusing on popularizing the technology to remote, no public network coverage areas of hydrological elements monitoring and data transmission.

4.4 Increasing Special Investment, the Introduction of Scientific Research Talent

Departments at all levels need to twin water resources as a key project to implement. The department should actively seek financial support from higher authorities to increase research efforts, and horizontally join with local governments to promote the establishment of digital twin systems, and increase the budget for R&D investment in government finances to provide financial support for electronic data and sustainable maintenance of the digital twin water system.

The Yangtze River Basin needs to focus on building a composite talent pool. With the continuous development of the information age, the water conservancy sector needs to enrich the type of talent, out of the traditional "strong professional, strong water conservancy" talent concept, expand the professional coverage of talent. It is conducive for creating a good development Environment, to promote talent familiar with the system construction, governance environment, the implementation of work, for the digital twin system of localized problem solving, model establishment, follow-up development to provide security.

5 Conclusion

DPSIRM model and entropy technology is applied to measure the level of water ecological civilization in the Yangtze River Economic Zone. During this process, the digital twin technology will make the assessment more efficient. We find that, On the one hand, digital twin technology can be used in "driving force—pressure—state—impact—response—management" to help the assessment of water ecological civilization construction. On the other hand, creating interconnected systems and sharing information achievements, building sound infrastructure and a decision-making system and increasing special investment, the introduction of scientific research talent are encourage to secure the use of digital twin technology in the field.

References

1. F.Q. Wang, *Thinking and Planning for the Protection of Yangtze River Water Resources Under the New Situation*, vol. 47. (People's Yangtze River, 2016), pp. 8–11+21.

2. Z.Y. Fang, R.C. Xiao, Supporting Sustainable Economic Development with Sustainable use of Water Resources, vol. 10. (People's Yangtze River, 2003), pp. 6–8+51
3. K.W. Tang, Discussion on the connotation and evaluation system of water ecological civilization. Water Resour. Prot. **29**, 1–4 (2013)
4. Y.Y. Dong, Construction of ecological security evaluation system based on ecological elements-DPSIRM. Soil Water Conserv. Res. **27**, 333–339 (2020)
5. Q. Guo, J.Y. Wang, B. Zhang, Comprehensive evaluation of regional water resources carrying capacity based on DPSIRM framework. J. Nat. Resour.Resour. **32**, 484–493 (2017)
6. Y.F. Wang, Y.S. Liu, Y.R. Li, Spatial and temporal characteristics of coordinated development of urbanization and rural areas in the Bohai Rim. Geogr. Res.. Res. **34**, 122–130 (2015)
7. M.D. Jiang, X.M. Shen, Y.Y. Wang, L. Wang, Evaluation of the effectiveness and spatial and temporal differences of the implementation of the river chief system in Jiangsu Province. South-North Water Diversion Water Conservancy Sci. Technol. **16**, 201–208 (2018)
8. X.J. Jiang, J. Jin, Digital technology enhances economic efficiency: service division of labor, industrial synergy and digital twin. Manage. World **38**, 9–26 (2022)
9. Y.L. Mou, Y.R. Liang, K. Wu, Design of a county water resources supervision and integration platform for intelligent water conservancy—taking Binhu District of Wuxi City as an example. Jiangsu Water Resour. **316**, 52–57 (2022)
10. Y. Li, J.W. Sun, Research on the technology of river and lake supervision system platform based on digital twin. Jiangxi Water Conservancy Sci. Technol. **49**, 108–111 (2023)
11. J. Zhou, Power displacement: a new paradigm of e-government management based on the perspective of digital twin technology. Off. Autom. **28**, 27–29 (2023)

Influence of the Shihutang Navigation-Power Junction Project Operation on Hydrodynamic Characteristics of the Ganjiang River

Qiang Zhang and Yuxuan Zhao

Abstract The construction of a reservoir on the river can meet the social requirements for irrigation, navigation, water supply, and power generation in the area. However, reservoir development also disrupts the natural flow continuity of the river, leading to fundamental changes in important water environmental factors such as water temperature, water quality, morphology, and water level. In this study, the Shihutang navigation-power junction project mainly affects the middle reaches of the Ganjiang River, the Wan'an to Xiajiang section is taken as the study area, and a one-dimensional hydrodynamic mathematical model of the middle reach of the Ganjiang River is established. The model is calibrated and validated based on measured water level and discharge data in order to study the influence of the Shihutang navigation-power junction project operation on the hydrodynamic characteristics of the middle reach of the Ganjiang River. The results show that the changes in water level, discharge, and velocity in the upstream and downstream areas are generally consistent before and after the operation of the Shihutang navigation-power junction project. The operation of the Shihutang navigation-power junction project has a significant influence on the hydrodynamic characteristics at the dam site, with the influence decreasing as the distance from the dam site increases.

Keywords Mathematical model · Saint–Venant equations · Calibrated and validated · Water level · Discharge

Q. Zhang (✉)
College of Hydrodynamic and Ecology Engineering, Nanchang Institute of Technology, Nanchang, China
e-mail: zhangqiang8812@163.com

Y. Zhao
POWERCHINA Chengdu Engineering Corporation Limited, Chengdu, China

W. Wang et al. (eds.), *Hydraulic Structure and Hydrodynamics*, Lecture Notes in Civil Engineering 608, https://doi.org/10.1007/978-981-97-7251-3_31

1   Introduction

With the rapid development of water resources in China, large-scale, highly efficient, and well-regulated water conservancy projects are being rapidly constructed.
However, if the management fails to adopt reasonable scheduling methods, it can
have a series of influence on the ecological environment of the river basin and the
overall effects of cascade reservoirs. For example, the eutrophication of river-type
and lake-type reservoirs is mainly influenced by topography and geography. Reservoirs in hilly lake areas, for instance, have relatively slow water flow. During the
dry season, the reservoirs are affected by the surrounding rivers, causing a decrease
in water velocity, leading to the growth of algae and plankton, and increasing the
pollution level of the reservoirs [1, 2]. At the same time, the "peak shaving" function of reservoirs directly influences the spawning of fish downstream. The Shihutang
navigation-power junction project is an important line of defense to promote sustainable and healthy economic development in China and ensure the safety of people's
property. It serves as a buffer against regional flood and water shortage in the
river basin, and plays a crucial role in achieving a sustainable and healthy development of water resources. Reasonable water resource allocation plans can not only
provide domestic water supply but also balance power generation tasks in the region,
while meeting production and living needs and driving local economic development
sustainably. Therefore, it is necessary to analyze the changes, trends, characteristics,
and laws of the hydrological and water quality situation of rivers, in order to scientifically, reasonably, and effectively allocate and utilize water resources, and reconcile
the conflicts between water resources, socio-economic development, and ecological
environmental protection [3–5].

   This study takes the Wan'an to Xiajiang section as the study area and establishes
a one-dimensional hydrodynamic mathematical model of the middle reach of the
Ganjiang River. The model is calibrated and validated based on measured water level
and discharge to study the influence of the Shihutang navigation-power junction
project on the hydrodynamic characteristics of the middle reach of the Ganjiang
River. The results can provide a scientific basis for the rational development and
utilization of water resources in the river basin and guide the planning, scheduling,
and evaluation of reservoirs in the Yangtze River system's structure and stability.

2   Model Establishment

In this study, the Mike11 software was utilized to establish a one-dimensional hydrodynamic mathematical model. The use of Mike11 software is advantageous due to its
recognized reliability, widespread application, user-friendly interface, practicality,
and convenience. Specifically, the controllable structure module of the software
allows for the automatic adjustment of gate openings, switch heights, and over-
current capacities based on predefined scheduling rules and control strategies. This

capability assists in effectively managing and controlling the hydraulic dynamics within the model.

2.1 The Saint–Venant Equations

The Saint–Venant equations consist of the continuity equation and the momentum equation [6–8].

$$\frac{\partial A}{\partial t} + \frac{\partial Q}{\partial x} = q \tag{1}$$

$$\frac{\partial Q}{\partial t} + \frac{\partial \left(\alpha \frac{Q^2}{A} \right)}{\partial x} + gA\frac{\partial h}{\partial x} + \frac{gQ|Q|}{C^2 AR} = 0 \tag{2}$$

In the equations: A refers to the cross-sectional area of flow; R is the hydraulic radius; Q represents the discharge; α is the momentum correction coefficient; h is the water level; g is the acceleration due to gravity; q is the incoming discharge from the side; x and t are the coordinates of the calculation point in space and time respectively; C is the Chezy coefficient.

2.2 Discretization and Solution of the System of Equations

To solve the Saint–Venant equations, it is not possible to obtain a general solution, so only an approximate solution can be obtained by prescribing boundary and initial conditions. In the solution process using Mike11 software, a 6-point Abbott-Ionescu difference format is used [9–11]. The model alternately arranges grid points in the order of water level, discharge, and water level, which are respectively called water level points (h) and discharge points (Q), as shown in Fig. 1:

Fig. 1 Schematic diagram of section calculation nodes

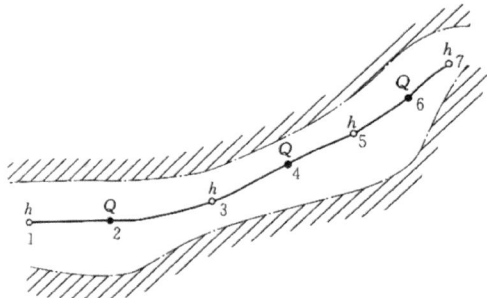

The water level calculation point is based on measured cross-sectional data. In each time step, the Q and h values are alternately calculated using an implicit format. This calculation is faster than the conventional 4-point difference format and can maintain stability even when the Courant number exceeds 250. The Abbott 6-point implicit format discretizes the above control equations. In this discretization format, water level and flow rate are not calculated simultaneously at each grid node. Instead, they are calculated in sequential order as h points and Q points. This format is unconditionally stable and can maintain stability under relatively large Courant numbers, allowing for longer time steps to save computation time.

2.3   *Boundary Conditions*

In this study, the non-steady flow and parameter file initiation method is used. The computational area is from Wan'an Reservoir on the main stem of the Ganjiang River to the Xiajiang Hydrological Station. The upstream boundary is defined by the discharge into Wan'an Reservoir, while the downstream boundary is simulated using the Xiajiang water level and flow relationship curve for the corresponding period. The inflow from tributaries is simplified based on the river network generalization and includes the following: Baisha inflow at 97.45 km (with corresponding flow process and point source distribution), Shangshalan inflow at 107.34 km (with corresponding flow process and point source distribution), and Xintian inflow at 129.97 km (with corresponding flow process and point source distribution).

3   Model Calibration and Validation

The model calibration period is from January 1, 2008, to December 31, 2008. After repeated adjustments to the Manning's roughness coefficient, a range of 0.03–0.05 is obtained. The calibrated water level values compared to the measured values have an error range of -0.03 to 0.04 m. Similarly, the calibrated discharge values compared to the measured values have an error range of -76.8 to 31.9 m^3/s. The calibration results show that the water level errors are within 0.1 m and the relative discharge errors are within 10%, which meet the requirements of relevant specifications. The model is suitable for hydrodynamic simulations in the middle reach of the Ganjiang River basin. Taking the Dongbei Station as an example, the calibration results are shown in Figs. 2 and 3.

The model validation period is from January 1, 2009, to December 31, 2011. Taking the Dongbei Station as an example, the validation results are shown in Figs. 4 and 5.

Validation was conducted using the measured water level and discharge. The results of the validation showed that the calculated water level matched the trend of the measured data, with an error within 0.1 m. The relative error of the flow was

Fig. 2 Calibration results of water level at Dongbei station

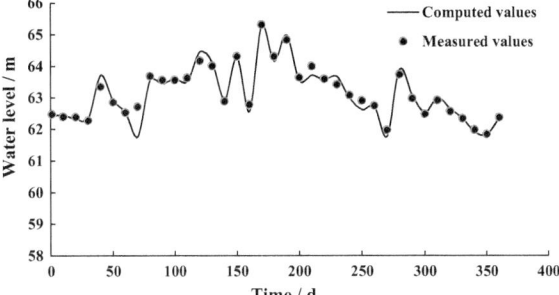

Fig. 3 Calibration results of discharge at Dongbei station

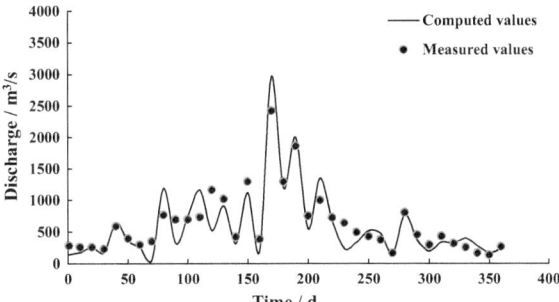

Fig. 4 Validation results of water level at Dongbei Station

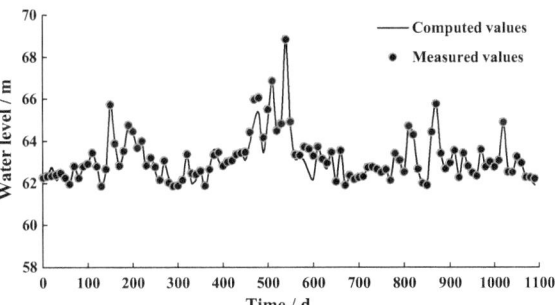

Fig. 5 Verification results of discharge at Dongbei station

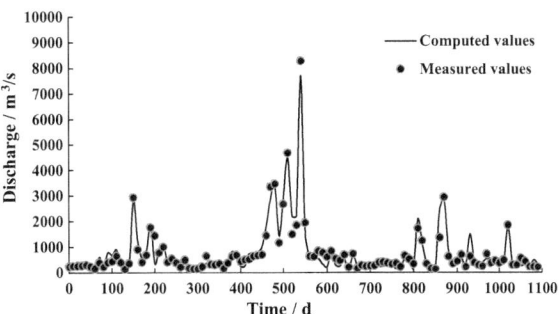

also within 10%, meeting the requirements of relevant specifications. Therefore, the model can be used for hydrodynamic simulation in the middle reaches of the Ganjiang River Basin.

4   Influence Analysis of the Shihutang Navigation-Power Junction Project

In order to study the influence of the operation of the Shihutang navigation-power junction project on the hydrodynamic characteristics of the middle reaches of the Ganjiang River, a one-dimensional hydrodynamic model was established using Mike11. The study area of the model is the river channel from Wan'an to Xiajiang in the middle reaches of the Ganjiang River. One-dimensional hydrodynamic models were created for both the natural flow conditions and the operational conditions of the Shihutang navigation-power junction project. The differences in discharge processes and water level variations were compared, and the extent of changes in the hydrodynamic conditions in the middle reaches of the Ganjiang River caused by the navigation-power junction project were analyzed.

4.1   Calculation Cases

Before the operation of the Shihutang navigation-power junction project

The study area for the natural runoff model is the main stream of the Ganjiang River from Wan'an to Xiajiang station. The inlet boundary is set as the discharge process at Wan'an from January 1, 2012, to June 30, 2013. The outlet boundary is defined by the water level and flow relationship curve of Xiajiang during the same period (Fig. 6).

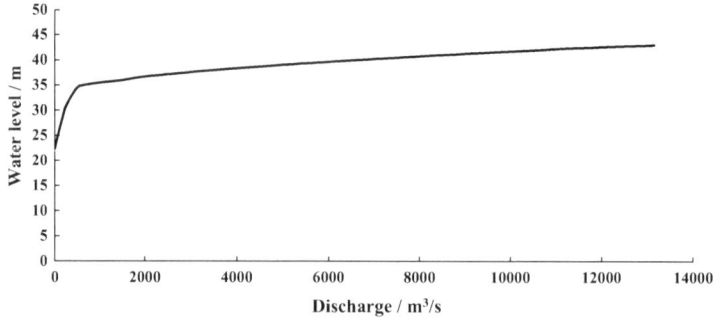

Fig. 6 Water level and flow relationship curve at Xiajiang station

After the operation of the Shihutang navigation-power junction project

After the operation of the Shihutang navigation-power junction project, the model is based on the discharge process of Wan'an Reservoir from January 1, 2012, to June 30, 2013, as the boundary condition for the inlet. The outlet boundary is determined by the water level and flow relationships curve at Xiajiang station during the same period. The inflow of tributaries is simplified based on the river network, with Baisha joining at the 97.45 km mark, Shangshalan joining at the 107.34 km mark, and Xintian joining at the 129.97 km mark. The flood dispatching rule for the Shihutang navigation-power junction project is based on the discharge at the Sihutang dam site. The content of the dispatching rule is as follows [12, 13]:

When the upstream flow at the Sihutang dam site exceeds 4700 m^3/s (the critical discharge for closing the sluice gate), water storage for power generation is halted, and flood discharge is initiated to restore the river flow to its natural state. This continues until the incoming discharge decreases below the critical discharge, at which point water storage and utilization operations resume.

When the discharge upstream of the Shihutang Dam is less than 4700 m^3/s (the critical discharge for closing the sluice gate), the sluice gates are gradually closed. Except for retaining a discharge of 187 m^3/s to ensure navigation for ships, all other incoming water is used for reservoir storage and utilization, while maintaining a normal water level of 56.5 m.

4.2 Calculation Results and Analysis

After the successful calculation of the model, water level and discharge data from three locations, namely Dongbei Station (at 20.5 km), Ji'an Station (at 111 km), and Xiajiang Dam Site (at 177 km), are obtained to analyze the influence of the operation of the Shihutang navigation-power junction project on the hydrodynamic characteristics of the middle reaches of the Ganjiang River.

Water level

The variation of water level in each section are shown in Table 1 and Figs. 7, 8 and 9. At Dongbei Station, the range of water level changes before and after the operation of the Shihutang navigation-power junction project is − 0.503 to 0.254 m (obtained by deducting the water level after the operation of the Shihutang navigation-power junction project from the natural condition, same for the following). The operation of the Shihutang navigation-power junction project affects the water level at Ji'an Station by − 0.212 to 1.364 m, and at Xiajiang Station by − 0.687 to 0.808 m.

Discharge

The variation of discharge in each important section can be seen in Figs. 10, 11 and 12 and Table 1. At Dongbei station, the range of discharge variation is − 300.9

Table 1 Range of hydrological element changes before and after the operation of the Shihutang navigation-power junction project

Hydraulic elements	Dongbei station	Ji'an station	Xiajiang station
Water level (m)	− 0.503~0.254	− 0.212~1.364	− 0.687~0.808
Discharge (m³/s)	− 300.9~311	− 730.5~1118	− 516.4~844.9
Velocity (m/s)	− 0.147~0.065	− 0.143~0.08	− 0.102~0.078

Fig. 7 Water level variation at Dongbei station

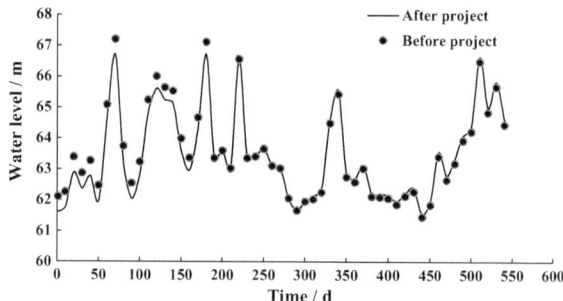

Fig. 8 Water level variation at Ji'an station

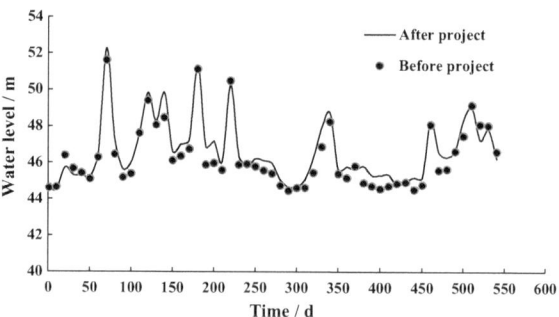

Fig. 9 Water level variation at Xiajiang station

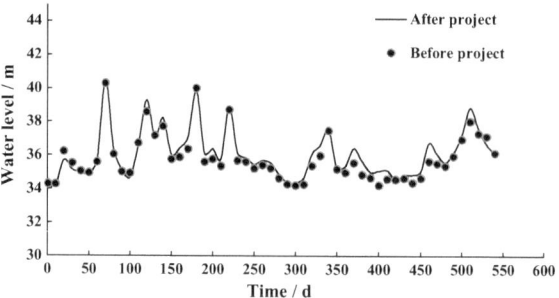

to 311 m³/s (obtained by subtracting the discharge after the Shihutang navigation-power junction project operation from the natural condition, the same applies below). The daily average discharge remains relatively stable. At Ji'an Station, the discharge variation ranges from − 730.5 to 1118 m³/s. Due to the regulation and flood control function of the reservoir, the flow in August and September notably increases. At Xiajiang Station, the dischargevariation ranges from − 516.4 to 844.9 m³/s.

Velocity

The variation of velocity in important sections are shown in Table 1. The velocity range at the Dongbei Station is − 0.147 to 0.065 m/s (obtained by subtracting the flow

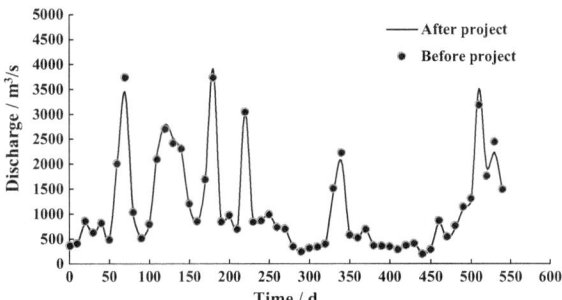

Fig. 10 Discharge variation at Dongbei station

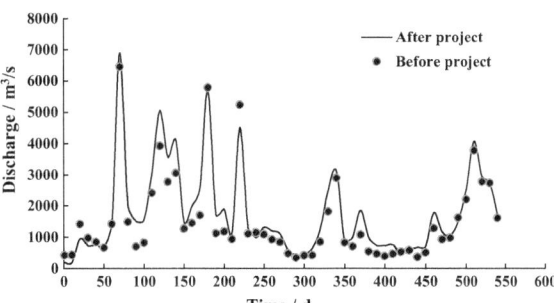

Fig. 11 Discharge variation at Ji'an station

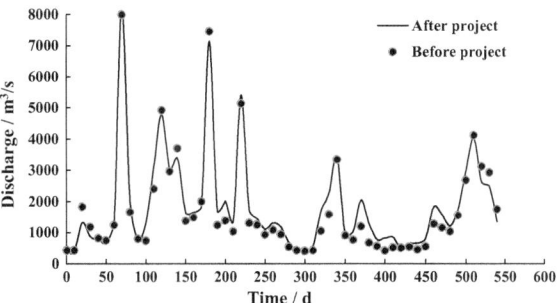

Fig. 12 Discharge variation at Xiajiang station

velocity after the operation of the Shihutang navigation-power junction project from the natural conditions, the same applies to the following). The downstream velocity at the dam site of the Shihutang navigation-power junction project has changed, with a flow velocity range at the Ji'an station of − 0.143 to 0.08 m/s. The velocity at the Xiajiang station has slowed down, with a range of − 0.102 to 0.078 m/s.

According to the above analysis, it can be seen that the variation trend of water level, discharge and velocity of each hydrologic station before and after the operation of the Shihutang navigation-power junction project is basically the same, and the variation range of water level, discharge and velocity is small. The operation of the Shihutang navigation-power junction project has little influence on the hydrodynamic characteristics of the middle reaches of the Ganjiang River. It can be seen from Table 1 that the closer the distance to the dam site of the Shihutang navigation-power junction project, the greater the influence of hydraulic characteristics, while the farther away from the dam site, the smaller the influence.

5 Conclusion

Reservoirs play a crucial role in the development of human society by addressing challenging water resource allocation issues such as flood control, aquaculture, and power generation. However, the construction of reservoir dams has also disrupted the stability and continuity of river ecosystems, posing a threat to the functionality and biodiversity of the original river ecosystem. It has also altered the pre-existing water cycle system in the region and influenced the frequency of local geological hazards. In order to study the influence of the Shihutang navigation-power junction project on the hydrological situation of the middle reaches of the Ganjiang River after implementing peak shaving and flood regulation measures, this study collected field measurements of water level and flow rate data in the middle reaches of the Ganjiang River. A one-dimensional hydrodynamic model was established for the section from Wan'an to Xiajiang, and this was used to analyze the influence of the operation of the Shihutang navigation-power junction project on water level, discharge, velocity, and other hydrodynamic factors in the river section. The main conclusions are as follows:

(1) A one-dimensional hydrodynamic mathematical model was established for the middle reaches of the Ganjiang River, and it was calibrated and validated using field measurements of hydrological data. The model calibration and validation results were satisfactory, demonstrating that the model is capable of accurately simulating the movement of water flow in the studied river section.

(2) Simulations were conducted for two cases: one with the operation of the Shihutang navigation-power junction project and another without. The results show that the water level, discharge, and velocity variations at each station before and after the operation of the Shihutang navigation-power junction project are generally similar. The magnitude of changes in water level, discharge, and flow

velocity is not significant. Overall, the operation of the Shihutang navigation-power junction project has a minimal influence on the hydrodynamics of the middle reaches of the Ganjiang River.

(3) The operation of the Shihutang navigation-power junction project has a significant influence on the hydrodynamic characteristics at the dam site. The influence becomes smaller as the distance from the dam site increases.

Acknowledgements The study is funded by the Jiangxi Provincial Department of Education Science and Technology Project (GJJ190968).

References

1. Q.H. Cai, Taking eco-system protection into fully consideration and improving reservoir regulation process. China Water Resour. **02**, 14–17 (2006)
2. D.M. Li, M.L. Hu, C. Miao et al., Influence of the hydro-junction on the nature reserve of four major chinese carps in Xiajiang section of Ganjiang river and its Countmeasures. Trans. Oceanol. Limnol. **147**(04), 77–82 (2015)
3. Y.Y. Chen, Y.D. Mei, H. Cai et al., Multi-objective optimal operation of key reservoirs in Ganjiang river oriented to power generation, water supply and ecology. J. Hydraul. Eng. **49**(05), 628–638 (2018)
4. S.L. Guo, J.H. Chen, P. Liu et al., State-of-the-art review of joint operation for multi-reservoir systems. Adv. Water Sci. **21**(04), 496–503 (2010)
5. J. Zhu, L.M. Wang, F.X. Jia et al., Calculation methods and case study of wetland ecological water demand in Northern China. Chin. J. Environ. Eng. **11**, 112–118 (2007)
6. H.H. Wang, G.H. Guan, C.C. Xiao, Comparison and optimization of sparse matrix solution methods in one-dimensional saint-venant equation difference numerical algorithm. J. Irrig. Drainage **40**(3), 116–124 (2021)
7. Y.F. Geng, X. Zheng, X. Ke, Urban flood process analysis with a 1D/2D coupling model: a case study of Quzhou city. J. Southeast Univ. (Nat Sci. Ed.) **49**(05), 1005–1010 (2019)
8. J.Y. Zhang, J. Xu, X.J. Yuan et al., Discussion on application of superposition of 1—D and 2—D coupling model for dike failure due to outside flood and water-logging model to flood control protection area. Water Resour. Hydropower Eng. **48**(05), 87–94 (2017)
9. Y.F. Xiao, M. Zhou, T. Hu, et al., Research on flood routing modeling and flood propagation law for the three gorges reservoir based on MIKE11. Water Resour. Power **40**(10), 74–77+194 (2022)
10. L.L. Zhao, Y.J. Zhang, Risk analysis of underground channel flood based on Mike11 one-dimensional river breach model. Henan Water Resour. South-to-North Water Diversion Proj. **49**(09), 24–25 (2020)
11. K.Z. Wang, Analysis of flow capacity of Liufangdi section of main stream of the Huaihe river based on MIKE 11 and 21 models. Pearl River **41**(08), 15–20 (2020)
12. C.H. Ling, Layout and optimization of navigation and power hub at Shihutang in the Ganjiang River. Pearl River **35**(03), 51–56 (2014)
13. M.X. Cao, B. Wu, C.L, et al., Experimental study on fall of stage downstream of Shihutang hydro-junction on the Ganjiang River. Hydro-Sci. Eng. **137**(01), 15–21 (2013)

Study on Linkage Alarm of Site Abnormality of Large Pumped Storage Power Station Under Video Monitoring

Feng Cao, Jishuang Han, Jing Li, and Guangyong Zeng

Abstract When alarming the abnormal state of the construction site, due to the lack of systematicness in the composition analysis of specific construction safety influencing factors at different stages, the accuracy of linkage alarm is relatively high. Therefore, the study on linkage alarm of abnormal state of large pumped storage power station under video monitoring is put forward. Combined with different stages of construction, the existing factors affecting construction safety are comprehensively analyzed from three aspects: construction preparation, foundation and main structure. Based on the image information collected by video surveillance, the existence state of safety influencing factors in the video surveillance picture is identified through the convolution layer including the upper and lower layers, and the operation content that needs to be alarmed at the construction site is determined by combining the correlation between safety influencing factors. In the test results, the design of linkage alarm method not only has high stability for the overall accuracy of alarm in different test conditions, but also keeps above 90.0% all the time.

Keywords Video surveillance · Large pumped storage power station · Abnormal site · Linkage alarm · Convolution layer · Safety influencing factors · Relationship

1 Introduction

No matter from the perspective of smooth construction of large-scale pumped storage power station site or from the perspective of ensuring the safety of construction personnel, it is extremely necessary to make effective alarm treatment in combination with the actual abnormal situation of the site [1–3]. In order to solve this problem,

F. Cao (✉) · J. Han · J. Li
Engineering Construction Management Branch of China Southern Power Grid Peak Shaving Frequency Modulation Power Generation Co., Ltd, Guangzhou, Guangdong, China
e-mail: 22823401@qq.com

G. Zeng
China Southem Power Grid Energy Storage Co., Ltd, Guangzhou, Guangdong, China
e-mail: 13926169267@139.com

W. Wang et al. (eds.), *Hydraulic Structure and Hydrodynamics*, Lecture Notes in Civil Engineering 608, https://doi.org/10.1007/978-981-97-7251-3_32

365

the linkage alarm method based on big data is widely used, but it needs a lot of actual data as the basis. When the analysis of the influencing factors of construction safety is not comprehensive enough [4–6], the corresponding alarm effect will be obviously affected. In addition, the linkage alarm method based on deep learning is also one of the more common methods [7–9], but this method has obvious limitations in the practical application stage, which requires higher application environment [10, 11]. Combined with the above analysis, it can be seen that there is still a huge space for the study of abnormal linkage alarm in the construction site, and it has extremely important practical significance [12, 13].

On this basis, this paper puts forward the study of abnormal linkage alarm of large pumped storage power station site under video monitoring, and analyzes and verifies the application effect of the designed alarm method through comparative testing.

2 Design of Linkage Alarm Method for Site Abnormality of Large Pumped Storage Power Station

2.1 Classification of Factors Affecting Site Construction Safety

In order to realize the linkage alarm of abnormal site, it is extremely necessary to comprehensively analyze the existing construction safety influencing factors [14]. The identification of safety influencing factors is defined and identified in an orderly and orderly manner according to the background reasons, performance characteristics and expected consequences of various hazards or uncertain factors in the project. Therefore, combining with different stages of construction, this paper has carried out specific research from three angles: construction preparation, foundation and main structure [15–17]. Among them, the specific construction safety influencing factors in the construction preparation stage are shown in Table 1.

Combined with Table 1, it can be seen that the analysis of safety influencing factors in the construction preparation stage is mainly divided into five aspects [18, 19]: safety sign setting, site enclosure placement, closed management implementation, site office and accommodation, and public signs setting, mainly focusing on construction personnel and construction environment.

The composition of specific construction safety influencing factors in the foundation stage is shown in Table 2.

Combined with Table 2, it can be seen that the analysis of the influencing factors of site safety in the foundation construction stage in this paper is mainly divided into three aspects [20, 21]: the implementation of border protection measures, the load state at the pit side and the earthwork excavation construction, involving construction personnel, construction environment, construction technology, construction equipment and construction materials.

Table 1 Analysis of influencing factors of site safety in construction preparation stage

Number	Project	Factors affecting safety after construction
1.	Safety signs	No major hazard warning signs and on-site safety signs have been set up
2.	Site fencing	Failure to set up enclosed enclosures and arrange enclosure heights as required
3.	Closed management	There is no gate set up at the entrance and exit of the construction site, and there is no guard room set up
4.	On site office and accommodation	Unclear division of construction operations, material storage areas, and office and living areas
5.	Publicity signs	No safety signs set up on the construction site

Table 2 Analysis of influencing factors of site safety in foundation construction stage

Number	Project	Factors affecting safety after construction
1.	Edge protection	Failure to set up edge protection measures that meet the protection requirements according to the depth requirements of the foundation pit
2.	Pit edge load	The minimum distance from the edge of the foundation pit to the edge of the groove, where building materials are piled up, does not meet the design requirements
3.	Earth excavation	Dump soil is piled up near the foundation pit, and the walking route of construction machinery is not executed according to the plan
4.	Auxiliary support frame	The installation of the construction elevator wall bracket and the connection method and angle of the structure do not comply with the specifications
5.	Protective devices	The top height of the crane is > 30 m and is higher than the surrounding buildings. No obstacle indicator lights are installed

The specific construction safety influencing factors in the main structure stage are shown in Table 3.

Combined with Table 3, it can be seen that the analysis of safety influencing factors in the construction preparation stage in this paper covers many aspects, such as vertical pole foundation, scaffold board and protective railing, vertical pole foundation, support stability, limb, hole opening, access protection, suspended operation, protective facilities, attached wall frame, protective device and multi-tower operation, and also involves construction personnel, construction environment, construction technology, construction equipment and construction materials [22, 23]. According to the above-mentioned methods, the comprehensive analysis of the factors affecting the site safety of large pumped storage power stations is realized, which provides a reliable basis for the subsequent abnormal linkage alarm.

Table 3 Analysis of safety influencing factors in main structure stage

Number	Project	Factors affecting safety after construction
1.	Pole foundation	Construction personnel entering the operating radius of construction machinery
2.	Scaffolding boards and protective railings	Scaffolding not equipped with vertical and horizontal sweeping poles according to standard requirements
3.	Pole foundation	No toe boards with a height greater than 180 mm have been installed on the scaffold board operation layer
4.	Support stability	The bottom of the formwork support pole is not equipped with a base or base plate according to the specification requirements
5.	Protection of edges, openings, and passageways	When the height to width ratio of the template support exceeds the specified value, the wall connecting rod is not set according to the regulations
6.	Suspended operation	No protective measures have been taken for the reserved openings, stairwells, and elevator wellheads of the construction in progress
7.	Protective facilities	No continuous edge protective railings are set up along the edge of the working face
8.	Wall bracket	No corresponding safety warning signs have been set up in the main construction areas and hazardous areas
9.	Protective devices	No protective railings or other reliable safety measures have been installed at the suspended operation site
10.	Multi tower operation	The construction elevator is not equipped with a ground protective fence or does not meet the specifications

2.2 Linkage Alarm of Abnormal Site Based on Video Monitoring

Combined with the analysis results of 1.1 on the safety influencing factors of large-scale pumped storage power station, when designing the corresponding linkage alarm mechanism of site abnormality, based on the image information collected by video surveillance, the corresponding alarm is made by identifying the existing state of the safety influencing factors in the video surveillance picture [24], combined with the analysis results of 1.1 on the relationship between the safety influencing factors in construction preparation, foundation and main structure stage. Among them, convolutional neural network is introduced to identify the existing state of security influencing factors in video surveillance pictures.

In order to ensure the comprehensiveness of the visual sensing domain analysis results of image information collected by video surveillance, a convolution layer with upper and lower layers is set up. Among them, the first convolution layer identifies the edge features and color features of the image information collected by video surveillance, and the extraction results of the corresponding image features can be expressed as

$$F(x) = \{S(x), C(x)\} \tag{1}$$

Among them, $F(x)$ represents the image information collected by the first convolution layer for video monitoring x its feature extraction results, $S(x)$ represents the image information collected by video surveillance x its edge features, $C(x)$ represents the image information collected by video surveillance x its color characteristics.

The second convolution layer identifies the texture features and surface features of the image information collected by video surveillance, and the extraction results of the corresponding image features can be expressed as follows

$$G(x) = \{W(x), B(x)\} \tag{2}$$

Among them, $G(x)$ represents the image information collected by the first convolution layer for video monitoring x its feature extraction results, $W(x)$ represents the image information collected by video surveillance x its texture features, $B(x)$ represents the image information collected by video surveillance x its surface features.

In this way, it is determined that the video surveillance has collected the image information x if it includes the factors affecting the site safety of large pumped storage power stations? It is assumed that the factors affecting the site safety of large pumped storage power stations are included, specifically y_i, and exists $Y = \{y_1, y_2, y_3, ..., y_i\}$, among which, Y shows the analysis results of the factors affecting the safety of large pumped storage power stations in Part 1.1, so the operations that need to be alarmed at the site under the linkage alarm mechanism at this time can be expressed as follows

$$h(X) = y_i \rightarrow con(\frac{d_i}{F(x) * G(x)}) \tag{3}$$

Among them, $h(X)$ indicates the operation content that the site needs to report to the police under the linkage alarm mechanism. *con* represents the correlation function, d_i represents characteristic parameter corresponding to site safety influence factor of large pumped storage power station.

According to the above-mentioned way, the linkage alarm of abnormal site of large pumped storage power station is realized.

3 Application Testing

3.1 Test Preparation

When analyzing the practical application effect of the design method, a comparative test was carried out based on the open video surveillance image data of the site scene, in which the control groups participating in the test were the commonly used linkage alarm method based on big data and the linkage alarm method based on deep learning. In the specific testing process, considering that there may be some differences in the quality of the video surveillance image data of the test site, the training data and the test data under different methods are set to be completely consistent, so as to reduce the problem of low reliability of the test result analysis caused by the difference of independent variables. Among them, the video surveillance image data of the training group is 5.6 GB, and the video surveillance image data of the test group is 0.8 GB. When analyzing the test results, this paper takes the alarm accuracy as the evaluation index, and its specific calculation method can be expressed as follows

$$\mathrm{Pr} = \frac{n_r}{n_r + n_w} * 100\% \tag{4}$$

Among them, Pr indicates the alarm accuracy, n_r indicates the number of accurate alarms, n_w indicates the number of missed or false alarms.

On this basis, the test results of the three methods are analyzed.

3.2   Test Results and Analysis

Note: The corresponding status of each alarm test number is as follows: "1" means that the height of the enclosed enclosure is lower than the safety requirements; "2" means that the construction personnel enter the working radius of the construction machinery, and "3" means that the edge of the working surface is not provided with a continuous edge protective railing; "4" means that the height of the footboard is less than 180 mm when the scaffold board is not set on the operation floor; "5" means that the elevator wellhead is not provided with protective measures.

Combined with the test results shown in Fig. 1, it can be seen that under three different linkage alarm methods, the corresponding accuracy shows obvious differences. Among them, under the big data linkage alarm method, its overall accuracy is stable in the range of 80.0–90.0%. In the test results of the deep learning linkage alarm method, the alarm accuracy for different States shows obvious fluctuations, with the maximum reaching 90.45%, but the minimum only 76.44%. In contrast, under the linkage alarm method designed in this paper, not only the overall accuracy has high stability, but also it is always stable above 90.0%. Based on the above test results, it can be concluded that the abnormal linkage alarm method designed in this paper under video monitoring has good application value.

Fig. 1 Comparison chart of test results of different methods

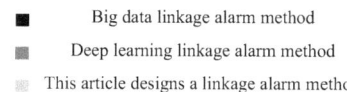

■ Big data linkage alarm method

■ Deep learning linkage alarm method

░ This article designs a linkage alarm method

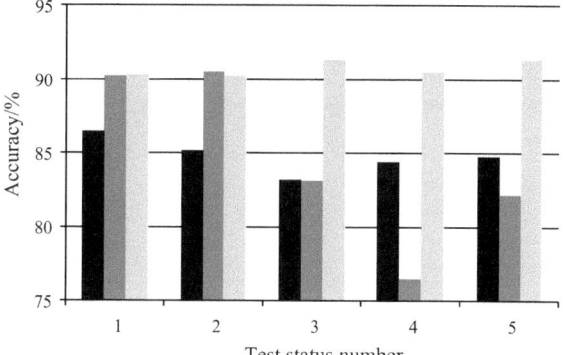

4 Conclusion

In order to ensure the safety of the construction site to the greatest extent, it is extremely important to give a comprehensive and accurate alarm for the existing safety hazards in combination with the correlation between different construction links and construction operations. In this paper, a study on the linkage alarm of large-scale pumped storage power station site anomalies under video surveillance is proposed. On the basis of a comprehensive analysis of the composition of site safety influencing factors, accurate alarm is realized by identifying the characteristic factors in the image information collected by video surveillance and combining the correlation between various safety influencing factors. With the help of the abnormal linkage alarm method designed in this paper, I hope it can provide valuable help for the safety management of the construction site.

References

1. K. Silva et al., Systematic approach to transboundary radioactivity monitoring for accidents in external nuclear power plants. J. Nuclear Eng. Radiation Sci. **7**(4), 041701 (2021)
2. Gao R, Wu F, Zou Q et al (2022) Optimal dispatching of wind-PV-mine pumped storage power station: a case study in Lingxin Coal Mine in Ningxia Province, China. Energy 243(15):123061.1–123061.14
3. A. Abdulkareem et al.,Development and construction of automatic three-phase power changeover control circuit with alarm, in *IOP Conference Series: Materials Science and Engineering*, vol. 1036, no. 1. IOP Publishing (2021)
4. X. Li et al., Research on false alarm detection algorithm of nuclear power system based on BERT-SAE-iForest combined algorithm. Ann. Nuclear Energy **170**, 108985 (2022)

5. N. Hemakumari, S. Bismillah Khan, C. Priya, M. Vidhyalakshmi, H. Peer oli, Internet of Things (IoT) based smart plant monitoring system. Int. J. Modern Trends Sci. Technol. **7**(5), 144–146 (2021)

6. C. Pan et al.Application of intelligent power consumption management and control technology in enterprise off-site enforcement. Meteorol. Environ. Res. **12**(3), 41–46 (2021)

7. F. Antonello et al., A novel association rule mining method for the identification of rare functional dependencies in complex technical infrastructures from alarm data. Expert Syst. Appl. **170**, 114560 (2021)

8. R. Carmona-Sánchez, M.A. Carrera-Álvarez, C. Peña-Zepeda, Prevalence of primary eosinophilic colitis in patients with chronic diarrhea and diarrhea-predominant irritable bowel syndrome. Revista de Gastroenterología de México (English Edition) **87**(2), 135–141 (2022)

9. E. Zhenwei et al., Transformer condition monitoring technology based on surface acoustic wave passive wireless sensing antenna, in *E3S Web of Conferences*, vol. 257 (EDP Sciences, 2021)

10. M. Uppalapati, K. Strohl, R. Sibilia, 854 sexsomnia in a divorce proceeding and its custody implications. Sleep **44**, A332 (2021)

11. L. Feng, L. Hou, Nursing intervention of cognitive impairment after cerebral infarction based on internet of things video monitoring. Microprocessors Microsystems **83**, 104013 (2021)

12. A. Soloy et al., A fully automated method for monitoring the intertidal topography using video monitoring systems. Coastal Eng. **167**, 103894 (2021)

13. A.J. McIvor et al., Unoccupied aerial video (UAV) surveys as alternatives to BRUV surveys for monitoring elasmobranch species in coastal waters. ICES J. Marine Sci. **79**(5), 1604–1613 (2022)

14. G. Simarro, D. Calvete, P. Souto, UCalib: cameras autocalibration on coastal video monitoring systems. Remote Sens. **13**(14), 2795 (2021)

15. H. Ghaffarian et al., Automated quantification of floating wood pieces in rivers from video monitoring: a new software tool and validation. Earth Surf. Dyn. **9**(3), 519–537 (2021)

16. T. Noda et al., Experience with the use of intraoperative continuous nerve monitoring in video-assisted neck surgery and external cervical incisions. Laryngoscope Investigative Otolaryngol. **6**(2), 346–353 (2021)

17. A. Callens et al., Automatic creation of storm impact database based on video monitoring and convolutional neural networks. Remote Sens. **13**(10), 1933 (2021)

18. X. Song,Intelligent video and image cloud operation monitoring center. J. Phys.: Conf. Ser. **1744**(2) (IOP Publishing, 2021)

19. G. Coro, M.B. Walsh, An intelligent and cost-effective remote underwater video device for fish size monitoring. Ecol. Inform. **63**, 101311 (2021)

20. B.J. Kolls, B.E. Mace, A practical method for determining automated EEG interpretation software performance on continuous Video-EEG monitoring data. Inf. Med. Unlocked **23**, 100548 (2021)

21. M. D'aponte, Remote monitoring and video surveillance in the workplace between Italian law and the case law of the European court of human rights. Courier of Kutafin Moscow State Law University (MSAL) **1**, 112–125 (2021)

22. U. Holder, T. Ehrmann, A. König, Monitoring experts: insights from the introduction of video assistant referee (VAR) in elite football. J. Bus. Econ. **92**(2), 285–308 (2022)

23. D. Gura et al, A complex for monitoring transport infrastructure facilities based on video surveillance cameras and laser scanners. Transp. Res. Proc. **54**, 775–782 (2021)

24. D. Zhang et al., Coastal fisheries resource monitoring through a deep learning-based underwater video analysis. Estuarine Coastal Shelf Sci. **269**, 107815 (2022)

Observation of Navigational Hydraulic Characteristics in the Downstream Entrance Area

Xin Wang and Shouyuan Zhang

Abstract Strong transverse flows and fluctuations often occur in the downstream approach channel and entrance area during flood discharge at high dam hubs. This study takes a power station as a research sample. Flow velocities, transverse flow intensities, fluctuation characteristics and ship navigation status are analyzed and studied through field observations. The results show that during flood discharge the entrance area experiences transverse short-wave fluctuations and significant transverse currents. The oscillation of hydraulic jumps is the direct factor causing fluctuations in the downstream entrance area. The fluctuation amplitude of the center and edge of the entrance area was generally large, and the fluctuation spread to the approach channel and continuous attenuation. The main fluctuation attenuation occurs at the transition from the broad riverway to the downstream restricted approach channel. As the flood discharge increases, the lateral rolling of the ships becomes more pronounced. The change in power station discharge has a relatively small effect on the ships' heeling and trim.

Keywords Entrance · Fluctuation · Flow velocity · Navigation data · Observation

1 Introduction

In inland navigation, navigable structures are vital structures for overcoming the hydraulic head difference after river channelization [1]. Favorable flow conditions in the entrance area are essential to ensure the safety of shipping and to increase shipping efficiency. The entrance area is a significant transition area connecting the approach channel and the downstream ship channel, which is located at the junction between still water and flowing water [2, 3]. Affected by the daily regulation of the

X. Wang (✉) · S. Zhang
National Key Laboratory of Water Disaster Prevention, Nanjing Hydraulic Research Institute, Nanjing, China
e-mail: xwang@nhri.cn

S. Zhang
e-mail: zhangsy@nhri.cn

© The Author(s) 2025
W. Wang et al. (eds.), *Hydraulic Structure and Hydrodynamics*, Lecture Notes in Civil Engineering 608, https://doi.org/10.1007/978-981-97-7251-3_33

power station and flood discharge of the hub, the flow pattern of the downstream riverway is complicated, including fluctuation and transverse flow, which affects the safety of ships entering and leaving the restricted waterways.

The impact of high dam discharges on navigation is significant. The flood discharge regulation of the hub and output modification of the power station can cause periodic changes in the hydraulic characteristics of the downstream riverway [4–6]. The surge wave and fluctuation induced by discharge directly affect the flow conditions downstream. In several navigation hubs with high hydraulic head difference in China, the problem of fluctuation in the entrance area has gradually emerged, which can cause severe heeling when ships passing through the entrance. The surge wave in the downstream entrance area of the Dajiang Ship Lock in Gezhou Dam has caused significant influence on navigation [7]. Ship test results show that the heaving amplitude of ship can reach 1 m, and the existence of surge flow poses a serious threat to navigation safety. The measured maximum navigable discharge in Dajiang Ship Lock is only 57% of the designed value. Likewise, during the discharge at the Wufengxi Ship Lock in the Wuqiang River in Hunan Province, the mainstream surges into the entrance area, resulting in significant transverse flow. The measured maximum navigable discharge is only 39% of the designed value [8]. Besides, when the power station is discharging, the fluctuation cannot attenuate sufficiently as it propagates downstream, resulting in transverse oscillation near the downstream entrance area [9]. When ships pass through this area, the hull experiences significant oscillation amplitudes, which persist for an extended duration, often causing seasickness among the majority of the crew members [10]. Therefore, it is necessary to study the navigational hydraulic characteristics of the entrance area in the approach channel for high dam hubs, which is of great significance to ensure the navigation safety and improve the navigational capacity.

The study sample is located in the southwest of China and a ship lift has been designed to meet Class IV waterway standards, which represents a design capacity for a fleet of 2×500-ton ships, while also accommodating 1000-ton bulk-cargo ship [9]. Due to the stringent requirements for water surface fluctuation amplitudes during the docking of ship compartments and the pronounced fluctuations in the entrance during actual operations, our research focuses on the fluctuation characteristics of the downstream entrance area.

Figure 1 illustrates the main components and methods used in our observations. This study mainly records and analyzes the flow velocities and fluctuations in the entrance area under varying overflow discharges by field observation. In addition, the study monitored the actual operational and navigational data of 44 transiting vessels for a total of 384 h. The research focuses on analyzing the hydraulic fluctuation characteristics in the downstream entrance area under different combinations of flood discharge and power station discharge. The study aims to investigate the navigational status and provide technical support to address safety concerns related to wave-induced vessel restraint and to further improve navigation flow conditions.

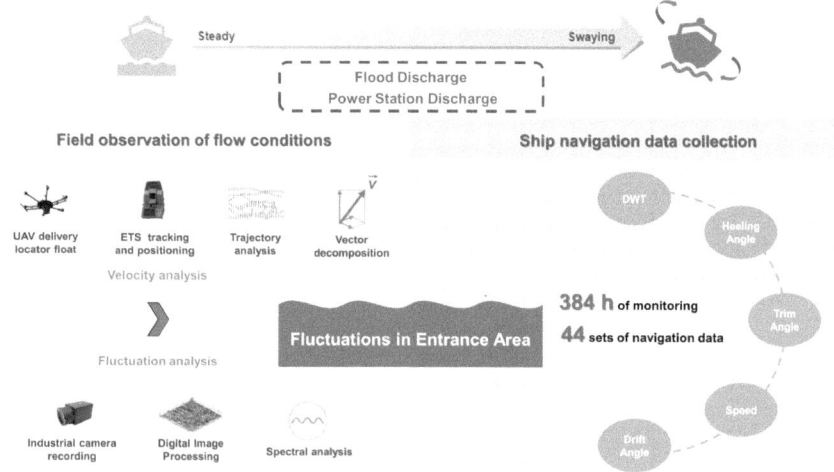

Fig. 1 The main components and methods of the study

2 Methodology

2.1 Flow Velocity

The entrance area of the downstream navigation channel is characterized by its vast area, measuring approximately 200 m in length and 60 m in width. During the flooding, the area experiences significant wave fluctuations. In previous observations, a fixed-point deployment of Acoustic Doppler Velocimeter (ADV) was used for flow velocity measurements. However, due to safety considerations and waterproofing requirements, this approach was discarded for the current study. After several site surveys, it was determined that the observation site offered ample space and favorable flying conditions. It was therefore decided to collect flow velocity data using a drop-and-drift method. Unmanned Aerial Vehicles (UAVs) were used to drop floating markers into the entrance area. These markers are lightweight and have excellent water tracking characteristics. High precision automatic tracking Electronic Total Stations (ETSs) were used to track the prisms embedded in the markers, providing accurate positional information. The marker positions were used to accurately trace trajectories and calculate flow velocities from the recorded time information and drift paths. Figure 2 shows the drop area of the floating markers.

Fig. 2 The layout of the fluctuation measuring points

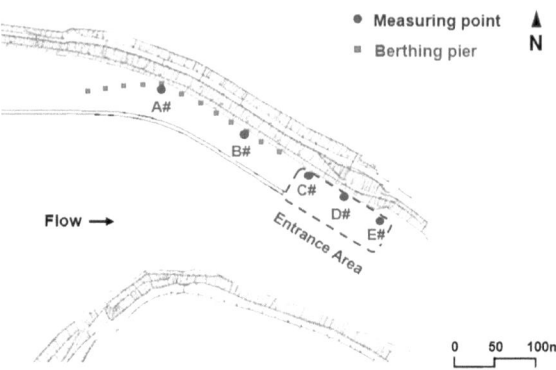

2.2 Fluctuation Data

The field observations were conducted during the flood season from July to October. Fluctuations in the downstream approach channel and the entrance area were recorded under spillway flood discharge conditions. Fluctuations measuring points 1# and 2# are located upstream of the entrance at a distance of 360 m and 210 m respectively. They are positioned to detect water level fluctuations near the auxiliary lock chamber and the interior of the approach channel. Points 3# to 5# are evenly spaced over a 200 m area in the entrance area, with approximately 50 m between measuring points. For fluctuation data collection and analysis, industrial cameras were used to continuously record water surface fluctuations within the range of the measuring points. Based on the principles of stereoscopic vision and digital image processing techniques, a three-dimensional water surface was constructed from a sequence of stereoscopic images taken at different time intervals, from which fundamental fluctuation param were extracted. In addition, dual-lens Particle Image Velocity Measurements (PIVs) were installed on the riverbank to compare and verify fluctuation data in critical areas of the entrance. This was done to ensure the accuracy and reliability of the data collection method mentioned above. Figure 2 shows the layout of the fluctuation measuring points.

2.3 Navigation Data

The navigation data collected includes a variety of parameters such as vessel speed, drift angle, heeling angle, trim angle and dead weight tonnage (DWT). Particular attention is paid to analyzing the heeling and trim amplitudes of ships within the entrance area. Navigation data is collected directly from the inertial navigation system. The drift angle is derived through post-processing and extensive analysis of the data. The inertial navigation system is a combined GPS/INS navigation device equipped with three gyroscopes, three accelerometers and a receiver. It combines

GPS and IMU technology to provide highly accurate navigation attitude information. To ensure real-time data acquisition and completeness, several base stations are strategically placed along the river bank in the downstream approach channel.

3   Results

3.1   Flow Velocity Distribution

The velocity distribution at the entrance was observed onsite. The downstream approach channel and entrance area velocity standards for longitudinal flow were established to align with national regulations, setting the maximum allowable value to 2.0 m/s. Transverse velocities were limited to 0.3 m/s, while backflow velocity had a cap of 0.4 m/s [9].

Under the conditions of full-load operation at the power station with no flood discharge, the flow velocity in the entrance area exhibited relatively modest values. The maximum recorded longitudinal flow velocity was approximately 1.75 m/s. The entrance area showed an extensive backflow. However, the intensity of backflow remained relatively modest. Only a few isolated flow velocity measuring points indicated backflow velocities exceeding the established limit, with the highest recorded backflow velocity reaching 0.51 m/s. The majority of measuring points recorded transverse flow velocities below the 0.3 m/s threshold. Nevertheless, it's worth mentioning that two measuring points located near the channel edges slightly exceeded the limit, registering at approximately 0.38 m/s.

During the full-load operation of the power station, with a flood discharge of 2000 m³/s, significant fluctuations were observed in the entrance area, accompanied by an expanded backflow region. Notably, specific measuring points within the upper right corner of the entrance area recorded transverse flow velocities exceeding the limit, with the highest transverse flow velocity reaching 0.61 m/s. Concurrently, longitudinal flow velocity maintained relative stability at around 1.0 m/s.

Upon further increasing the flood discharge to 2500 m³/s, significant transverse fluctuations within the entrance area were evident. The longitudinal flow velocity showed minor variations compared to the previous two sets of observations. The maximum longitudinal flow velocity was 1.12 m/s, with the highest transverse flow velocity reaching 0.92 m/s. The extent and intensity of backflow increased, with the maximum backflow velocity recorded at 1.43 m/s. Furthermore, additional flow velocity measurements were carried out when the power station was operating below full capacity. Detailed velocity data are given in Table 1.

Table 1 The flow velocity in the entrance area

Total discharge (m³/s)	Power station discharge (m³/s)	Flood discharge (m³/s)	Flow velocity (m/s)		
			Trans-verse	Longi-tudinal	Back-flow
1700	1700	0	0.26	0.32	0.46
3800	3800	0	0.25	0.68	0.67
4600	4600	0	0.28	0.72	0.70
6400	6400	0	0.38	1.75	0.51
8400	6400	2000	0.61	1.10	1.18
8900	6400	2500	0.92	1.12	1.43

3.2 Entrance Area Fluctuations

Field observations showed that during flood discharge conditions, there were notable fluctuations in the entrance area, while the water surface remained calm when the power station discharging flow independently. The discharge of floodwater from the surface spillway channel results in a hydraulic jump once it flows over the auxiliary weir of the stilling pool. A larger discharge induces more intense hydraulic jumps with greater wave amplitudes. The oscillation of hydraulic jumps is the direct factor causing fluctuations in the downstream entrance area. The fluctuations in the approach channel are a consequence of entrance area waves propagating upstream in the direction of the navigation wall and gradually diminishing in intensity. As the waves travel downstream of the stilling pool, they follow the longitudinal contours of the riverbed, influenced by the topographical features of the riverbed and the revetments on the right bank of the power station. These complicated boundary conditions cause fluctuations in the entrance area, perpendicular to the embankment.

Table 2 presents the maximum wave heights at different wave measuring points. The maximum wave height (MWH) effectively illustrates the time-averaged characteristics of maximum wave amplitudes. Fluctuations in the entrance area are random waves, with the significant wave height (SWH) being a primary factor influencing the safety of navigation.

Table 2 Fluctuations in the downstream entrance

Power Station discharge (m³/s)	Flood discharge (m³/s)	Maximum wave height (m)				
		Point A#	Point B#	Point C#	Point D#	Point E#
5711	1042	0.39	0.34	0.43	0.54	0.70
5749	1563	0.35	0.32	0.37	0.47	0.58
5731	2044	0.48	0.40	0.55	0.63	0.76
5750	2500	0.51	0.44	0.65	0.71	0.84
6400	2500	0.51	0.58	0.59	0.93	0.85
6400	3000	0.54	0.61	0.61	1.05	0.85

When the power station was functioning with a discharge of 5750 m³/s, and the discharge via the spillway was 2500 m³/s, the SWH in the approach channel was 0.5 m whilst the SWH at the entrance equaled 0.65 m. The entrance area edge had a height of 0.84 m. Under full-load operation of the power station, with a flood discharge of 3000 m³/s, the MWH generated in the center of the entrance area was 1.05 m. As the surface spillway discharge increases, the MWH also increases sharply. With an increase of 500 m³/s in flood discharge, the MWH rises by about 10%.

4 Discussion

While the power station operating independently, it will not have a negative effect on the flow condition in entrance area and will not induce significant transverse flows. This perspective is consistent with the partial-load power station operation observed in Gao's study [10]. When the power station operates at full-load with a flood discharge rate of 3000 m³/s, the maximum transverse velocity of the entrance area was 0.38 m/s. Combined with the observed navigation results, fully loaded ships could overcome the transverse flow within 0.40 m/s. The operation of the power station did not affect the navigation safety. When the surface spillway was discharging, the intensity of the transverse flow and the backflow increased obviously, but the longitudinal flow velocity remained unchanged. Most of the measuring points with the largest transverse flow intensity were located at the end of the backflow in the entrance area, and the transverse velocity in the core area of the ship channel was not more than 0.4 m/s. The ship operator could overcome the partial strong transverse flow by increasing the speed, so the effect of the transverse flow on navigation was relatively limited, and the observational navigation data also supported this view [9].

Figure 3 shows the spectrum analysis results of the measuring points in the center of the entrance area at different flood discharges. The transverse fluctuation in the entrance area was particularly pronounced during surface spillway discharge. Based on the measured time series data of wave characteristics, Fourier transform is applied for further spectral analysis of the fluctuations. When the power station operates at full-load with 2500 m³/s flood discharge capacity, the main fluctuation frequency was 0.127–0.176 Hz, and the period was approximately 5.7–7.9 s. When the flood discharge was 3000 m³/s, the main frequency of fluctuation was between 0.156 and 0.176 Hz, and the period was about 5.7–6.4 s. Based on the fluctuation characteristics of different discharge combinations, the fluctuation of the entrance area was mainly a short-period oscillation, with a period of 6 s. In the other studies of fluctuations during partial-load operation, it was confirmed that there were short-period characteristics in the entrance area [9]. However, the specific wave frequencies are not reported in the study. In addition, the wave height in the center of the entrance area was higher than that in the entrance, and the fluctuation amplitude in the center of the entrance area was prominent when the power station was operating at full capacity.

In the entrance area, both at the center and the edges, the wave amplitudes are more pronounced. These fluctuations move towards the still water region along the

Fig. 3 Spectrum analysis of
the measuring points. (Red
dotted line is the main
frequency position)

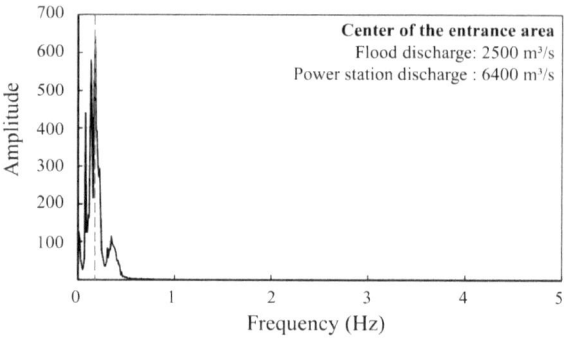

(a) 2500 m³/s Flood discharge capacity.

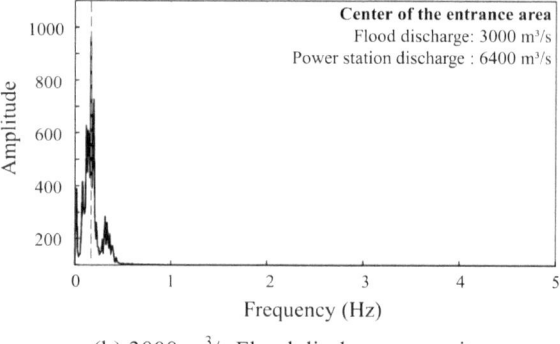

(b) 3000 m³/s Flood discharge capacity.

approach channel, gradually attenuating. When the power station operates at partial-load, the fluctuations in the approach channel remain relatively stable, even with increased flood discharge. A 140% increase in flood discharge results in a 30% increase in MWH in the center of the approach channel, with an average value of 0.40 m. Under full-load operation with flood discharge, the average MWH within the approach channel reaches 0.56 m. Analysis of 6 sets of fluctuation data shows that the attenuation rate of fluctuations from the entrance to the channel is consistently between 55.7 and 63.5% across different discharge combinations.

Figure 4 illustrates the complete process of fluctuation attenuation. Figure 4a indicates that during the partial-load operation, the greatest fluctuation attenuation occurs within the region from the entrance to the approach channel, with an attenuation rate of 20.0–32.3%. Figure 4b shows that under full-load operation, the highest attenuation rate occurs from the center of the entrance area to the entrance, approximately 36.5–41.9%. The attenuation capacity within the approach channel is relatively poor, resulting in a reduction in wave height of 5–8 cm. The main fluctuation attenuation occurs at the transition from the broad riverway to the downstream restricted approach channel.

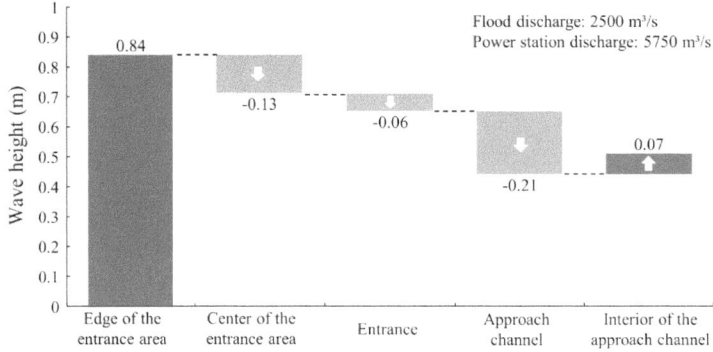

(a) Partial-load operation.

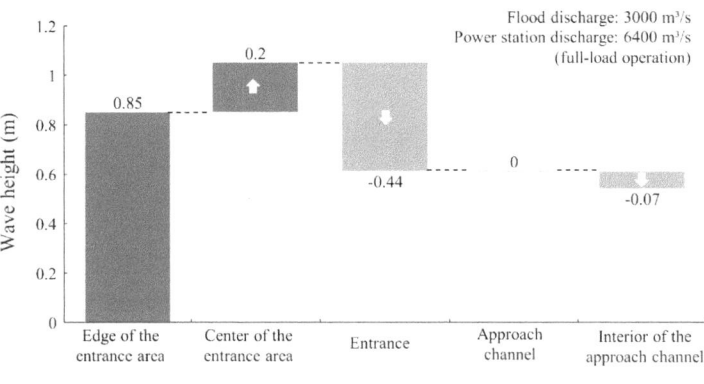

(b) Full-load operation.

Fig. 4 The complete process of fluctuation attenuation

Navigation data is a direct reflection of water flow characteristics. Based on 44 sets of data, it is evident that the drift angle of the ship is predominantly in the range of 20–35°. No significant increase in drift angle was observed with increasing flood discharge. The speed measurements also show that the ship's speed remains within the range of 3.0–7.0 m/s for different flood discharge. However, as the flood discharge increases, there is a slight increase in the actual navigation, with an increase of 14–19%. The transverse fluctuation also induces rolling motions in ships [9]. At a flood discharge of 1400 m³/s, the navigation data showed that a 500 DWT ship experienced a heeling angle of up to 6.2°. Lighter loaded ships are confronted with excessive heeling amplitudes when passing through areas of pronounced transverse fluctuations in the entrance area. The violent heeling amplitudes pose a significant threat to the stability and navigational safety, with a risk of capsize [10].

Figure 5 shows the heeling angles for the same DWT ships passing through the entrance area at full-load operation. The heeling amplitudes increase as the flood discharge increases. It is noteworthy that 89% of the ships passing through the entrance area are able to maintain heeling angles within 3.5°. With no flood discharge, ships experience minimal rolling, not exceeding 1°. However, as the flood discharge increases, Fig. 5 shows that the envelope area gradually increases. When the flood discharge reaches 2200 m³/s, there is a significant range of heeling angles from 2.11 to 5.09°. This reflects the increased difficulty in controlling, which is closely related to the navigational skills of the operators and the navigation routes.

Figure 6 shows the trim angles for the same load at full load. The trim angles of the sips are directly proportional to the flood discharge. With no flood discharge, the trim angle is less than 0.2°. When the flood discharge reaches 2200 m³/s, the maximum pitch angle is 0.83° and for all discharge combinations the trim angle remains below 1°. Comparing the heel and trim data, it is clear that the ships experience much larger heeling angles than trim angles as they pass through the entrance area.

Figure 7 illustrates the effect of different power station discharge on the heeling and trimming. For a constant flood discharge, the heeling angle of the vessels fluctuates around 2.5° as the power station discharge increases from 3780 to 5980 m³/s, indicating that the power station discharge has a relatively small effect on the heeling of the ships. In addition, the trim angle remains at 0.5° and is independent of the power station discharge. In summary, it can be concluded that variations

Fig. 5 Heeling angles in entrance area

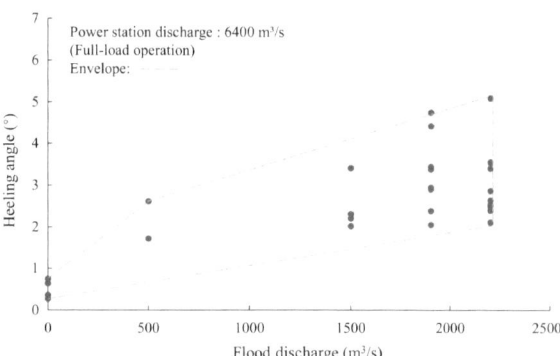

Fig. 6 Trim angles in entrance area

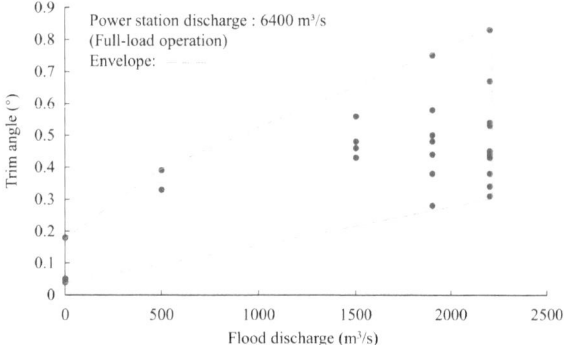

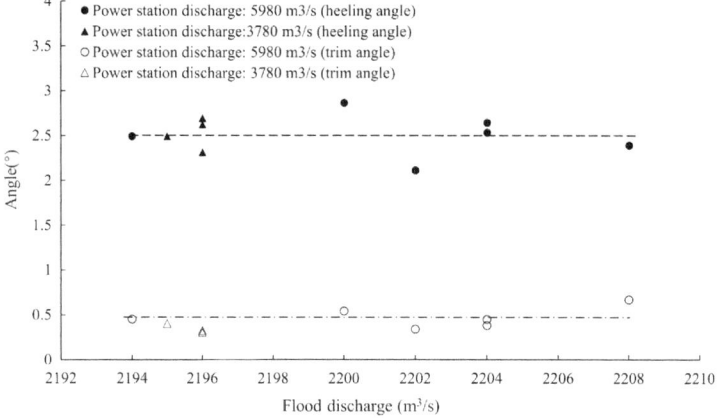

Fig. 7 Influence of different power flows on heeling and trimming

in power station discharge have negligible impacts on navigation conditions in the downstream entrance area. The physical model experiments in Zhang's study provide some support for this view. When the power station discharge increases by 90% (from 3253 to 6181 m³/s), the variations of the unsteady flow parameter and the abating hydraulic index indicate that the power station regulation has a minimal effect on the navigation and berthing [11]. The downstream river channel in the study sample is generally straight, with a U-shaped cross-section [12]. Navigational characteristics in this type of channel are not highly sensitive to the power station regulation [12, 13].

5 Conclusions

The study investigated the fluctuations of the entrance area during the flooding period with a long-term observation of the flow velocity, fluctuations and actual navigation situation, and the collected data are systematically analyzed. The main conclusions were as follows:

- The oscillation of hydraulic jumps downstream of the auxiliary weir is the primary source of short-period fluctuations in the downstream entrance area, with the fluctuations being directly proportional to the flood discharge.
- Short-period fluctuations propagate from the entrance area to the approach channel, and the main attenuation occurs at the transition from the broad riverway to the downstream restricted approach channel.
- The flood discharge induces the transverse flow intensities within the entrance area, which also increases the intensity and extent of the backflow.

- The power station discharge exerts minimal influence on navigation flow conditions, and changes in discharge do not result in alterations to vessel heeling or trim.

The study provides direct data support for the downstream navigation conditions, providing valuable insights for the navigational hydraulic characteristics. The results will not only provide valuable guidance for the operational management of the power station, but will also be an important reference for ensuring the safety of navigation. In future studies, the relationship between reservoir regulation and navigational status can be further explored to improve navigation safety and efficiency.

References

1. Z.Y. Wan, Y. Li, X.G. Wang, J.F. An, B. Dong, Y.P. Liao, Influence of unsteady flow induced by a large-scale hydropower station on the water level fluctuation of multi-approach channels: a case study of the three gorges project, China. J. Water **12**(2922) (2020)
2. Z.Y. Wan, Y. Li, L. Cheng, X.G. Wang, B. Wang, J.F. An, Investigating hydraulic operational schemes of a large-scale multi-lane lock group concerning water-level fluctuations in a branched approach channel system. J. Ocean Eng. **260**(111758) (2022)
3. U. Ji, E.-K. Jang, G. Kim, Numerical modeling of sedimentation control scenarios in the approach channel of the Nakdong river estuary barrage. J. Sediment. Res. **31**, 257–263 (2016)
4. M. Xie, J. Zhou, C. Li, and P. Lu, Daily generation scheduling of cascade hydro plants considering peak shaving constraints. J. Water Resour. **142**(04015072) (2016)
5. G. Ardizzon, G. Cavazzini, G. Pavesi, A new generation of small hydro and pumped-hydro power plants: advances and future challenges. J. Renew. Sustain. Energy Rev. **31**, 746–761 (2014)
6. T.B.A. Couto, J.D. Olden, Global proliferation of small hydropower plants-science and policy. J. Front. Ecol. Environ. **16**, 91–100 (2018)
7. Y.K. Zhou, The problem of the channel connecting section outside the entrance of Dajiang ship lock in Gezhou Dam. J. Port Waterw. Eng. **9**, 27–30 (1996)
8. L.F. Lu, N. Wang, X.G. Pu, Study on improvement measures of navigation conditions of ship locks constructed under complex conditions in mountainous areas. J. Waterw. Harbor. **5**, 579–583 (2018)
9. Y.A. Hu, J.F. An, J.J. Zhao, X. Wang, Prototype observation of hydraulic fluctuation characteristics at entrance of approach channel downstream of Xiangjiaba ship lift. J. Port Waterw. Eng. **12**, 22–26 (2020)
10. C. Gao, Research on flow conditions at lower reach outlet areas of approach channels for ship lifts at Xiangjiaba Dam. J. Shanxi Archit. **48**, 181–183 (2022)
11. Z. Zhang, Y.H. Liu, L.H. He, X.J. Zhang, Influence of Xiangjiaba daily regulation on operation of Zhongzui Wharf. J. Port Waterw. Eng. **9**, 145–150 (2023)
12. X.J. Zhang, Z.Z. Hu, Y.H. Liu, C.L. Huang, Research on propagation characteristics of unsteady flow caused by daily regulation of Xiangjiaba Hydro-power station. J. Waterw. Harb. **5**, 414–418 (2015)
13. M.X. Cao, X.S. Pang, Progress in studies of hydropower station discharge influence on the downstream navigation channel. J. Hydro-Sci. Eng. **6**, 94–104 (2011)

Research on the Optimization Layout of the Connection Layout of the Multi-lock Dam Plain River Network and Pond River System in Langfang City, China

Shuo Feng, Jinyong Zhao, Jing Zhang, Yicheng Fu, and Wenqi Peng

Abstract In view of the uneven distribution of water resources in the multi-lock and dam plains, the insufficient connection and layout of river systems such as isolated wetlands and pond barriers in floodplains, resulting in insufficient water resources allocation capacity, poor flood storage and drainage, and serious water ecological damage. Taking Langfang City as an example, based on graph theory and complex network theory, considering the flow capacity of DAMS and gates and human demand for water landscape, the importance of rivers and canals in river network—optimizing river network layout—establishing comprehensive evaluation index system of river network layout—comprehensive evaluation is taken as the main line to evaluate and determine the importance of rivers and canals in river network. The overall layout of river network connectivity in the north, middle and south regions of Langfang is optimized and evaluated. The results show that there are large differences in the current layout of waterways in the three regions, with poor river network connectivity in the southern part of the Langfang. The overall layout of the optimized river network was improved by 13.0%, 3.0% and 12.4% compared to the overall layout in Langfang. The optimized scheme provides technical support and theoretical basis for the optimal implementation of the regional river and lake connectivity project in Langfang.

Keywords River system connectivity · Complex networks · Graph Theory · Landscape ecology · Langfang City

S. Feng (✉) · J. Zhao · J. Zhang · Y. Fu · W. Peng
China Institute of Water Resources and Hydropower Research, Beijing 100038, China
e-mail: fengshuo@edu.iwhr.com

© The Author(s) 2025
W. Wang et al. (eds.), *Hydraulic Structure and Hydrodynamics*, Lecture Notes in Civil Engineering 608, https://doi.org/10.1007/978-981-97-7251-3_34

1 Introduction

The pattern of river and lake system and their connectivity have an important impact on regional water resources allocation, water storage, flood control and disaster reduction, ecological protection and restoration, and landscape improvement [1–3]. With the rapid development of economy and society, human activities have gradually affected the pattern of natural river system, especially the construction of water conservancy projects such as locks and dams, blocking the longitudinal and horizontal connectivity of regional rivers and changing the pattern of regional river network, resulting in uneven distribution of regional water resources, isolated wetlands of floodplain and barrier of ponds and other problems. The river and lake water system connectivity adjusts and optimizes the river and lake water system pattern by maintaining, repairing and constructing the hydraulic connection between rivers, lakes and reservoirs, improving the uneven spatial and temporal distribution of water resources, and improving the coordination between the water resources distribution pattern and the economic and social pattern [4].

The study of river system pattern mainly evaluates rivers of different levels according to the indicators of river geomorphology, which is the basis for the analysis of river network morphology and function [5]. River system development theory, the law of river number and river length, Strahler river classification method, river fractal characteristics and other theories have promoted the development of the study of river network pattern evolution to a large extent [6–9]. In order to quantitatively describe the distribution pattern and spatiotemporal evolution of river network, many characteristic indexes of river system have been proposed successively. Based on historical topographic maps and remote sensing images, river network density, water surface rate, river network development coefficient, trunk stream area to length ratio, river network structure stability and box dimension are used to analyze the change trend of regional river system in spatial distribution and time evolution [9–13]. Among them, indicators of both hydrologic pattern and structural connectivity are more frequently applied, while hydraulic connectivity is relatively less studied. Huang Cao et al. established the evaluation index system of river system pattern and connectivity, and analyzed and proposed the optimization plan of river system connectivity planning for different areas of Dongting Lake [3]. Zuo Qiting et al. believe that the connectivity of river and lake system includes not only the physical connection and pattern distribution of natural water bodies such as rivers, lakes and wetlands and artificial water bodies such as reservoirs, artificial rivers and lakes, but also the "quantity and quality exchange" between water bodies [14]. Phillips et al. established a hydraulic connectivity evaluation system for urban rivers based on flow resistance and hydrological processes [15]. Liang Xiao et al. constructed a hydrological connectivity evaluation system for urban river networks in plains cities, which contains three criteria: water system pattern, structural connectivity and hydraulic connectivity [16]. Different scholars have different emphases on the evaluation of river network connectivity. In view of the dense river system in the plain river network, under the control of water conservancy projects such as gates and pumps, the water flow movement is complicated and the water system connectivity is not smooth, which affects the self-purification ability of the river, so it is necessary to consider the influence of different sluice and dam flow capacity.

This paper takes Langfang city as an example, focusing on the current problems of water scarcity, spatial and temporal distribution imbalance, insufficient ecological water use in the water system, and ecological fragility of the Yongding River flood-plain, combining with the strategic objectives of the Beijing-Tianjin-Hebei synergistic development strategy, vertically taking into account the influence of human activities such as locks and dams, horizontally taking into account the role of pits, ponds, and puddles in storing water and preventing floods, and at the same time taking into account the hydrophilic capacity of the people, and establishing a multifactor The evaluation index system of water system pattern, water system connectivity and hydrophilic ability is established. Based on graph theory and complex network theory, the importance of rivers and canals between backbone rivers is determined through weighted connectivity to optimize the river network connectivity layout. Provide technical support for the implementation of the regional river and lake water system connectivity project in Langfang.

2 Study Area

Langfang City is located in the northeast of the North China Plain, flat terrain, dense river network, ponds and dams, a total of 5 ponds, 275 locks, 7 rubber dams. There are 112 river channels in Langfang, including 10 main flood channels and 10 main drainage channels. There are 5 ponds in Langfang City, and the rivers connected with them are: Liusong million mu Lotus pond—Qinglongwan Reduction River, East Zhangwu Wetland—Long River, Dongdian—Zhongting River, Wenanwa—Tanli Trunk Canal, Jiakouwa—Ziya River. Langfang river network river system is shown in Fig. 1. This paper takes the river network of Langfang City as an example, takes 19 major rivers as the framework (except Zi Ya new River), and artificial ditches and dams as the connecting objects (see Table 1), optimizes the connectivity layout of the river network of Langfang City, and makes a quantitative evaluation of the northern river network of Langfang City.

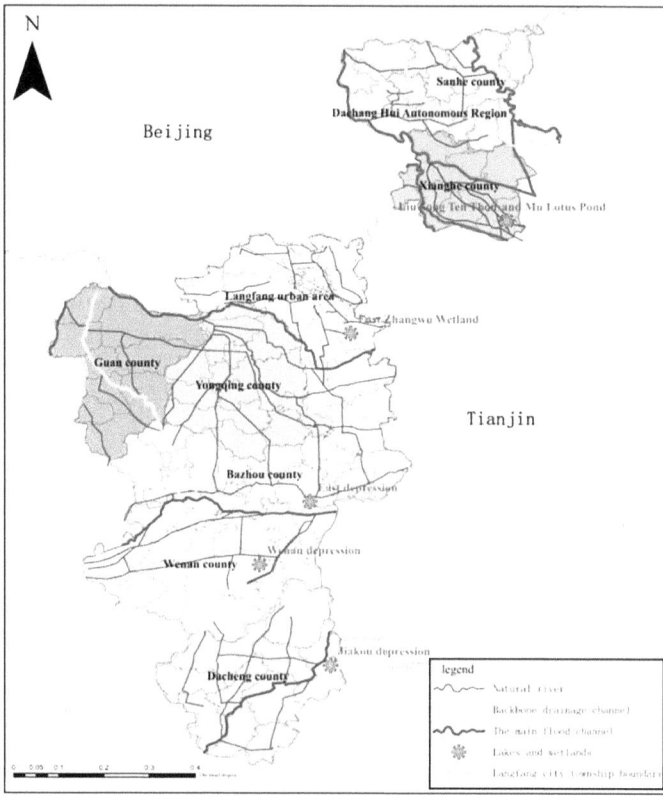

Fig. 1 Langfang river network river system

3 Data Sources and Evaluation Indicators

3.1 Data Sources

Through the "Langfang Hydrology and Water Resources Use Guidelines" to obtain the Langfang City, the locks and dams of the state overwater capacity and the length of the river, of which the length of the river section in the accuracy of 0.1 km requirements, the application of resolution of 1 m Google Earth imaging data resolution and measurement of the length of the river section, and with the Langfang City, the average difference of the statistical length of the river compared with the average difference of less than 0.1 km. other data in combination with the collection of the vector map as well as the site The other data are combined with the collected vector map and field measurement data for data calibration.

Table 1 Status of main rivers and canals in Langfang river network

River channels	Channel function	Gate and dam (number)			Whether the channel is silted	Gate and dam scheduling mode
		Regulating gate	Drainage sluice	Rubber dam		
Chaobai river	The main flood channel	2	5	4	No	F/T
Ju river		2	3		No	F
Take Ju into the Chao			1		No	F
Qinglong Bay river			2		No	F
Beiyun river		1	1		No	F
Yongding river			5	1	Yes	F/T
Daqing river			5		No	F/T
Baigou river			1		No	F
Ziya river		2	2		Yes	Y
Ziya new river			/		No	Y
Boqiu river	Backbone drainage channel	4	7		No	F/T
Phoenix river					No	Y
Long river		4	3	2	No	F/T
Tiantang river		1	1		No	F
Mangniu river		4	1		Yes	F
Paisan main canal			2		Yes	Y
Xiong GUba new river		1			No	F
Ren River drainage Canal		2	1		Yes	Y
Ren Wen trunk Canal		2	1		No	Y
Black dragon River west branch					Yes	Y

Note F is flood season opening; *T* is water transfer opening; *Y* is year-round opening

3.2 Graph Model Generalization

Mapping the current river system based on the current river system in Langfang city. Identify the river system of Langfang City through Google earth, totaling 96 rivers and canals, and a total of 19 backbone rivers in Langfang City (excluding Ziya New River). Arcgis is used to extract the river system of Langfang river network, and the edges indicate the backbone rivers and artificial rivers and canals, and the locks and dams, river intersections, and ponds and precipitates are all regarded as nodes. Among them, the wetland of the pit pond is directly generalized to the point on the corresponding connecting river and canal, and there is no role of dividing the river and canal, but it has the role of water storage and flood prevention. Langfang is located in the northeastern part of the North China Plain and has a small difference in elevation, so it is generalized to an undirected graph model.

3.3 Judgment of the Importance of River Canals in the River Network

In complex networks, the degree of a node can be described as the importance of that node in the complex network, while in real network systems, the individual edge energy flow capacity is not the same and varies greatly, so entitled to the degree of the node in the network Fi is more reflective of the importance of the node in the complex network. In this paper, the maximum overflow capacity of the gates and dams on the river section is used as the side weights considering the flood control and mitigation needs [17].

$$wil = \text{MAX}(Qup, Qdown)$$

Standardized treatment of marginal weights: For quantitative indicators, the standardization method (difference method) is often adopted, calculated as follows.

$$wil = \begin{cases} \frac{wil'}{\max(wil') + \min(wil')} \\ 1 - \frac{wil'}{\max(wil') + \min(wil')} \end{cases}$$

where, wil and wil' are the standardized and actual values of the quantitative factors.

Importance of nodes in complex networks:

$$Fi = \sum_{j=1, j \neq i}^{n} wij$$

The importance of a river and canal in a river network is mainly related to the importance of the nodes connecting the river and other rivers, so the importance Fl of the river and canal in the river network is mainly expressed by the sum of the node importance of all nodes connected with other rivers and canals on the river and canal.

$$Fl = \sum Fi$$

3.4 Selection of Connectivity Layout Optimization Scheme

Considering that the river network of Langfang Plain is composed of main river channels and artificial ditches, the connectivity of artificial ditches is intermittent. In order to control flood control and diversion of water during the period of flood and drought, the water allocation is realized through sluice and dam regulation., Based on the main rivers in the river network ($e1$, $e2$, …, en), By selecting k scheduling paths composed of important rivers and canals between any two major rivers, $Lk(ei, ej)$ where k = 0, 1, 2.

3.5 Comprehensive Evaluation Method

(1) Evaluation index system

The network connectivity of regional river system originates from the concept of landscape ecological network connectivity in landscape ecology. For the evaluation of the connectivity layout of river network in plain area, multi-index evaluation method is commonly used at present, and the index system usually includes river frequency, river network density, water storage, structural connectivity, river system ring degree, node connection rate and other indicators [18, 19]. As Langfang is a plain area, the difference in topography is small, and there are densely covered rivers and canals, and numerous pits and dams, the changes in river velocity and water dynamic force are small and can be ignored, and the impact of dams and dams on the connectivity of river network is great. Urban water system is different from general water system, so the problems of flood control and drainage of urban water system should be paid attention to in the process of urban development [16]. Therefore, considering the influence of different types of dams and dams, the structural connectivity and river network connectivity index are optimized. In this paper, river network density and water storage capacity are selected to measure the river system pattern of Langfang City; structural connectivity and river network connectivity are selected to measure the river system connectivity of Langfang City; and river network connectivity and layout index system of river network in plain region are established by measuring the water philicity of Langfang city through the coverage rate of cities and towns (see Table 2).

(2) Comprehensive evaluation method

Table 2 Evaluation index system of river network connectivity layout

Target layer	Criterion layer	Indicator layer	Calculation formula	Remarks
Indicator system for evaluating river network system connectivity	Drainage layout	Density of river network	$D = L/A$	L: Total length of river network
		Quantity of water stored	$Q = (Ah - Dg)LW$	A_h: Normal water level; D_g: River depth; L_i: River length; W: River width
	Drainage connectivity	Structural connectivity	$N = nc/n$	n_c: The number of cut points in the river map model, in which the cut point coefficient of large sluice is 0.8, the cut point coefficient of medium sluice is 0.5, and the cut point coefficient of small sluice is 0.2. n is the total number of vertices in the drainage diagram model
		River network weighted connectivity	$C = \dfrac{\sum\limits_{i=1,i\neq j}^{n} \sum\limits_{j=1}^{n} dij}{n \cdot n - 1}$	d_{ij}: Maximum overflow capacity of the river is the value of the side weights, The edge weights corresponding to the shortest distance traveled from v_i to v_j
	Hydrophilic ability	River network town coverage	$P = Pi/P$	P_i: Number of river networks passing through human settlements

First, the difference method is used to standardize each index, and then the comprehensive evaluation index is calculated. The formula is as follows:

$$F = k1x1 + k2x2 + k3x3 + k4x4 + k5x5$$

where, ki is the weight. $x1$, $x2$, $x3$, $x4$, $x5$ is the five preferred index values corresponding to each river section.

The analytic hierarchy process (AHP) is often used to assign weights to the preferred 5 indicators. Finally, the weight coefficients of river network density, water storage, structural connectivity, river network water connectivity and river network town coverage are 0.11, 0.22, 0.11, 0.22 and 0.33, respectively. The CR of the judgment matrix is less than 0.1, and the consistency is acceptable.

$$F = 0.11x1 + 0.22x2 + 0.11x3 + 0.22x4 + 0.33x5$$

4 Analysis and Discussion of Research Results

According to the geographical characteristics of Langfang at the present stage, Langfang intersects with Beijing and Tianjin, so that the main body of Langfang's border area is not connected, and the North three counties are not connected with other regions. Moreover, the lower reaches of Yongding River in the middle of Langfang is the Yongding River pan area, which is located between Beijing and Tianjin and is responsible for the task of flood detention and sediment sediment removal of Yongding River. Therefore, Langfang is divided into three regions as the research object.

4.1 Evaluation of Langfang's Current River System Layout

Based on the current layout of the river network in Langfang (some sections of the river are silted and blocked) compared to the layout of all water systems in Langfang (see Figs. 2, 3 and 4), the results of the evaluation of the river network connectivity layout in Langfang are shown in Fig. 5. The overall composite scores for river network connectivity in the northern, central, and southern portions of Langfang City decreased by 11.0%, 13.6%, and 15.8%. However, river network connectivity increased by 147% in the northern region and river structure connectivity increased by 21.37% in the southern region. All other indicators decreased or remained unchanged. Due to the siltation and blockage of some river sections in Langfang's current river network layout, the number of river system nodes and the number of river chains are reduced, and the river system pattern and the index of the criterion layer of hydrophilic ability are related to the number of river chains. For example, the density of river network is calculated by the ratio of the total length of river network to the regional area, and the coverage rate of river network towns is related to the number of towns through which river network passes. The connectivity degree of the river network is related to the water transfer capacity of the water nodes in the river network, and the connectivity affects the water transport capacity of the river network. Therefore, on the basis of improving the connectivity of river network, it is also necessary to improve the river system pattern and hydrophilic ability of river network. It is necessary to judge the importance of artificial canals between major river channels and select the optimal connectivity layout of river network.

4.2 Determination of the Optimization Scheme for the Overall Layout of the Langfang River System Connection

The network structure of water system is relatively complex. With the increase of nodes, the maximum number of possible connecting river sections is also increasing,

Fig. 2 River network river
system in the northern part of
Langfang City

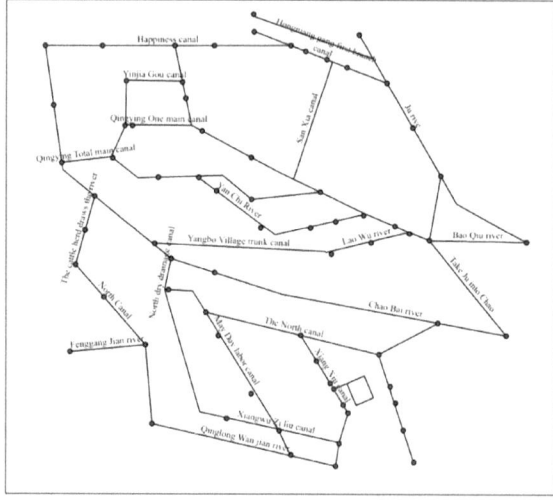

(a) Layout of all river systems

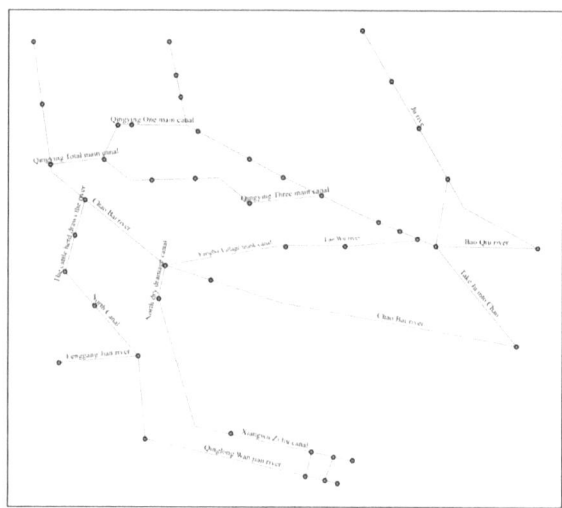

(b) Layout of the current river system

which leads to the failure of nodes to give full play to the ability of connecting river sections [16]. Based on the judgment of the importance of each river in the river network, the 19 main rivers and canals in the river network of Langfang are taken as the framework, and the canals with higher importance among the main rivers and canals in Table 3 are selected and connected to form an optimized overall layout plan of the river network connectivity of Langfang (See Fig. 6). Among them, the river network in the south of Langfang City between Zi Ya River-Black dragon River west branch is the Jiakouwa stagnant flood storage area, with the right embankment of

Fig. 3 River network river system in central Langfang City

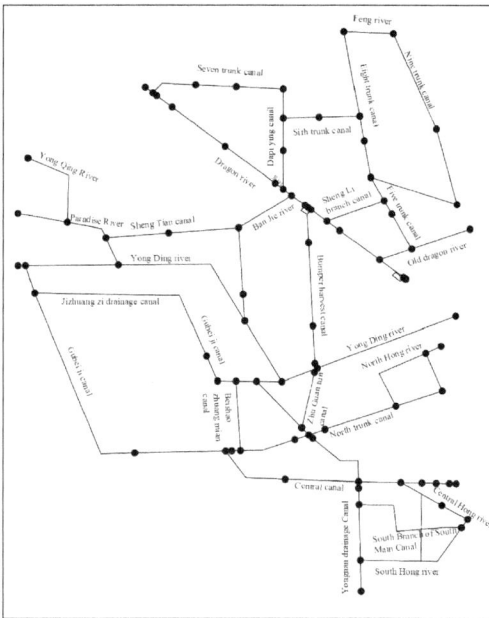

(a) Layout of all river systems

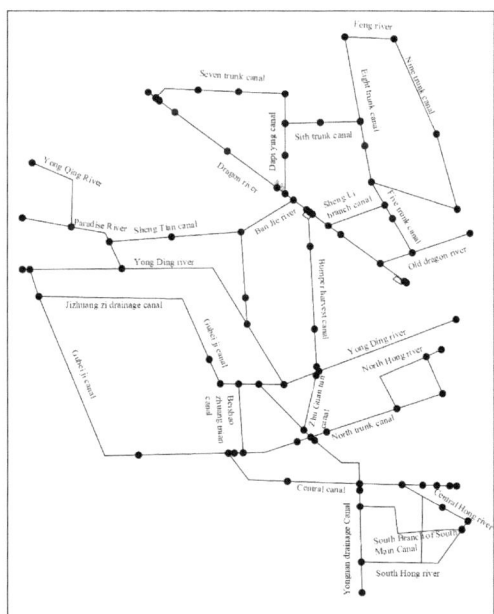

(b) Layout of the current river system

Fig. 4 River network river
system in the south of
Langfang City

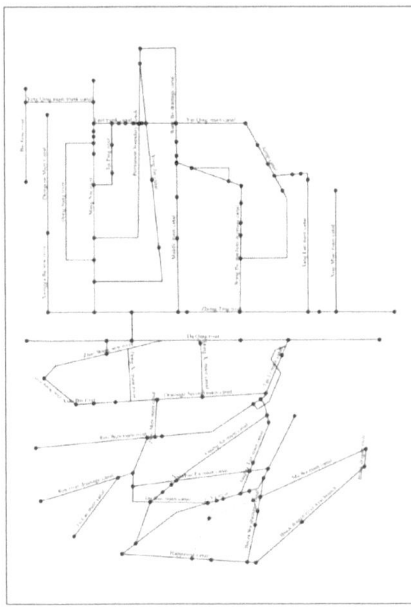

(a) Layout of all river systems

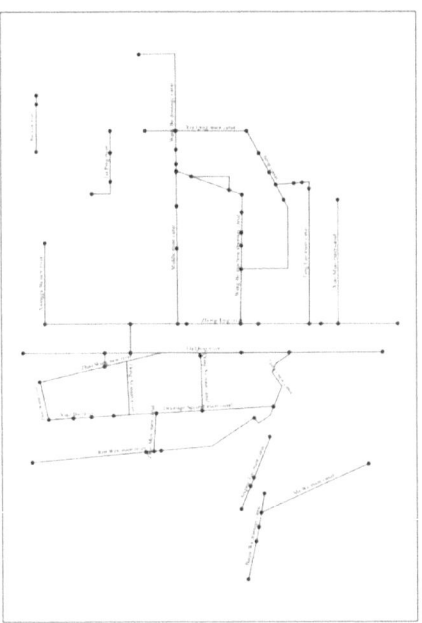

(b) Layout of the current river system

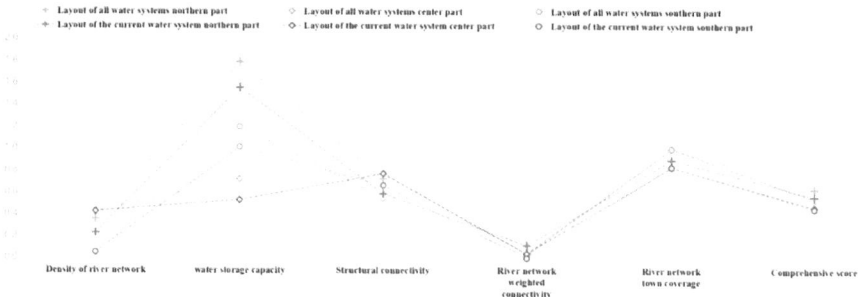

Fig. 5 Evaluation of the current river system layout in Langfang City

Zi Ya River and the left embankment of the South Canal as the puddle precipitation embankment, so this paper does not consider the connecting river and canal between the Zi Ya River and the Black dragon River west branch.

Table 3 Importance of rivers and canals in the river network of Langfang City

Northern part		Center part		Southern part	
Name of the river canal	Importance of river canal	Name of the river canal	Importance of river canal	Name of the river canal	Importance of river canal
Chao Bai river	13.17	Yong Ding river	18.48	Zhao Wang new river	5.94
May Day labor canal	5.04	Dragon river	5.07	Da Qing river	3.08
Yinjia gou	3.14	Bumper harvest canal	4.93	Zhon Ting river	1.54
The cattle herd draws the river	2.98	Gubei ji canal	4.34	Dong Xi Trunk Canal	1.30
Qinglong Bay river	2.91	Paradise river	4.25	Zhao Wang river	1.29
⋮	⋮	⋮	⋮	⋮	⋮
North dry drainage canal	0.03	Nine trunk canal	0.18	Ren Wen trunk canal	0.20
Yin Jia ditch canal	0.02	North trunk canal	0.09	Happiness Canal (South)	0.19
East trunk canal	0.02	Sixth trunk canal	0.07	Longmen kou trunk canal	0.18

Fig. 6 Overall layout of
optimized river system
connectivity in Langfang
City

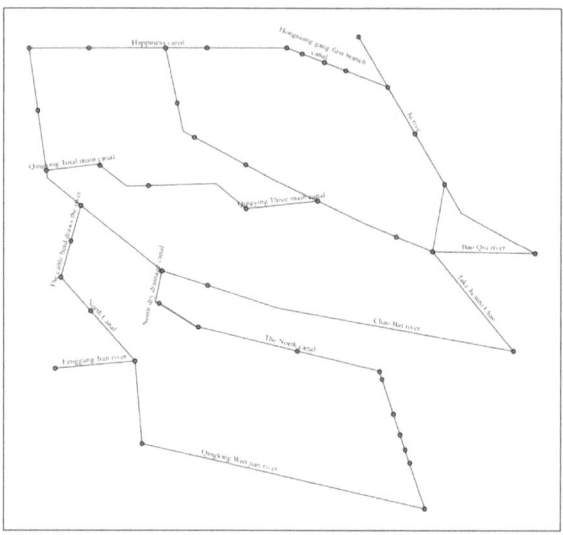

(a) River system in the north of Langfang City

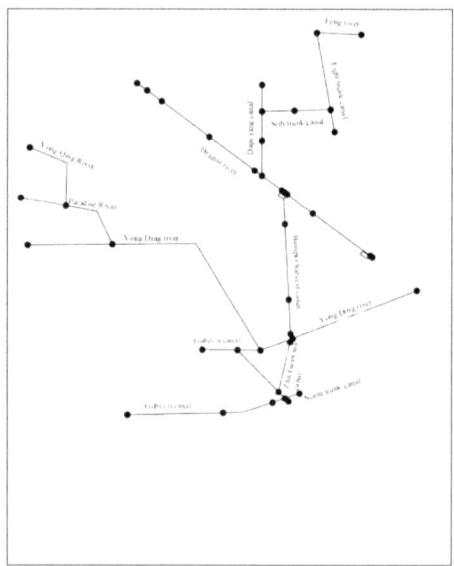

(b) River system in the central of Langfang City

4.3 Optimization Evaluation of Overall River System Connectivity in Langfang City

The overall layout of river network connectivity in Langfang City was optimized by selecting key river channel connectivity that improved the water transport capacity of

Fig. 6 (continued)

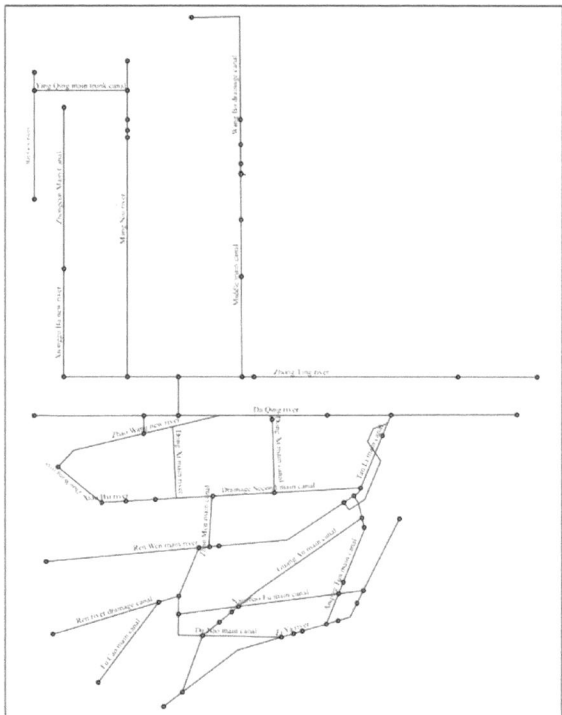

(c) River system in the south of Langfang City

the river network, as shown in Fig. 6. According to the analysis in Fig. 7, compared with the current river network layout of Langfang City (some river sections are silted and blocked), the overall comprehensive scores of river network connectivity in the northern, central and southern regions of the optimized scheme increase by 13.0%, 3.0% and 12.4%, in which the structural connectivity index decreases in the northern, central and southern regions, and the connectivity degree of river network improves. The connectivity of the north and middle river networks in Langfang City increased by 11.27% and 84.13%, and the connectivity of the south river network from disconnected to 0.0381. The optimization scheme further improves the capacity of water diversion or flood discharge in Langfang City, and improves people's water-loving comfort through most human gathering places.

By comparing the evaluation results of the optimized river network connectivity layout, the current river network layout (some river sections are silted and blocked) and all water system layout in Langfang City, it can be seen that there are differences in the three conditions in the three criterion layers of water system pattern, water system connectivity and hydrophilic capacity. It should be pointed out that the connectivity layout of regional river network is not the best when all water systems are connected, and the connectivity layout of regional river network is not the best when a single index of a certain criterion layer is not the best, and the structural connectivity and

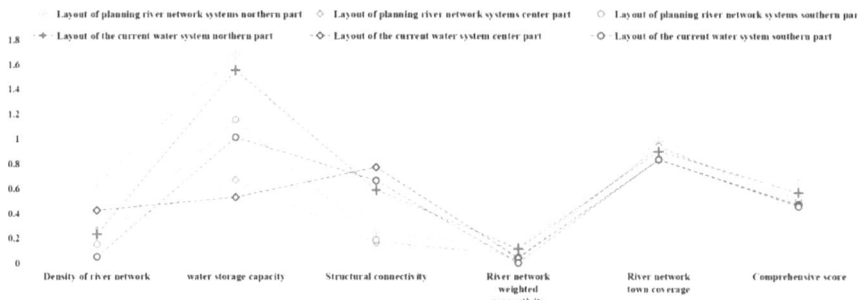

Fig. 7 Evaluation results of indicators for optimizing river network connectivity layout in Langfang City

river network density of the optimal water system connectivity scheme obtained in this study are also not the best. This also shows that it is not possible to rely only on the size of a single index to determine the connectivity of water systems in each scheme, and it is necessary to select relevant indicators from multiple criterion layers to construct the overall layout evaluation system of river network connectivity. By comparing the Comprehensive Water Conservancy Plan of Langfang City, the planning and design of the surrounding water system of the counties of Langfang City, the 13th Five-Year Plan of Langfang City and the Comprehensive Ecological Planning of Langfang City, the optimized overall layout plan of the river network connectivity of Langfang City covers the connecting routes of the transport corridor and the diversion corridor involved in the planning and design of Langfang City, and the connecting route of the surrounding water system of the counties of Langfang City.

5 Conclusions

This paper focuses on the optimization of river network connectivity layout in plain river network, and establishes the evaluation process of optimizing river network connectivity layout, which includes "Determining the importance of nodes and canals in river network—optimizing the connectivity layout scheme of river network—comprehensive score of the overall connectivity layout of river network".

(1) Taking major rivers as the framework, ditches, ponds and dams in the river network as the connecting objects, considering the needs of flood control and disaster reduction, taking the maximum overflow capacity of sluices and dams in the river reach as the weight, and determining the importance of nodes and canals in the river network through the obtained weighted adjacency matrix of the shortest path of the river network, so as to determine the optimal water system connection scheme. However, it is necessary to consider the control ability of dam and gate in future research.

(2) Aiming at the optimization of water system connectivity layout of plain river network, based on the water system connectivity evaluation system, a comprehensive evaluation method was adopted to evaluate different water system connectivity schemes from three aspects: water system pattern, water system connectivity and hydraulic connectivity. The final evaluation scores of water system connectivity layout schemes under different scenarios were obtained, and the optimization schemes were evaluated. It provides reasonable planning suggestions for the optimization and implementation of the plain river network connectivity project.

(3) In this paper, by optimizing the evaluation process of river network connectivity layout, it is determined that the comprehensive scores of each region in Langfang City after optimizing the overall layout of river network connectivity increase by 13.0%, 3.0% and 12.4%. However, due to the limited water quality data in this paper, the water quality of the area before and after connectivity has not been simulated. Therefore, in future studies, water quality analysis indexes can be further added to comprehensively evaluate the overall layout optimization scheme of river network connectivity.

References

1. Z. Li, C. Liu, X. Hao et al., Theoretical basis and priority areas of the interconnected river system network research. Acta Geogr. Sin. **76**(03), 513–524 (2021)
2. C. Liu, Z. Li, Z. Wang et al., Key scientific issues and research directions of the interconnected river system network. Acta Geogr. Sin. **76**(03), 505–512 (2021)
3. C. Huang, Y. Chen, Z. Li et al., Optimization of water system pattern and connectivity in the Dongting Lake area. Adv. Water Sci. **30**(05), 661–672 (2019)
4. Z. Dong, J. Zhao, J. Zhang, Three types flows via four dimensional connectivity ecological model. Water Resour. Hydropower Eng. **50**(6), 134–141 (2019)
5. M. Xia, Z. Zhou, H. Zhao, Evaluation of water system connectivity of the district around Chaohu Lake based on comprehensive indexes. Geogr. Geo-Inf. Sci. **33**(01), 73–77 (2017)
6. W.S. Glock, The development of drainage systems: a synoptic view. Geogr. Rev. **21**(3), 475–482 (1931)
7. A. Strahler, Hypsometric (area-altitude) analysis of ersional topography. Geol. Soc. Am. Bull. **63**(11), 1117–1142 (1952)
8. M. Veltri, P. Veltri, M. Maiolo, On the fractal description of natural channel networks. J. Hydrol. **187**, 137–144 (1996)
9. R.W. Phillips, C. Spence, J.W. Pomeroy, Connectivity and runoff dynamics in heterogeneous basins. Hydrol. Process. **25**(19), 3061–3075 (2011)
10. V. Jain, S.K. Tandon, Conceptual assessment of (dis)connectivity and its application to the Ganga River dispersal system. Geomorphology **118**(3–4), 349–358 (2010)
11. G. Xu, Y. Xu, L. Wang. Temporal and spatial changes of river systems in Hangzhou Jiaxing Huzhou Plain during 1960s–2000s. Acta Geogr. Sin. **68**(7), 96–104 (2013)
12. Z. Lin, Y. Xun, X. Dai et al., Effect of urbanization on the plain river network structure in the lower reaches of the Yangtze River. Resour. Environ. Yangtze Basin **11**, 2612–2620 (2019)
13. Y. Gao, Y. Liu, G. Yan et al., Analysis on variation of water system structure and connectivity of Qinhuai River Basin. Adv. Sci. Technol. Water Resour. **40**(5), 32–39 (2020)

14. Q. Zuo, G. Cui, Quantitative evaluation of human activities affecting an interconnected river system network. Acta Geogr. Sin. **75**(7), 1483–1493 (2020)
15. R. Hillips, C. Spence, J. Pomeroy, Connectivity and runoff dynamics in heterogeneous basins. Hydrol. Process. **25**(19), 3061–3075 (2011)
16. J. Liang Xiao, S.B. Wenhui et al., Water system connectivity evaluation in plain urban river network based on AHP-entropy weight method: A case study of Langfang City. South-to-North Water Transfers Water Sci. Technol. **2**, 352–364 (2022)
17. X. Gao, Z. Hu, C. Yan et al., Construction and application of water system connectivity evaluation index system considering hydraulic connectivity. Water Resour. Prot. **38**(02), 41–47 (2022)
18. S. Si, X. Sun, *Complex Network Algorithms and Applications* (National Defense Industry Press, Beijing, 2015)
19. C. Fu, C. Chen, Longbing Li et al., Research on the connotation and evaluation index system of interconnected river system network. Hydroelect. Power **42**(07), 2–7 (2016)

Research on the Whole Life Cycle Management of Water Conservancy Project Based on K-means Algorithm

Jiabao Zhu

Abstract Research on full life cycle management in water conservancy engineering is currently limited, resulting in issues like improper project siting, high construction costs, and insufficient risk management capabilities. To address these challenges, this paper proposes an innovative management approach using the K-means clustering algorithm. This method leverages the K-means algorithm's simplicity and fast convergence to cluster and analyze diverse data from different stages of water conservancy projects. It then presents these clustering results visually to assist project managers in making informed decisions. Using the K-means algorithm in the planning and design stage of water conservancy projects as an example, we demonstrate its feasibility, efficiency, and scientific rigor in managing the entire project life cycle. The research findings indicate the following: The research results show that (1) In the planning and design stage, the K-means algorithm aids in making informed decisions about engineering construction sites by clustering and analyzing relevant data. (2) The improved K-means clustering algorithm exhibits high processing efficiency and accuracy when handling various types of water conservancy project data. (3) The K-means algorithm's application can span the construction, operation, and maintenance stages. It can cluster and analyze data related to material consumption, equipment usage, construction efficiency, and quality, ensuring compliance with quality standards, cost control, and project optimization. Additionally, it can be used to cluster and analyze data akin to water conservancy projects, like historical risk factors, aiding in risk evaluation and prediction.

Keywords Water conservancy project · Total life cycle management · K-means · Scientific decision-making

J. Zhu (✉)
School of Resources, Shandong University of Science and Technology, Shandong Province, Tai'an City 271000, China
e-mail: 2836126205@qq.com

© The Author(s) 2025
W. Wang et al. (eds.), *Hydraulic Structure and Hydrodynamics*, Lecture Notes in Civil Engineering 608, https://doi.org/10.1007/978-981-97-7251-3_35

1 Introduction

Since the 18th National Congress of the Communist Party of China (CPC), the government has emphasized the significance of water control and issued crucial directives for the scientific management of the entire life cycle of water conservancy projects. However, a comprehensive review of water conservancy literature reveals a scarcity of studies addressing the importance of scientific management in practice, particularly in the context of full life cycle management for water conservancy projects. This has led to a series of problems, including inappropriate site selection, high construction costs, and insufficient levels of risk management and dataization [1–3].

Aiming at the current situation of the whole life cycle management of water conservancy projects, some scholars have carried out in-depth research. For example, Zhao Nan put forward the concept of "BIM+", which introduces the Internet and big data into project management and promotes the digitalization and systematization of water conservancy project management [4]. Sun Shaonan improved the informatization level of the water conservancy project by expressing information about each stage of the project in a coordinated way through BIM+GIS technology [5]. Li Sha, on the other hand, innovatively combined the concept of a full life cycle with EPC mode and achieved benefits superior to traditional management methods [6].

Although Previous research in water conservancy project management has limitations, such as the complexity and cost of handling multi-source data in BIM, technical challenges and high costs in "BIM+GIS" with poor compatibility and efficiency, and potential risk and change management issues when combining the whole life cycle concept with the EPC model. This paper introduces a novel full life cycle management approach for water conservancy projects using the K-means clustering algorithm. The method offers several advantages, including simplicity, low technical requirements, fast data processing, decision support, optimized resource allocation, enhanced construction efficiency, and capabilities for risk assessment and early warning.

The paper details how the K-means algorithm is applied in the planning and design phases and extended to the construction, operation, and maintenance phases, aiding in identifying data relationships and supporting informed decision-making by project managers.

2 Methods

This paper reviews the current state and challenges of whole life cycle management in water conservancy engineering based on literature research. The original concept is to utilize the K-means clustering algorithm in the stages of design and planning, construction, and operation and maintenance of water conservancy projects. This approach aims to address issues related to site selection, cost, data processing, and risk management, offering a novel and effective method for comprehensive life cycle management.

2.1 K-means Clustering Algorithm

The K-means clustering algorithm is a commonly used unsupervised learning algorithm for dividing a dataset into K non-overlapping clusters. The algorithm is based on the idea of a distance metric, usually using Euclidean distance, and assigns sample points to the nearest clusters by iterating and updating the center of mass of the clusters until the convergence condition is reached [7].

Assume that the given dataset $X = [x_1, x_2, x_3, \ldots, x_n]$, Divide the samples into clusters to give the centroids corresponding to each sample data, the steps of the algorithm are shown below:

(1) Choose K centers of mass at random such that:

$$K = [\mu_1^{(0)}, \mu_2^{(0)}, \ldots, \mu_k^{(0)}] \tag{1}$$

(2) Define the loss function:

$$J(c, \mu) = \min \sum_{i=1}^{M} \|x_i - \mu_{ci}\|^2 \tag{2}$$

(3) Pin any x_i and match to the nearest center of mass:

$$c_i^t < -\arg \min_{\mu} \|x_i - \mu_k^t\|^2 \tag{3}$$

(4) For each category of center of mass μ_i, recalculate:

$$\mu_k^{(t+1)} < -argmin_k \sum_{i:c_i^t=k}^{b} \|x_i - \mu\|^2 \tag{4}$$

(5) Make the number of iterations $t = 0, 1, 2, \ldots, m$ and repeat steps 3 and 4 until the loss function J converges.

After reviewing the information, the K-means clustering algorithm exhibits various advantages and disadvantages. The following table compares the key strengths and weaknesses of the algorithm and suggests corresponding enhancement strategies for its shortcomings.

2.2 Application of K-means Algorithm in Total Life Cycle Management

Table 1 shows that K-means clustering algorithm has the following advantages: simple principle, fast convergence, and applicability to large-scale data. Although there are some disadvantages, such as subjectivity of center of mass selection, easy to fall into local optimal solutions, and sensitivity to outliers, they can be overcome by elbow method, random initialization and data preprocessing.

The whole life cycle management of water conservancy projects includes the comprehensive management of planning, design, construction, operation, maintenance and disposal [8]. The process involves planning, implementation, and monitoring at all stages and emphasizes continuous improvement and sustainable development. Under the concept of total lifecycle management, water conservancy projects become complex, long-term and integrated, thus generating a large amount of data. The K-means clustering algorithm can efficiently process various water conservancy project data, analyze the relationship between the data, and provide data support for total lifecycle management. The following Fig. 1 briefly shows how to use K-means clustering algorithm to manage water conservancy projects under the concept of whole life cycle management.

In summary, the K-means clustering algorithm in the water conservancy project life cycle management has a great advantage of data processing and results analysis ability, can effectively solve the problems in the water conservancy project life cycle management.

3   Data Sources and Analysis of Results

To fully demonstrate the feasibility and advantages of the K-means clustering algorithm in the entire life cycle management of water conservancy projects, this paper specifically focuses on the scientific site selection during the design and planning phase of water conservancy projects in Sichuan Province [9]. Sichuan, with its intricate topography and abundant water resources, serves as the research subject. When embarking on water conservancy projects, the initial and foremost consideration is annual surface water storage, and the region's rainfall volume directly influences the amount of surface water storage, which, in turn, indirectly affects the profitability

Table 1 Characteristics of K-means clustering algorithm

Vintage	Drawbacks	Improved methodology
Simple principle	Determine the value of K	The law of the Elbow
Fast convergence	Local optimum solution	Multiple random initializations
Suitable for large-scale datasets	Sensitive to outliers, noise	Data preprocessing

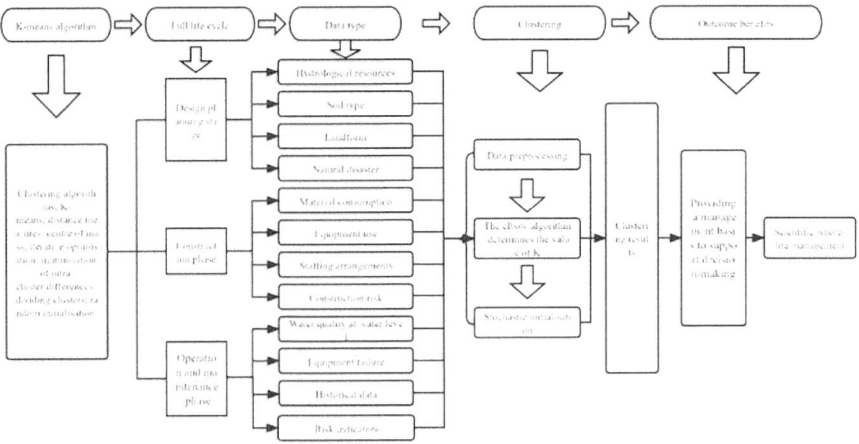

Fig. 1 The realization path of the whole life cycle of water conservancy project under the concept of K-means clustering algorithm

of constructing water conservancy facilities in the area [10]. Consequently, the data chosen, from a whole life cycle perspective, primarily encompasses rainfall and groundwater storage during the design and planning stage. To ensure the data's scientific rigor, accuracy, and official sources, the author meticulously collected and organized data from two reliable sources: the Global Statistical Data Analysis Platform (EPS data platform) and the official website of the Sichuan Provincial Water Resources Department.

3.1 Data Processing and Analysis

Rainfall data for major cities in Sichuan Province from 2006 to 2020, along with surface water and groundwater reserves for previous years, were gathered from the EPS data platform. Excel software was employed to calculate the maximum and minimum values, mean, median, and variance of the collected data. The analysis revealed that Ya'an City exhibited the highest mean annual precipitation, maximum annual precipitation, and precipitation variance. This indicates that Ya'an City receives substantial annual precipitation, but it experiences significant fluctuations and variations. On the contrary, Panzhihua has the lowest precipitation among these cities, with minimal annual fluctuations and a relatively stable pattern, despite the lowest annual precipitation being 537.70 mm.

Furthermore, concerning groundwater storage data for major cities in Sichuan Province (2020–2022), Chengdu City registered the highest average groundwater storage at 17.509 billion m³, while Neijiang City had the lowest average groundwater storage at 9.167 billion m³. It's important to note that Chengdu City displayed substantial interannual fluctuations in groundwater storage, whereas Suining City exhibited the smallest fluctuation, reflecting a more stable groundwater level. The study's findings were processed in line with the operational steps of the K-means clustering algorithm as outlined in the research methodology for the collected rainfall and surface water storage data.

3.2 Analysis of Results

Rainfall data clustering results. Introducing the rainfall dataset of each city from 2006 to 2020, the idea of elbow rule is introduced for the shortcomings of K-means clustering algorithm's subjectivity in the selection of K-value, and the following Fig. 2, Fig. 2a, is the folding diagram of determining the K-value of the number of rainfall clusters with elbow rule:

Fig. 2 Elbow fold plot for determining the value of K

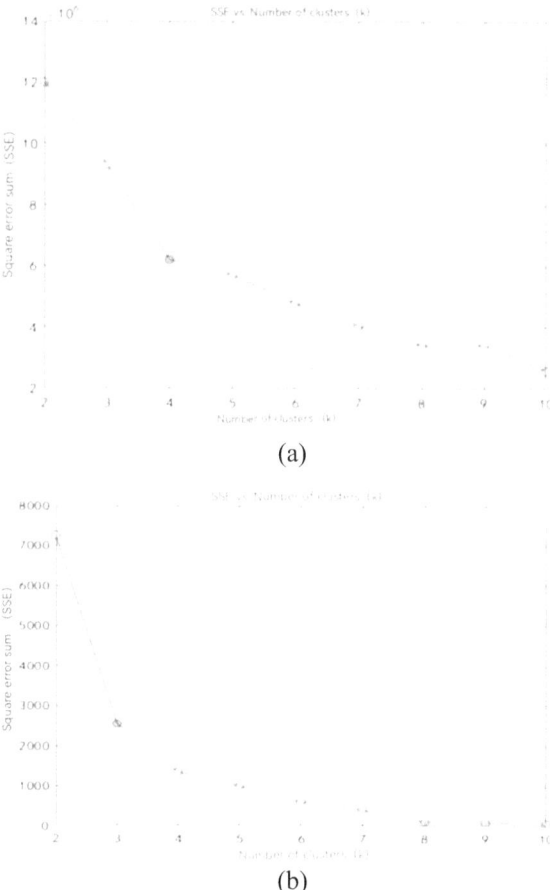

(a)

(b)

Combined with the results of the operation of the elbow rule, this paper makes $K = 4$ and $K = 3$ to analyze the clustering of rainfall and surface water storage in each city in the past years, and transforms the algorithmic solving steps above into MATLAB code for output solving, and the final output of the K-means clustering results are shown in Tables 2 and 3:

Table 2 Rainfall clustering results

Form	Municipalities	Average (mm)	Centre-of-mass distance
Analogy 1	Luzhou	1157.66	579.042
	Leshan	1247.45	435.442
	Zhibin	1046.83	907.368
Analogy 2	Yaan	1603.11	546.887
Analogy 3	Chengdu	955.93	650.573
	Zigong	939.94	649.975
	⋮	⋮	⋮
	Meishan	996.50	578.405
	Ziyang	898.89	507.617
Analogy 4	Nanchong	1058.91	582.703
	Guangan	1102.20	0
	Dazhou	1228.20	752.776
	Bazhong	1157.74	498.923

Table 3 Surface water clustering results

Form	Municipalities	Average surface water	Centre-of-mass distance
Analogy 1	Chengdu	203.51	0
Analogy 2	Zigong	11.65	22.90
	Nanchong	6.21	5.19
	Meishan	11.35	4.35
	Zhibin	18.61	17.26
	Guangan	4.39	8.41
	⋮	⋮	⋮
	Ziyang	2.07	12.34
	Dazhou	15.93	11.98
	Deyang	13.49	7.91
Analogy 3	Mianyang	28.77	3.86
	Leshan	25.40	9.72
	Yaan	38.73	13.57

In rainfall category 3, in addition to those listed in the table above, Panzhihua City, Deyang City, Mianyang City, Guangyuan City, Suining City, Neijiang City are also included. Surface water category 2 also includes Panzhihua City, Luzhou City, Suining City, Guangyuan City, Neijiang City, and Bazhong City.

Analysis of clustering results. The application of the K-means clustering algorithm has led to the categorization of cities into four distinct groups based on their rainfall patterns. These categories exhibit significant differences in mean annual rainfall, while cities within the same category display relatively consistent rainfall patterns. Based on this clustering outcome, specific interpretations can be assigned to each category. For example, Category 2 is characterized as the high rainfall layer, Category 4 as the very high rainfall layer, Category 1 as the medium rainfall layer, and Category 3 as the low rainfall layer. This categorization facilitates efficient site selection for hydraulic projects, especially when the desired reservoir capacity is known, or it provides essential data support for matching reservoir design capacity when the construction area is specified.

Likewise, the clustering results obtained from surface water storage data enable the division of cities into three layers. For instance, Category 1 represents the water resource-rich layer, Category 3 signifies the water resource-medium layer, and Category 2 indicates the water resource-scarce layer. This categorization empowers managers to make informed decisions regarding the construction location or reservoir capacity of water projects based on surface water storage, enhancing the efficiency of site selection and planning.

Having successfully applied the K-means clustering algorithm in the planning and design phase, particularly for rainfall data, we are now extending this method to various stages, including construction operation, and maintenance. In the construction phase, K-means can play a pivotal role in cost control by clustering data related to material consumption, equipment usage, and the allocation of construction personnel. Furthermore, it can analyze project cost, schedule, and risk data, enabling managers to develop scientifically sound construction plans that meet requirements for shorter durations, lower costs, and higher quality. During the operation and maintenance phase, the K-means clustering algorithm can cluster and analyze data concerning water level, water quality, and flow rate, assisting managers in evaluating the performance of water conservancy projects. Additionally, by conducting a clustering analysis of historical risk factors, it can forecast potential issues, providing a basis for contingency plans to prevent operation and maintenance problems. Simultaneously, the algorithm can cluster malfunctioning equipment, enhancing equipment operation efficiency and safety while enabling predictive maintenance.

4 Conclusion and Outlook

Using Sichuan Province as a case study, this research applies the K-means clustering algorithm to categorize the primary factors relevant to the planning and design phase of water conservancy projects, specifically rainfall and surface water storage. The study yields the following conclusions:

(1) During the planning and design stage, the K-means clustering algorithm classifies eighteen cities into four distinct rainfall categories, each with significantly varying annual precipitation levels. This enables engineering managers to swiftly determine a city's rainfall category, preliminary reservoir design capacity, and site selection scope. Similarly, the categorization of cities into three groups based on surface water storage facilitates the more precise determination of reservoir design capacity and site selection range, aiding managers in promptly identifying suitable construction locations. Therefore, the K-means clustering algorithm proves useful for tasks like site selection and engineering program design in water conservancy projects when sufficient data is available. (2) In terms of algorithms, this study employed data processing techniques to remove outliers when applying the K-means clustering algorithm. It used the elbow rule to determine K-values for rainfall and surface water storage. Additionally, random initialization was implemented to prevent the algorithm from converging to local optima. Based on the balance of data processing speed and accuracy, it is evident that the K-means clustering algorithm, after addressing its own limitations, offers excellent data processing capabilities, rapid convergence, and precise processing results. (3) The experience gained from applying the K-means algorithm in the planning and design phase can be extended to other phases, including construction and operation and maintenance. In the construction phase, the algorithm can be employed to process data such as material consumption, equipment usage, and personnel allocation, aiding managers in progress control. During the operation and maintenance phase, it can be utilized for predictive maintenance plans by clustering water level, water quality, and historical risk factors. This proactive approach helps reduce project attrition rates and enhances overall efficiency.

This study introduces a novel concept of comprehensive life cycle management for water conservancy projects, specifically emphasizing full life cycle management facilitated by the K-means clustering algorithm. The K-means clustering algorithm, known for its simplicity, robust data processing capabilities, and fast convergence, holds significant potential to advance scientific management and contribute to the high-quality development of water conservancy projects in China.

However, this study has two noteworthy limitations: first, it primarily demonstrates the feasibility of the K-means clustering algorithm in the planning and design stage, with less in-depth research on other project phases. Second, while the K-means clustering algorithm exhibits superior data processing capabilities compared to the BIM+GIS system, it may require additional time when handling extensive datasets. While the K-means clustering algorithm demonstrates great promise in the overall life cycle management of water conservancy projects, further research is essential to address these limitations effectively.

References

1. X. Chunyue, J. Wei, Problems and improvement measures in water conservancy project design. Urban Constr. Theory Res. (Electron. Ed.) **31**, 177 (2018). https://doi.org/10.19569/j.cnki.cn1 19313/tu.201831150
2. S. Jianzi, C. Zan, X. Min, et al., Solution to construction problems of water conservancy project. Water Conservancy Technol. Supervision (04), 105–106+189 (2019)
3. C. Shiyong, Exploration of risk management problems of water conservancy project. Bus. News **22**, 146–147 (2019)
4. Z. Nan, L. Wanqu, C. Yanping, Research on the whole life cycle management of water conservancy project based on BIM technology. Sichuan Water Resour. **43**(04), 116–121 (2022)
5. S. Shaonan, S. Yichang, Research on the whole life cycle construction management of water conservancy project based on BIM+GIS. China Rural Water Conservancy Hydropower, (10), 131–137+142 (2022)
6. L. Sha, Z. Yu, Application of full life cycle management concept in project cost management under EPC mode of water conservancy project. Small Hydropower (01), 6–7+23 (2022)
7. W. Sen, L. Chen, X. Shuaijie, A review of research on K-means clustering algorithm. J. East China Jiaotong Univ. **39**(05), 119–126 (2022). https://doi.org/10.16749/j.cnki.jecjtu.202209 14.001
8. X. Liu, T.Y. Zhang, Y. Sun, Analysis of full life cycle management of water conservancy projects in Shanghai. Pearl River Water Transp. (05), 53–55 (2023). https://doi.org/10.14125/ j.cnki.zjsy.2023.05.001
9. H. Pei, Spatial and temporal distribution of surface water resources in Sichuan Province. Chengdu Univ. Technol. (2011)
10. Y. Yu, Z. Wen, J. Shang et al, Identification of strong runoff zone of karst water in Mingshui spring area based on the correlation between spring flow and precipitation. Water Resour. Conserv. **37**(03), 56–60 (2021)

Analysis of the Construction of Flood and Drought Disaster Prevention System in the Yellow River Delta

Miaomiao Li and Qiang Zhang

Abstract By considering factors such as topography, meteorology, and hydrology, this paper examines the causes of flood and drought disasters in Binzhou City, located in Shandong Province, which serves as the central city of the Yellow River Delta, and summarizes recent flood and drought events. The analysis highlights weaknesses in the current flood and drought prevention system, particularly regarding systematic river management and flood control informatization. To address these issues effectively, this study proposes a set of measures to enhance Binzhou's flood and drought disaster prevention system, including prioritizing improvements in flood control and drainage standards, expediting key flood control projects' implementation, strengthening river management, investigating potential risks of water conservancy projects, improving sea embankment construction while developing an intelligent platform for water affairs. These countermeasures aim to bolster Binzhou's comprehensive disaster prevention and mitigation capacity while reinforcing its water security.

Keywords Binzhou City · Flood and drought disaster prevention · Water security · River management · Informatization

1 Introduction

Binzhou City is situated in the hinterland of the Yellow River Delta, located in the northern part of Shandong Province and on the southwestern bank of Bohai Bay. The city's territory spans across both sides of the Yellow River, positioned at $36°41'19''$–$38°16'14''$ north latitude and $117°15'27''$–$118°37'03''$ east longitude. The region comprises mountainous areas in the south, yellow floodplains in the central region, and coastal beaches in the north [1]. Binzhou falls within East Asia warm temperate subhumid continental monsoon climate zone with an average annual precipitation of 573.1 mm (1956–2016 series). Because of highly concentrated intense rainfall events, from June to September, these months contribute to about 75 0.6% percent of

M. Li · Q. Zhang (✉)
Binzhou Municipality's Water Resources Bureau, Binzhou 256600, Shandong, China
e-mail: 314632983@qq.com

© The Author(s) 2025

W. Wang et al. (eds.), *Hydraulic Structure and Hydrodynamics*, Lecture Notes in Civil Engineering 608, https://doi.org/10.1007/978-981-97-7251-3_36

417

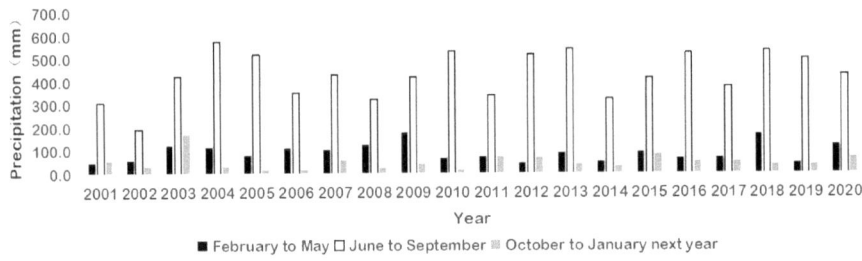

Fig. 1 Precipitation distribution from 2001 to 2020 in Binzhou City

the annual precipitation (as shown in the Fig. 1). Factors such as geography, geology, hydrology and extreme weather conditions, play a role in the high frequency and rapid transition of flood and drought disasters within Binzhou City.

In recent years, the flood and drought disaster prevention system in Binzhou has faced numerous challenges (as shown in the Fig. 2). In August 2010, it experienced the most severe flood disaster with a direct economic loss of 4,024 billion yuan. The average rainfall in August reached 340 mm. In August 2012, Binzhou was struck by Typhoon "Dawei", which marked the strongest typhoon causing an average rainfall of 108 mm during this period, affecting approximately 1.4 million people. From January to June 2019, the city witnessed an unprecedented drought occurrence with an average rainfall of 61 mm. According with the research that water-related disasters will negatively affect agricultural areas and crop production and threaten food security [2], the drought leaded to a reduction of 243,400 tons in grain production for the city. In August of the same year, Typhoon "Lekima" struck the city continuously for three days, leading to an average precipitation of 268 mm, which is equivalent to a volume of 2.5 billion cubic meters of water. The direct economic losses incurred from this calamity amounted to a total of 118.19 billion yuan [3]. The current period is of utmost importance for the implementation of the national strategy of ecological protection and high-quality development in the Yellow River Basin. It is imperative to prioritize flood and drought disaster prevention.

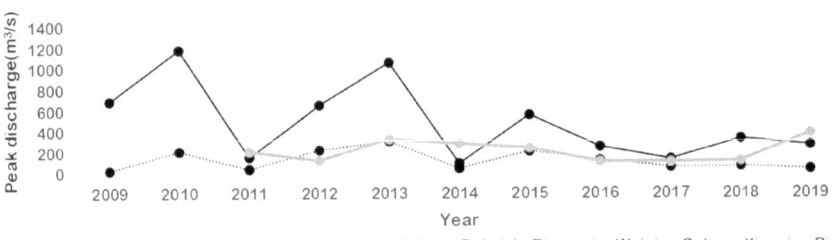

Fig. 2 Peak flow in Binzhou City

2   The Challenges in Flood and Drought Disaster Mitigation

2.1   The River Management System Lacks Robustness

The river system of Binzhou city is well-developed, as shown in the Table 1, belonging to the Huaihe River basin, Yellow River basin and Haihe River basin from south to north. The Huaihe River basin comprises Xiaoqing River, Xinghua River, Shengli River, Xiaofu River, Zhimai River, Beizhixin River, Sanhaozhi River and Yubei River among others. Meanwhile, the Haihe River basin has four transiting channels including Tuhai River, Majia River, Dehuixin River and Zhangweixin River with two main drainage channels in the city namely Qinkou and Chao rivers. Among them are Zhangweixin river, Tuhai River, Chao River which originate in Henan Province flowing through many cities in Shandong before entering into sea in Binzhou City. As important water-logging channels in Shandong Province, these rivers gather flood of a dozen cities and counties in upper reaches to Binzhou during flood season every year. The city often suffers from triple pressure of upstream flood, local waterlogging and tidal backwater [4]. At present, the backbone rivers have been basically treated. However, the small and medium size rivers still face problems such as low treatment rate, low flood control standards, asynchronous upstream–downstream management, and insufficient maintenance of embankment buildings, resulting in inadequate performance of their capacity for flooding prevention, drought relief, and becoming weak links in the disaster prevention system.

Table 1 Length and catchment area of the main rivers in Binzhou City

Rivers	Total length (km)	Length within the city (km)	Catchment area (km^2)
Yellow River	5464	94	0
Tuhai River	436.35	177.5	2129
Majia River	480	57.6	254
Dehuixin River	172.6	73.3	135.4
Xiaoqing River	233	74.9	1468.9
Chao River	75.5	75.5	1241.3
Qinkou River	124.8	124.8	2500.72
Xisha River	41	41	171
Xinli River	31	31	136

2.2 The Level of Support for Flood Control Information is not Sufficiently High

The traditional flood and drought disaster prevention system based on engineering projects has historically struggled to meet requirements, prompting the Ministry of Water Resources to propose the implementation of a smart water conservancy construction [5]. In this context, Binzhou urgently needs to prioritize the city's specific needs and leverage advanced technologies such as Internet of Things (IoT), video collection, artificial intelligence (AI), and big data. This will enable the development of an intelligent flood and drought disaster defense system equipped with functions like intelligent perception, watershed big data analysis, intelligent simulation modeling, and intelligent decision-making [6]. Currently, the digital and intelligent level of water affairs in Binzhou is relatively low, and the construction of intelligent defense system is still in its early stages. The development of flood and drought disaster monitoring, warning, and command platforms at both city and county levels needs improvement, particularly in terms of urban water-logging monitoring as well as urban flood prevention and emergency event management. There has been multiple investments from city, county, township, and functional departments resulting in independent constructions with single functions. Additionally, the basic data is incomplete with a low degree of automatic intelligent collection. The issues of information decentralization, fragmentation, and isolation are prominent, and the integration and analysis of historical data are low, which cannot provide effective support for later flood scheduling and flood/drought disaster prevention.

3   Countermeasures and Suggestions

The construction of flood and drought disaster prevention system in Binzhou City should fully implement the national security strategy, focusing on the principle of risk prevention and security guarantee. It is crucial to establish a solid bottom line thinking, enhance risk awareness, strengthen research and judgment on water security risks, improve coordination for prevention and control, establish mechanisms for prevention and resolution, and enhance capacity building. The aim is to prevent and minimize the damage caused by flood and drought disasters. We will strive to establish a "safe water conservancy" system that meets safety standards while effectively controlling flood damage through the construction of robust disaster prevention barriers.

Table 2 Flood control and drainage standards(m^3/s/km^2)

Drainage area (km^2)	30	40	50	70	100	500	1000	3000
64-year rain type	0.746	0.664	0.607	0.530	0.460	0.287	0.225	0.157
61-year rain type		1.213	1.108	0.969	0.840	0.510	0.425	3.310

3.1 Upgrade the Flood Control and Disaster Reduction Systems Through Enhancing the Standards for Flood Control and Drainage

Consolidate and enhance the flood control capacity of river basins, regions, and cities to achieve the same frequency resonance of the flood control system across these areas. The flood control and drainage standards for the backbone river should basically meet the "64-year rain type" and "61-year rain type" requirements specified by Shandong Province and Shandong Peninsula Basin Flood Control Plan (as shown in the Table 2). We are actively improving urban infrastructure and key water-logging areas to effectively prevent and mitigate floods while strengthening our ability to manage flood and drought risks. With all hazardous reservoirs are already eliminated, we aim for a compliance rate of over 85% for grade 5 or higher river and lake embankments during the "14th Five-Year Plan" period.

3.2 Eliminate Safety Hazards in Engineering Through Strongly Promoting Risk Elimination and Reinforcement

The investigation of hidden danger and safety appraisal engineering facilities such as reservoirs and sluices, as well as the reinforcement or scrapping of dangerous reservoirs and sluices, should be carried out dynamically. Emphasis should be placed on source prevention to eliminate potential risks in engineering safety. We should flexibly use methods such as county-district mutual inspection and surprise inspections to conduct comprehensive inspections of reservoirs, sluices etc., and establish a record of identified issues related to engineering operation and management. Then, the responsible person, rectification time limit, and measures must be implemented to ensure that the rectification is in place. Due to technical limitations and funding constraints, priority should be given to promoting the reinforcement projects for large-, medium- and small-sized dangerous reservoirs and sluices, as listed in Table 3.

Table 3 Reinforcement projects for dangerous reservoirs/sluices in Binzhou City

Management year	Scale type	Dangerous reservoirs/sluice	County district
2019	Large	Fanqiao Sluice, Nianlibao Sluice	Huimin, Bincheng
2019	Medium	Daxiao Sluice, Wangwen Sluice, Xingfu Sluice	Boxing, Huimin
2020	Medium	Dayuzhang Qushou Reservoir, Qintai Reservoir	Boxing, Bincheng
2020	Small	Shihe Sluice, Xihe Sluice, Juqun Sluice	Zouping
2021	Medium	Huangjiajing Sluice, Houzhou Sluice	Yangxin
2021	Small	Shanghe Sluice, Yazhuang Sluice	Zouping
2022	Large	Huangguangling Rubber Dam	Wudi
2022	Small	Beiyuan Sluice, Taitou Sluice, Yuyin Sluice	Zouping

3.3 Crack the Weak Links in Disaster Prevention and Mitigation Through Consolidating the Backbone River Channels and Small and Medium-Sized Rivers

Firstly, we need to focus on completing the comprehensive management project for flood control in Xiaoqing River and its important tributaries. Table 4 gives detail information of the future construction plan. Secondly, an overall plan for rivers management needs to be prepared at a municipal level with emphasis on key sections and regions requiring unified management while appropriately improving governance standards. Thirdly, incorporating small and medium-sized rivers into the integrated urban–rural water information platform will enable digitalized networked intelligent management. Fourthly, it's essential to establish joint dispatching mechanism between the upstream and downstream [7], especially for the passing rivers through multiple cities.

Table 4 Rivers planned to be treated before 2035 in Binzhou City

Basin	Rivers	Management year	Construction demand type
Haihe River Basin	Sha River	2023–2035	Improve standard construction
Haihe River Basin	Fujia River	2023–2035	Improve standard construction
Haihe River Basin	Maxin River	2023–2025	Meet standard construction
Huaihe River Basin	Yubei River	2026–2035	Information construction
Huaihe River Basin	Xingfu River	2026–2035	Meet standard construction
Huaihe River Basin	Dayuzhang River	2026–2035	Meet standard construction
Huaihe River Basin	Sanhaozhi River	2023–2035	Improve standard construction

3.4 Strengthen the Existing Disaster Management System Through Enhancing Disaster Prevention and Control Measures in Mountainous and Coastal Regions

Zouping City is situated in the southwestern part of Binzhou City, containing three distinct landforms: low mountain hills, gently sloping plains at the foot of mountains, and yellow pan plains [8]. To enhance the construction of flood prevention and control projects in Zouping's southern mountainous area, it is imperative to establish a centralized platform for monitoring and issuing. In contrast, Zhanhua District and Wudi County are located in the northern part of Binzhou City bordering Bohai Bay, and bear the responsibility of tide prevention. The repairment and upgrading of tidal dikes in Zhanhua and the construction work on Lubei embankment in Wudi must be proceeded.

3.5 Enhance the Level of Information-Based Support for Disaster Prevention Through Developing a Comprehensive Intelligent Platform for Water Affairs

During the "14th Five-Year Plan" period, it is imperative to establish and enhance the city's integrated monitoring and information network for water affairs, covering "four pre" parts, including prediction, pre-warning, preview and pre-arranged planning [9]. Additionally, constructing big data centers and intelligent application platforms that support crucial water applications is of great importance. Monitoring facilities for rainfall, water level, flow rate, water quality, and water conservancy project safety should be built. Furthermore, we will establish a Water Conservancy Internet of Things platform for real-time monitoring of river-related information such as riverside conditions or reservoirs' status.

4 Conclusion

In addition to enhancing engineering measures, material support, and information platform construction, it's necessary to strengthen information dissemination [10]. Personnel reorganization, increased interdepartmental cooperation, changes in working methodologies further impose higher demands on institutional development. To improve disaster prevention system continuously, Binzhou should persistently enhance standards while diligently enforcing responsibilities and clarifying tasks.

References

1. M.M. Li, M. Cao, H.M. Zhang et al., Discussion on soil and water conservation plan review system under the background of reforms to delegate power, streamline administration and optimize government services. China Water Resources **14**, 50–52 (2020)
2. E.P.A. Pratiwi, E.L. Ramadhani, F. Nurrochmad et al., The impacts of flood and drought on food security in Central Java. J. Civil Eng. Forum **2020**, 69 (2020)
3. The Compilation Committee of Binzhou Water Conservancy Annual: Water Conservancy Records of Binzhou City, pp. 111–119 (2021)
4. D. Li, L. Zhao, H.W. Li, Causes and countermeasures of farmland waterlogging. China Water Resources **07**, 41–42 (2014)
5. The Ministry of Water Resources of the People's Republic of China: Guiding Opinions on the Proactive Promotion of Smart Water Conservancy, pp. 1–9 (2022)
6. Xin, D.M.: Analysis on the intelligent construction of flood and drought disaster prevention in Benxi City of Liaoning Province. China Flood Drought Manage. **33**(8), 62–64 (2023)
7. J.J. Xu, Analysis on the construction of flood and drought disaster prevention engineering system in Qingdao. Shandong Water Resources **1**, 35–36 (2023)
8. J.G. Hu, Y.N. Cui, X.N. Li, Challenges and countermeasures of soil and water conservation construction in Zouping City. Shandong Water Resources **02**, 76–78 (2022)
9. X. An, Y. Yang, J.T. Wang et al., Development and application of "four pre" platform for Pearl River flood and drought disaster prevention. Water Resources Informatization **04**, 14–19 (2023)
10. Sotto, Jr., R.: Flood control measures in one municipality in Camarines Sur, Philippines: bases for community-based flood control interventions. J. Educ. Manage. Dev. Stud. **2**(2), 30–39 (2022)

Analysis of the Influence of River Boundary Water Level on the Drainage Capacity of Pipe Network—A Case Study of Jiangnan Area in Yiwu City

Penghui Li, Yantan Zhou, Yiwei Zhen, Qiannan Jin, and Xiangyong Meng

Abstract In order to ensure the safety of urban drainage and alleviate urban waterlogging, this study focuses on the impact of dynamic river water levels on pipeline drainage. The region south of Yiwu River in Yiwu City was selected as the study area. Utilizing the Mike modeling platform, simulations were conducted to assess pipeline drainage outlets under heavy rainfall conditions with return periods of 2-year, 3-year, and 5-year. These simulations considered connections to both the river's normal water level and various flood scenarios based on different design frequencies. The study also analyzed the influence of changes in river water levels on the drainage capacity of the pipeline network and their impact on urban waterlogging. This research provides a reference basis for understanding the causes of urban waterlogging in the study area and its mitigation.

Keywords Urban waterlogging · Drainage pipe network · River water level boundary · Mike model

1 Background of the Study

Climate change and urbanization have led to the increasing prominence of the intensity frequency and extent of urban storm flooding [1–4]. Recent flood disasters in cities such as Zhengzhou, China (July 2021), Hagen, Germany (July 2021), New York, USA (September 2021), Queensland, Australia (February 2022), and Seoul,

P. Li · Y. Zhou · Q. Jin · X. Meng
Zhejiang Institute of Hydraulics and Estuary (Zhejiang Institute of Marine Planning and Design), Zhejiang Province, Hangzhou 310020, China

Y. Zhen (✉)
China South-to-North Water Diversion Limited Corporation, Beijing 100036, China
e-mail: zhenyiwei@tongji.edu.cn

College of Civil Engineering, Tongji University, Shanghai 200092, China

© The Author(s) 2025
W. Wang et al. (eds.), *Hydraulic Structure and Hydrodynamics*, Lecture Notes in Civil Engineering 608, https://doi.org/10.1007/978-981-97-7251-3_37

South Korea (August 2022) have confirmed the severity of urban flood disasters [5–9]. Urban waterlogging has become an important problem facing the development process of urbanization in China and has received widespread attention.

Urban flood water is mainly discharged through the drainage pipe network. Ensuring urban drainage safety is contingent on two key factors. Firstly, the size and dimensions of the pipelines must meet design standards. Secondly, the effective connection between the urban river network and the drainage pipelines plays a crucial role in guaranteeing urban drainage safety [10]. Zhou Hong et al. pointed out that in China, the separate management of flood disasters and waterlogging disasters has artificially severed their close connection, resulting in a lack of effective connection between urban drainage and river drainage, further aggravating the degree of damage caused by waterlogging disasters [11].

2   Method

In the urban drainage and flood control planning pipeline scale design, pipeline drainage outlet often use constant water level (pipe free outflow) or set a fixed water level boundary, the above articulation relationship does not take into account the dynamic impact of river flooding on pipeline drainage, and the actual occurrence of heavy rainfall, the river water level remains high often affects the pipeline normal drainage, resulting in pipeline water flow jacking or even backing up, and then triggered the urban flooding. In view of the urban flooding problem, this paper selects the urban area south of Yiwu River in Yiwu City as the study area, and uses the Mike model to construct the hydrological and hydrodynamic one and two-dimensional coupling model of urban rainfall production—surface inundation—drainage network—water conveyance channels, etc., to simulate and analyze the impact of the dynamic water level of the river on the pipeline drainage capacity.

3   Model Building

3.1   Overview of the Study Area

The study area is the area south of Yiwu River in the planned central city of Yiwu, with an area of 52.3 km², and the land type classifications are mainly roads, green spaces, open spaces, buildings, and water surfaces, as shown in Table 1. The main rivers within the study area are Qingkou River, Nian River, Chitang River, and Shi River, in addition to water level stations on Qingkou River and Nian River. The study area is shown in Fig. 1.

Table 1 Table of classification of land types in the study area

Study area	Surface type (subsurface) area statistics (km²)					
	Roads	Green area	Clearing	Constructions	Water surface	Total
Area	5.3	25.9	2.8	14.5	3.8	52.3
Proportions (%)	10.1	49.5	5.4	27.7	7.3	100

Fig. 1 Map of the study area south of the Yiwu River in Yiwu City

3.2 Model Building

Upstream of the urban area and the surrounding mountainous areas of long calendar time (24 h) design rain type and rainfall is determined in accordance with the "short calendar time rainstorm in Zhejiang Province", and the design flood is inferred by the inference formula method of Zhejiang Province. The maximum 1 h rainfall in urban area in short calendar time is calculated by Yiwu City urban rainstorm formula and assigned by Chicago rain type; the rainfall in any calendar time in this rain type is equal to the design rainfall, if the rainstorm formula is: $a = \frac{S_p}{(t+b)^n}$, then the rain intensity process is:

$$\text{Pre-peak}: i = \frac{S_p}{\left(\frac{t}{r}+b\right)^n}\left[1 - \frac{nt_1}{t_1 + rb}\right]$$

$$\text{Post-peak}: i = \frac{S_P}{\left(\frac{t}{1-r}+b\right)^n}\left[1 - \frac{nt_2}{t_2 + (1-r)b}\right]$$

where a is the average rainfall intensity in the calendar time t; i is the instantaneous rainfall intensity; t_1 is the rainfall calendar time before the peak; t_2 is the rainfall calendar time after the peak; r is the relative position of the rainfall peak; and S_p, b, and n are the parameters of the rainstorm formula.

Taking the main stream of Yiwu River as the boundary, we constructed a hydrological and hydrodynamic simulation model with the coupling of "rainfall production—surface inundation—drainage network—flood channel" for the whole process of flood control in the whole area. The modeled catchment area includes the urban area and the mountainous areas upstream and around the urban area. MIKE Urban and MIKE 11 are used to construct a one-dimensional model of urban flood control and drainage, including urban drainage pipes, inland rivers, reservoirs, mountain ponds, drainage and flood control gates and pumps, etc., and a two-dimensional hydrodynamic model of the urban surface is constructed by MIKE 21 to simulate the evolution of surface inundation in the study area, and MIKE FLOOD is used to simulate and calculate the coupling of the one-dimensional model with the two-dimensional model.

The model generalizes a total of 8 river channels with a total length of 18,508 m, 135 sections, 5130 check well nodes, 5124 drainage pipes with a total pipe length of 120.1 km, and 114 drainage outlets. The river is coupled with the 2D terrain in the form of lateral connection, the pipe check well nodes are coupled with the 2D terrain, and the drainage outlets are coupled with the river. The model construction process is shown in Fig. 2, the one-dimensional river network generalization, stormwater pipe network and two-dimensional terrain generalization are shown in Figs. 3, 4 and 5, respectively, and the model coupling generalization is shown in Fig. 6.

3.3   Model Validation

The urban area of Yiwu City experienced a heavy rainfall on May 12, 2021, with a 24 h surface rainfall return period of about 1 year; the maximum 1 h surface rainfall return period of about 3 years, resulting in flooding in many places in the urban area. This rainstorm, the river water level station and the flooding point measured data on the model built in this paper to verify. The simulated water level at the river water level station is compared with the measured water level as shown in Figs. 7 and 8, and the actual investigation of flood-prone points is compared with the simulated inundation map as shown in Fig. 9. The results show that the simulation results match well with the measured river water level process, waterlogged points and waterlogged range.

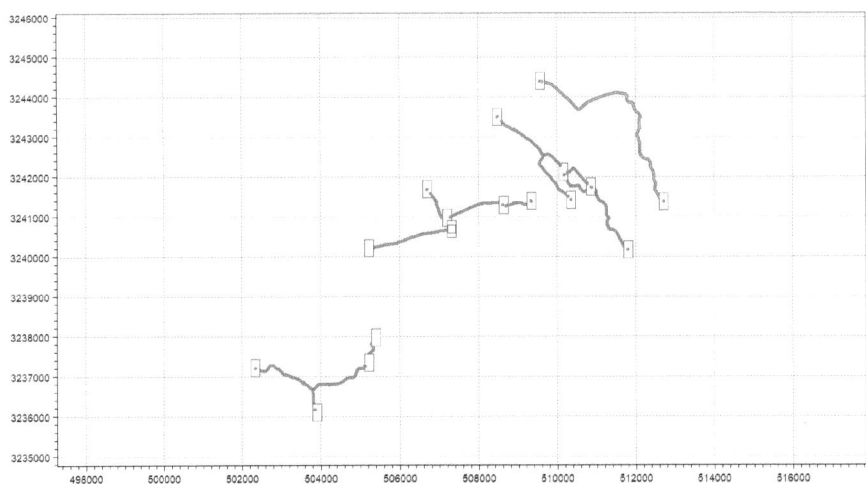

Fig. 2 Flow chart of constructing hydrological and hydrodynamic model for urban flood control

Fig. 3 Generalization of one-dimensional river network model in Jiangnan area

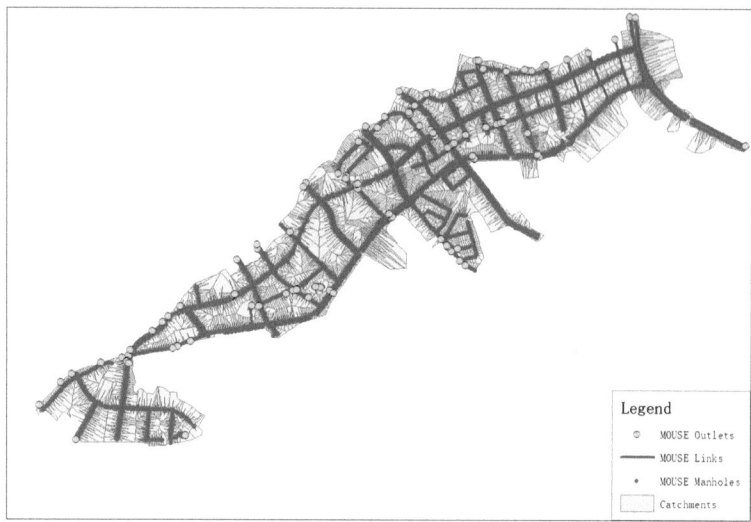

Fig. 4 Stormwater network model generalization for Jiangnan area

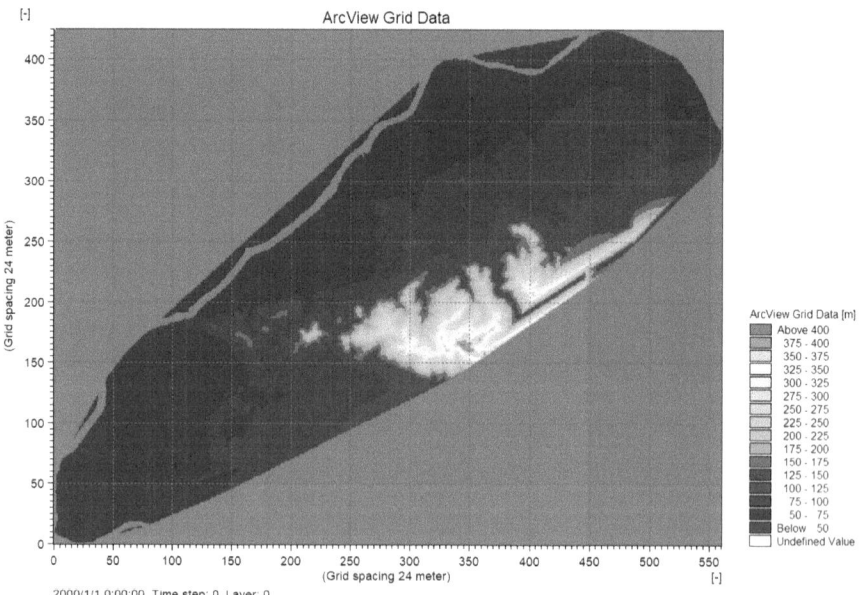

Fig. 5 Two-dimensional topographic generalization of the Jiangnan Area

Fig. 6 Generalized hydrodynamic coupling model for urban flood control in the Jiangnan area of Yiwu City

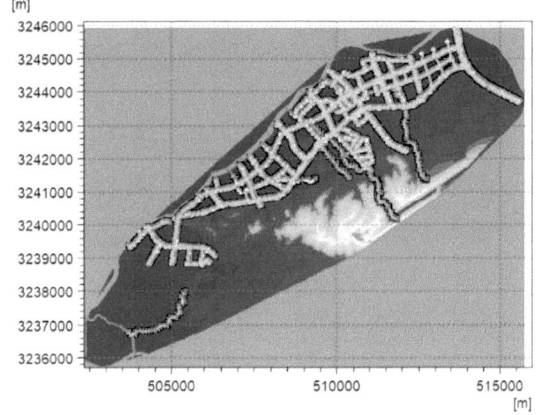

Fig. 7 Comparison of calculated water level and measured water level at Qingkouxi station during the "2021.5.12" rainstorm

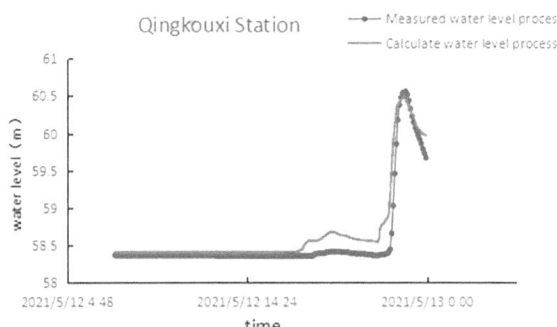

Fig. 8 Comparison of the calculated water level and measured water levels at Nianxi Station during the "2021.5.12" rainstorm

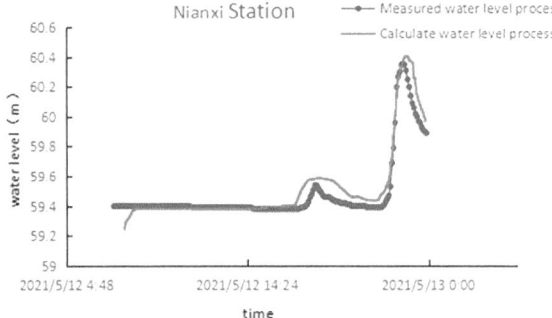

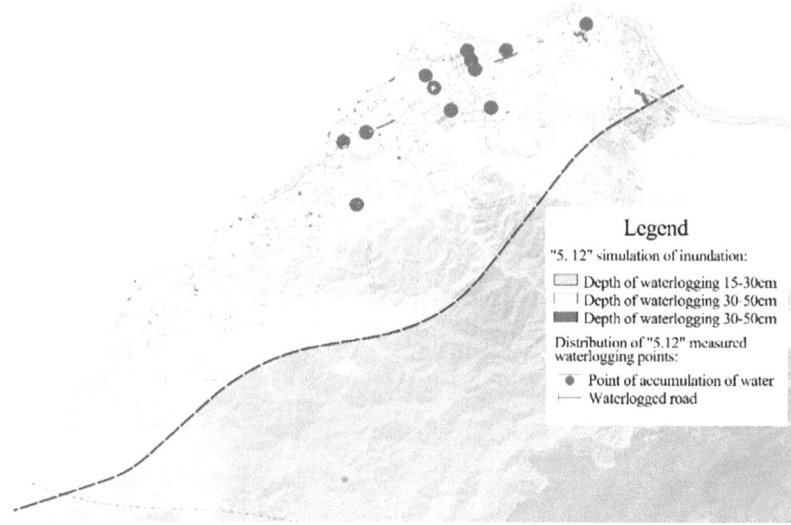

Fig. 9 Comparison of measured and modelled water accumulation points during the "2021.5.12" rainstorm

4 Results

4.1 Calculation Schemes

According to "Outdoor Drainage Design Standards" (GB50014-2021) and "Technical Specification for Prevention and Control of Urban Flooding" (GB51222-2017), the calculation standard for this assessment is: the design standard for pipelines in important areas of the central urban area is one in five years, in the central urban area is one in three years, and in the non-central urban area is one in two years. Urban flood prevention and control standards for the central city for one in 50 years, non-central urban areas for one in 20 years. Considering the richness of the comparative analysis of the program, the development of 2a, 3a, 5a storm return period conditions under the pipeline outfall were connected to the river standing water level (pipeline free outflow), 5 years, 20 years and 50 years of the design of the flood of 12 sets of combinations of calculation programs, as shown in Table 2.

Table 2 The basic situation table of each calculation scheme

Calculation schemes	Pipeline design rainfall return period			River design water level standard			
	2a	3a	5a	Standing water level	5a	20a	50a
Scheme 1	✓			✓			
Scheme 2	✓				✓		
Scheme 3	✓					✓	
Scheme 4	✓						✓
Scheme 5		✓		✓			
Scheme 6		✓			✓		
Scheme 7		✓				✓	
Scheme 8		✓					✓
Scheme 9			✓	✓			
Scheme 10			✓		✓		
Scheme 11			✓			✓	
Scheme 12			✓				✓

4.2 Calculation Results

According to the results of the model calculations, analysis the full pipe (pipe fullness ≥ 1.0 is full) length of the pipeline under each calculation scheme, and count the area of the flooded area in the study area where the depth of waterlogging reaches more than 0.15 m, as shown in Table 3.

The full pipe length and inundation area of pipes for different schemes are shown in Figs. 10 and 11. Schemes 1~4 (pink), 5~8 (blue) and 9~12 (orange) in Fig. 10 are divided into three groups of schemes, and with the same short-calendar storm return period, as the design water level at the pipe outfall increases, the full pipe length increases from 92.32 to 103.92 km, 93.71 to 104.42 km, 97.2 to 105.11 km, the full pipe length increased significantly. Schemes 1, 5, and 9 (vertical columns), Schemes 2, 6, and 10 (horizontal columns), Schemes 3, 7, and 11 (diagonal columns), and Schemes 4, 8, and 12 (brick columns) were divided into four groups of schemes, and the four schemes increased with the short-calendar-time storm return period from 2 to 5a, with the same design water level at the outfall, the full pipe length increased from 92.32 to 97.2 km, 97.4 to 102 km, 102.9 to 104.29 km, and 103.92 to 105.11 km, respectively, and compared with the first three groups of schemes where the increase in the full pipe length is obvious, the latter four groups of schemes where the increase in the full pipe length is much less variable.

Schemes 1–4 (dark red), 5–8 (green) and 9–12 (purple) in Fig. 11 are divided into three groups of schemes, and with the same short-calendar storm return period, the surface inundated area of the three schemes increases with the increase of the design water level at the pipeline outfalls from 0.5 to 2.7 km^2, from 0.6 to 2.8 km^2, and from 1.1 to 3.0 km^2, and the inundated area increased significantly. Schemes 1, 5, and 9

Table 3 Table of pipeline drainage and ground flooding in each calculation scheme

Calculation schemes	Pipe drainage condition		Ground flooding conditions	
	Full pipe length (km)	Full pipe rate (%)	Flooded area (km²)	Flooded rate (%)
Scheme 1	**92.32**	**76.9**	**0.5**	**1.0**
Scheme 2	97.4	81.1	0.6	1.1
Scheme 3	102.9	85.7	1.8	3.4
Scheme 4	**103.92**	**86.5**	**2.7**	**5.2**
Scheme 5	93.71	78.0	0.6	1.1
Scheme 6	98.92	82.4	0.7	1.3
Scheme 7	103.73	86.4	2.0	3.8
Scheme 8	104.42	86.9	2.8	5.4
Scheme 9	**97.2**	**80.9**	**1.1**	**2.1**
Scheme 10	102	84.9	1.2	2.3
Scheme 11	104.29	86.8	2.3	4.4
Scheme 12	**105.11**	**87.5**	**3.0**	**5.7**

The significance of bold corresponds to the subsequent figures, for the convenience of viewing

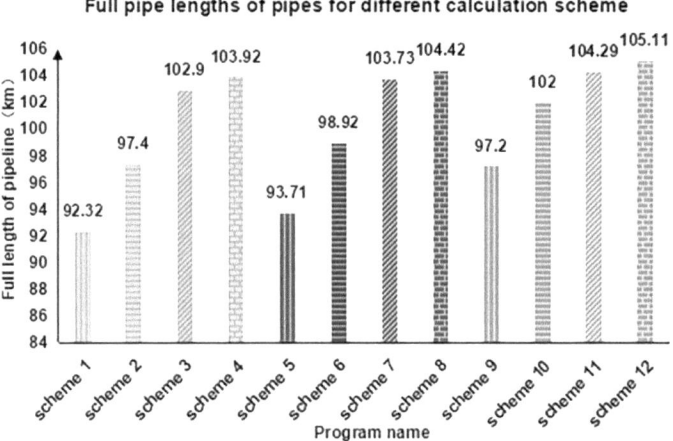

Fig. 10 Graph of full pipe lengths for different calculation schemes

(vertical column), Schemes 2, 6, and 10 (horizontal column), Schemes 3, 7, and 11 (diagonal column), and Schemes 4, 8, and 12 (brick column) were divided into four groups of schemes, and the four groups of schemes had the same design water level at the outfalls, and with the increase of short-calendar-time rainstorm return period from 2 to 5a, the inundated area of ground surface increased respectively from 0.5 to 1.1 km² The surface flooded area increased from 0.5 to 1.1 km², 0.6 to 1.2 km², 1.8 to 2.3 km², and 2.7 to 3.0 km², respectively with the increase of short-calendar

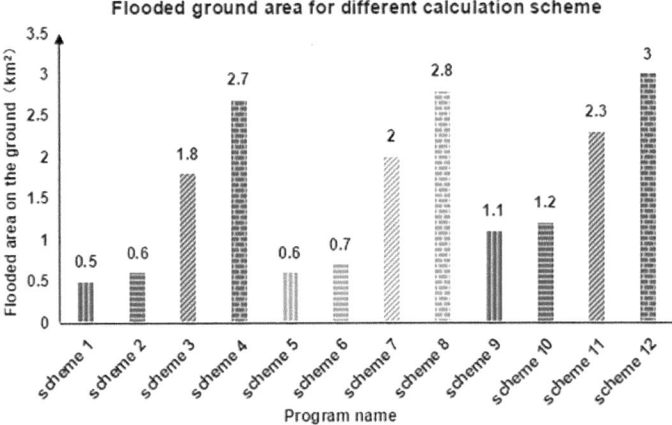

Fig. 11 Graph of flooded ground area for different calculation schemes

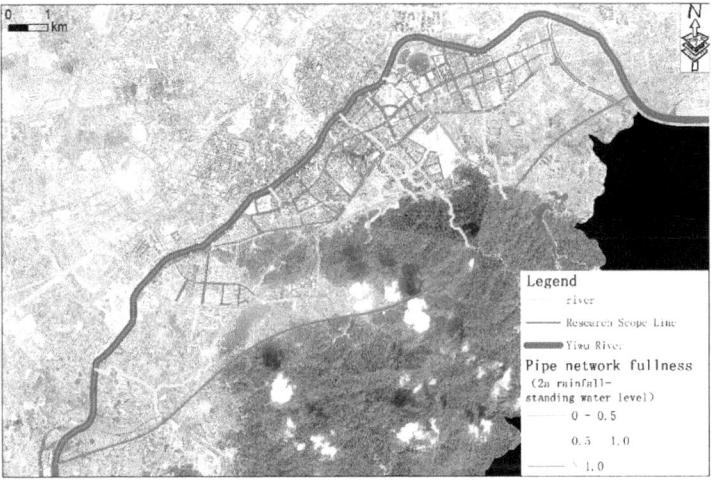

Fig. 12 Pipe network drainage filling classification diagram of calculation scheme 1

storm return period from 2 to 5a, and the change of flooded area in the last four schemes was smaller compared to the obvious change of flooded area in the first three schemes.

The pipe filling and the graded depth of inundation of the two-dimensional ground for typical comparison conditions (calculation schemes 1, 4, 9, and 12) are presented as shown in Figs. 12, 13, 14, 15, 16, 17, 18 and 19.

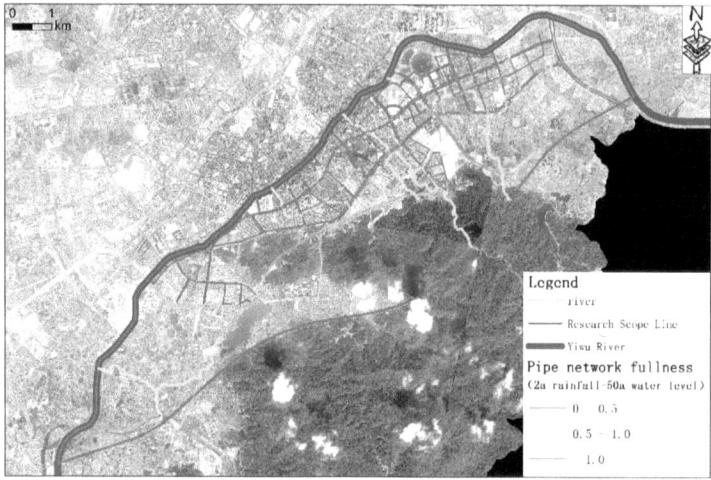

Fig. 13 Pipe network drainage filling classification diagram of calculation scheme 4

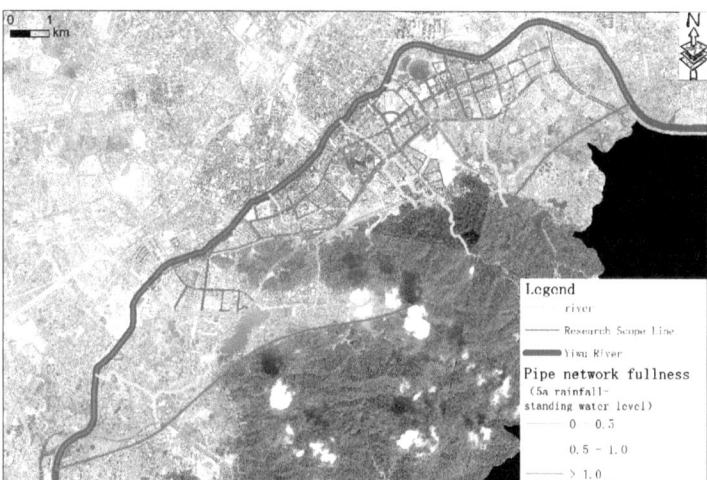

Fig. 14 Pipe network drainage filling classification diagram of calculation scheme 9

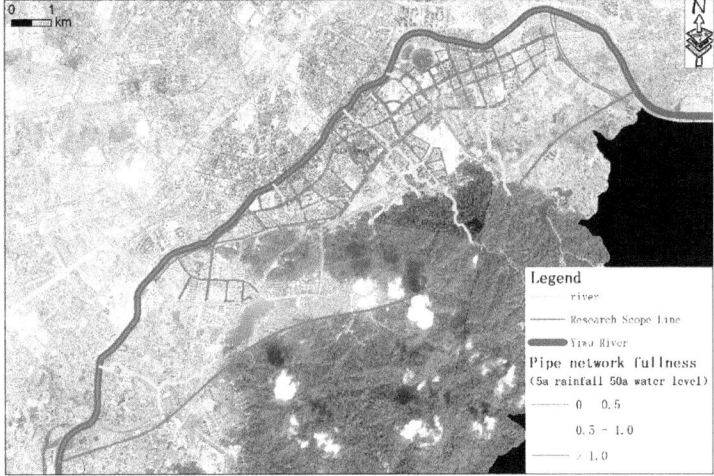

Fig. 15 Pipe network drainage filling classification diagram of calculation scheme 12

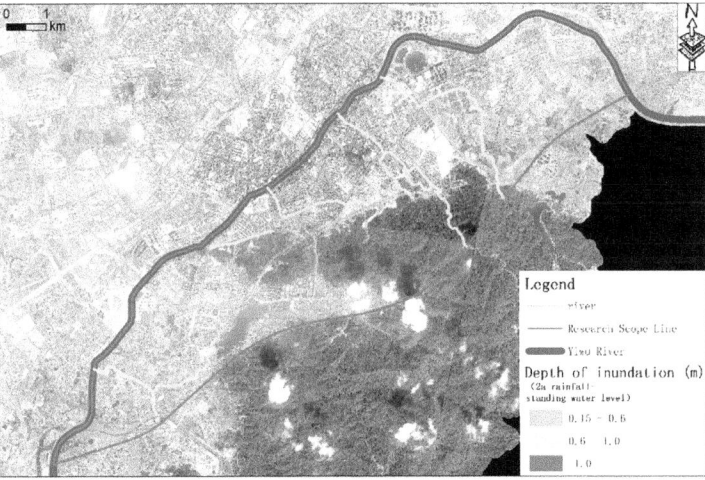

Fig. 16 Study area inundation map of calculation scheme 1

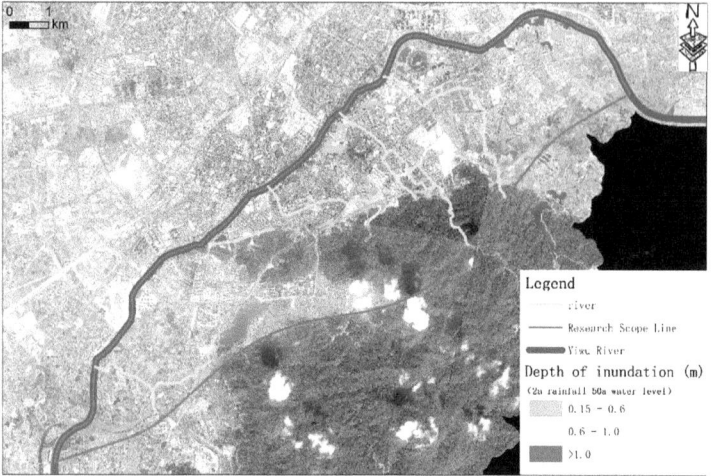

Fig. 17 Study area inundation map of calculation scheme 4

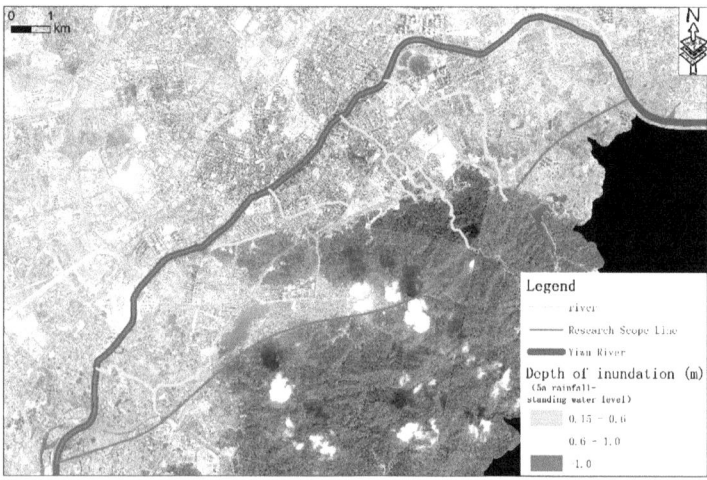

Fig. 18 Study area inundation map of calculation scheme 9

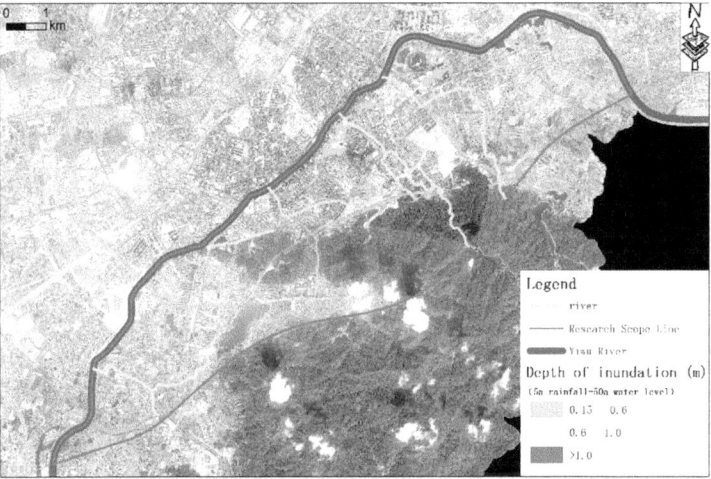

Fig. 19 Study area inundation map of calculation scheme 12

5 Conclusion

The paper takes the urban drainage area south of Yiwu River in Yiwu City as an example. Based on the MIKE model, the paper simulated and evaluated the drainage capacity of the pipeline system under different river water levels for various design rainfall return periods. The main findings are as follows.

(1) The stormwater pipeline system in the research area was constructed relatively early, with lower design standards, resulting in suboptimal overall drainage capacity. Under the free outflow condition of the articulation outlet of the urban area for the 2-year return periods of rainstorm, 76% of the pipeline reach full capacity.

(2) Based on scenario analysis, it was observed that, compared with the different design storm return periods of 2-year, 3-year, and 5-year, the influence of different boundary water levels, such as the normal water level of the river and the design floods for 5-year, 20-year and 50-year, had a more significant effect on the pipeline drainage capacity in the research area.

(3) For the evaluation and design of urban drainage pipe network, based on the simulation and calculation of free outflow condition of pipe drainage outlets, urban rainfall model can be constructed to simulate the impact of river design flood on the drainage network. This approach facilitates an effective connection between pipe drainage and river flooding, allowing for the optimization of drainage and flood prevention planning.

Acknowledgements The authors gratefully acknowledge the financial supports by the Joint Funds of the Zhejiang Provincial Natural Science Foundation of China (Grant No. LZJWZ23E090007);

Zhejiang Water Conservancy Science and Technology Program (RC2217); Zhejiang Water Conservancy Science and Technology Program (RB2202).

References

1. H. Kreibich, A.F. Van Loon, K. Schroter et al., The challenge of unprecedented floods and droughts in risk management. Nature **608**(7921), 80–86 (2022)
2. B.J. Van Ruijven, M.A. Levy, A. Agrawal et al., Enhancing the relevance of Shared Socioeconomic Pathways for climate change impacts, adaptation and vulnerability research. Clim. Change **2**(3), 481–494 (2014)
3. J.Y. Zhang, Y.T. Wang, R.M. He et al., Analysis of urban flooding problems and causes in China. Adv. Water Sci. **27**(4), 485–491 (2016)
4. X.B. Miao, M. Lü, F.C. Liang et al., An analysis of urban water logging and water quality control simulation in low-impact development(LTD). China Rural Water Hydropower **1**(87–91), 96 (2019)
5. China State Council Disaster Investigation Group, Investigation Report on the "7–20" Extremely Heavy Rainstorm Disaster in Zhengzhou, (Henan Province, 2022). https://www.mem.gov.cn/gk/sgcc/tbzdsgdcbg
6. Q.F. Hu, Y. Zhang, L.J. Li et al., Comparative analysis of GPM near-real-time inversion data for the "7-20" extreme rainstorm in Henan Province in 2021. Adv. Water Sci. **33**(4), 14 (2022)
7. W. Cornwall, Europe's deadly floods leave scientists stunned. Science (New York, N.Y.). **373**(6553), 372–373 (2021)
8. M. Hemmati, K. Kornhuber, A. Kruczkiewic, Enhanced urban adaptation efforts needed to counter rising extreme rainfall risks. Npj Urban Sustain. **2**, 16 (2022)
9. L.L. Lan, Z.B. You, Analysis and evaluation of the response to the major flooding process in Australia in February-March 2022. China Disaster Reduction **15**, 34–37 (2022)
10. M. Lou, X.C. Ye, Y.C. Wang, C.L. Zhu et al., Influence of river boundary water level on drainage capacity of pipeline. J. Water Resour. Water Eng. **29**(3), 169–174 (2018)
11. H. Zhou, J. Liu, C. Gao, S.F. Ou, Analysis on current situation and problems of urban waterlogging control in China. J. Disaster Sci. **33**(3), 147–151 (2018)

Study on Hydraulic Characteristics of Rock Plug Inlet/Outlet Blasting Process in a Power Station

Shaojia Yang, Yongqing Wang, and Tao Zhang

Abstract Hydraulic characteristics of rock plug blasting are the key to successful construction of rock plug inlet/outlet. In this paper, the hydraulic characteristics of rock plug blasting at the inlet/outlet of the lower reservoir of a pumped storage power station are studied by means of physical model test. The characteristic laws of water hammer pressure, surge of gate shaft and accumulation effect during blasting are obtained. The results verify the rationality of the design type of rock plug inlet, and demonstrate the feasibility of using gate direct water blocking scheme and high water level blasting during rock plug blasting. The research results can provide technical support for the design and construction of this project, and provide reference for similar projects.

Keywords Rock plug blasting · Model test · Hydraulic characteristics · Gate well surge · Water hammer pressure

1 Introduction

Underwater rock plug blasting is usually simulated by physical or numerical simulation to obtain information about blasting effect, rock fragmentation distribution, underwater pressure and its impact on the surrounding environment. This model experimental research is helpful for us to deeply understand the principle and law of underwater rock plug blasting, so as to optimize the blasting scheme in practical engineering applications.

With the development of mathematics, physics and computer technology, more theoretical methods and models have been developed in the experimental research of underwater rock plug blasting model, such as fluid-structure coupling model and

S. Yang (✉) · T. Zhang
PowerChina Huadong Engineering Corporation Limited, Hangzhou 311122, China
e-mail: yang_sj1@hdec.com

Y. Wang
Shangrao River and Lake Management and Protection Center, Shangrao 334000, China

W. Wang et al. (eds.), *Hydraulic Structure and Hydrodynamics*, Lecture Notes in Civil Engineering 608, https://doi.org/10.1007/978-981-97-7251-3_38

441

continuum breaking model. These theoretical methods and models are helpful to improve the accuracy and rationality of the experiment and deepen the understanding of underwater rock plug blasting.

In the experimental research of underwater rock plug blasting model, the continuous improvement of experimental devices and methods has also been widely concerned. For example, through the selection of simulated materials, the setting of experimental conditions and other means, the experimental results are more close to the actual situation. Through the data and knowledge obtained from the experimental study of underwater rock plug blasting model, the existing blasting design can be improved and verified to optimize the blasting effect in practical engineering applications.

A pumped storage power station uses the built reservoir as the lower reservoir, and the inlet/outlet of the lower reservoir is constructed by open air cushion underwater rock plug blasting. A ballast collecting pit [1–3] is arranged near the water intake, a temporary construction plug is set downstream of the sluice well to block water (or directly use the sluice gate to block water), and an air cushion room is set above the sluice collecting pit to form an air cushion by filling the sluice well with water. The air cushion reduces the blasting pressure and fills with water to block the movement of rock ballast [3–6]. See Fig. 1 for the schematic diagram of the open air cushion rock plug blasting method. Combined with the model test, the effects of filling water level, internal and external water head difference and air cushion volume change on reducing blasting effect were studied. To verify the design type of rock plug water inlet, select the optimal external water level with minimal blasting effect and the water filling level of gate well, and select the optimal air cushion volume to provide safety guarantee for engineering construction.

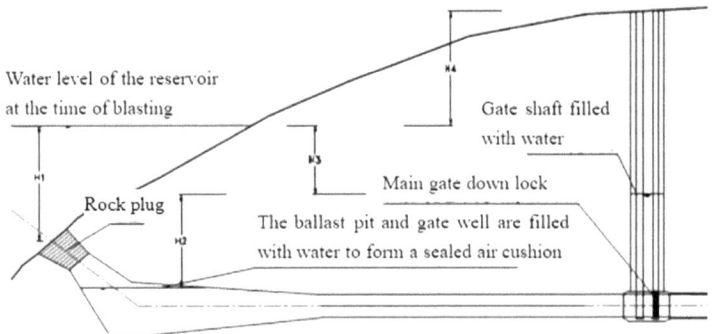

Fig. 1 Schematic diagram of open air cushion rock plug blasting

Fig. 2 Photos of the overall model

2   Model Test

2.1   Model

The model is designed according to the gravity similarity criterion. Considering the similarity of rock plug blasting model and test requirements comprehensively, the geometric scale of the model is set at 1:20, and the hydraulic element scale of the model is determined under the gravity similarity criterion.

As shown in Fig. 2, the simulation range is 80 m upstream and downstream of the center line of the rock plug inlet, and the height of the bottom top of the bank slope is simulated to 159.24 m. The lower boundary of the channel is taken 100 m downstream of the gate well. According to the scale calculation, the total length of the model is 25 m, the total width is 8 m, of which the reservoir is 9 m long and 8 m wide. The boundary center elevation of the model tunnel is 55.88 m, and the model is properly elevated to facilitate outflow.

2.2   Measurement Point Layout and Measurement Content

The pressure measuring point arrangement of rock plug blasting test is shown in Fig. 3. During the test, if the gate is used to block water, the 13# pressure measuring point will not be used, and the 14# pressure measuring point will be moved to the gate to measure the force of blasting shock wave on the gate. The scale grid is pasted on the side wall of the ballast collecting pit model, and the surface line coordinates of the ballast collecting pit are drawn to measure the shape of the ballast collecting pit.

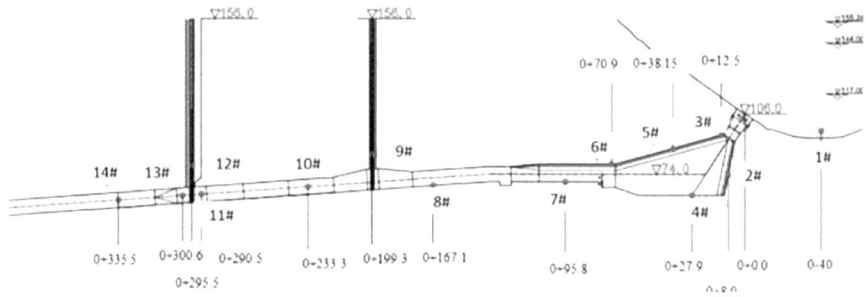

Fig. 3 Pressure measuring point arrangement of rock plug blasting test

2.3 Test Conditions

Rock plug blasting tests under various test conditions in Table 1 were carried out to study the effects of reservoir water level and gate well filling water level on blasting pressure, rock ballast distribution and surge of gate well, and to verify the rationality of the design type of ballast pit.

3 Rock Plug Blasting Test Simulation Scheme

During the test, dynamic and hydraulic processes are combined to measure the impact pressure produced by blasting and the dynamic water pressure produced by water flow. The scheme is as follows:

(1) Selection of explosive charge;
(2) Head water blocking blasting test;
(3) Gate water blocking blasting test;
(4) Comparison of two water retaining schemes.

Table 1 Rock plug blasting model test conditions

Reservoir level (m)	Gate well water level (m)			
120	105	110	115	/
125	105	110	115	120
130	115	120	125	/
140	125	130	135	140

4   Comparative Analysis of Rock Plug Blasting Test Results

4.1   Ballast Pile Shape of Ballast Collecting Pit

As shown in Fig. 4a, under the water blocking condition, the distribution range of ballast varies slightly with the difference of water level between reservoir and gate well. The smaller the difference of water level between reservoir and gate well, the closer the ballast is to the upstream. With the increase of the difference of water level between reservoir and gate well, the ballast moves downstream, the distribution range expands, and the top elevation of ballast decreases slightly. In general, the piles are distributed in a peak shape, and the peak position is basically near the extension line of the center line of the rock plug mouth, and the piles are mainly concentrated in the range of pile number $0 + 25{\sim}0 + 40$ m.

As shown in Fig. 4b, under the condition of gate water blocking, the maximum pile size reaches $0 + 54$ m. The maximum height of the ballast pile peak is 69.2 m. The distribution law is basically the same as that of water stopper. The difference of water level between reservoir and gate well is large, and the pile debris moves to gate well side, while the difference of water level between reservoir and gate well is small.

Under all test conditions of the two schemes, the pile piles after blasting are mainly concentrated in the range of pile number $0 + 25{\sim}0 + 40$ m, all of which do not reach the end of the ballast pit, and there are plenty of them. The peak height is also below the elevation of 69.2 m, and does not exceed the bottom elevation of the downstream flat cave section (70 m). Compared with the scheme of gate retaining water, the distribution law of ballast in ballast collecting pit is basically the same, the shape is still hump distribution, and the peak value of ballast is close to the height. The length and depth of the two schemes are reasonable.

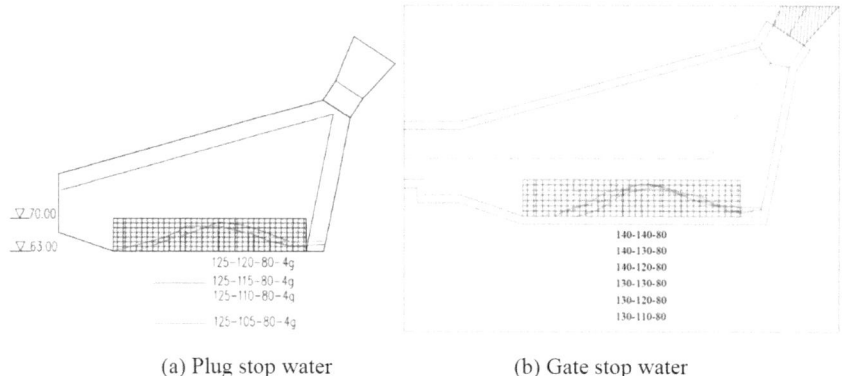

| (a) Plug stop water | (b) Gate stop water |

Fig. 4 Pile shape of ballast pit

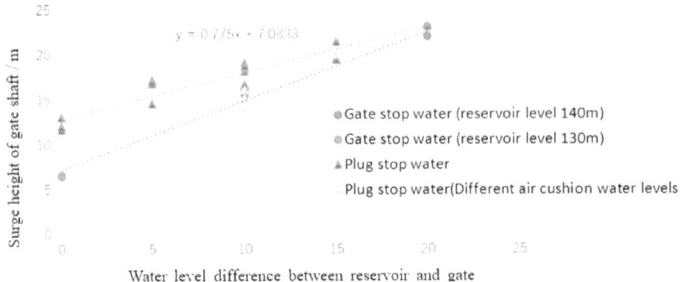

Fig. 5 Relation between surge water level of gate well and water level difference between reservoir and gate well

4.2 Surge Water Level of Gate Well

After the rock plug detonates, the rock ballast falls into the ballast collecting pit, and part of the reservoir water flows into the tunnel. At the same time, the gas in the inflatable air bag at the top of the ballast collecting pit quickly overflows to the reservoir. After the ballast collecting pit, the pressure in the tunnel near the gate well drops sharply, and the water level of the gate well drops rapidly, and then the water level of the gate well rises rapidly to the highest level with the water in the lower reservoir.

There is a linear relationship between the swell height of gate well and the inflated air cushion level of ballast collecting pit. The higher the inflated air cushion level, the higher the surge level of gate well. Under the same water level difference (15 m) between reservoir and gate well, the inflatable air cushion level increases from 80 to 85 m, and the maximum surge of gate well increases from 19.5 to 23 m.

As shown in the Fig. 5, the surge height of gate well is directly proportional to the difference of water level between reservoir and gate well. Compared with the stopper scheme, the surge height of the gate well is small when the water level difference between reservoir and gate well is low, but the surge height of the gate well increases with the water level difference between reservoir and gate well at a higher rate than that of the stopper scheme. When the water level difference between reservoir and gate well is 20 m, the surge height of the two schemes is basically the same.

4.3 Stress Analysis of Tunnel and Gate

The pressure distribution along the tunnel is shown in Figs. 6 and 7. The peak pressure of the gate water retaining tunnel appears at the bottom of the ballast collecting pit (measuring point 4#) and the gate position (measuring point 14#), and the gate pressure is obviously higher than the bottom pressure of the ballast collecting pit. Under the same reservoir water level, with the increase of water level difference between reservoir and gate well, the impact pressure of gate increases, and the relationship is basically linear. Under the same water level difference between reservoir and gate well, the impact pressure of gate tends to decrease with the increase of reservoir water level, which is mainly related to the larger air cushion pressure at the top of ballast collecting pit under high water level and the enhanced buffering effect. Under the same reservoir water level and the difference between reservoir and gate well water level, the gate pressure increases significantly compared with the stopper scheme.

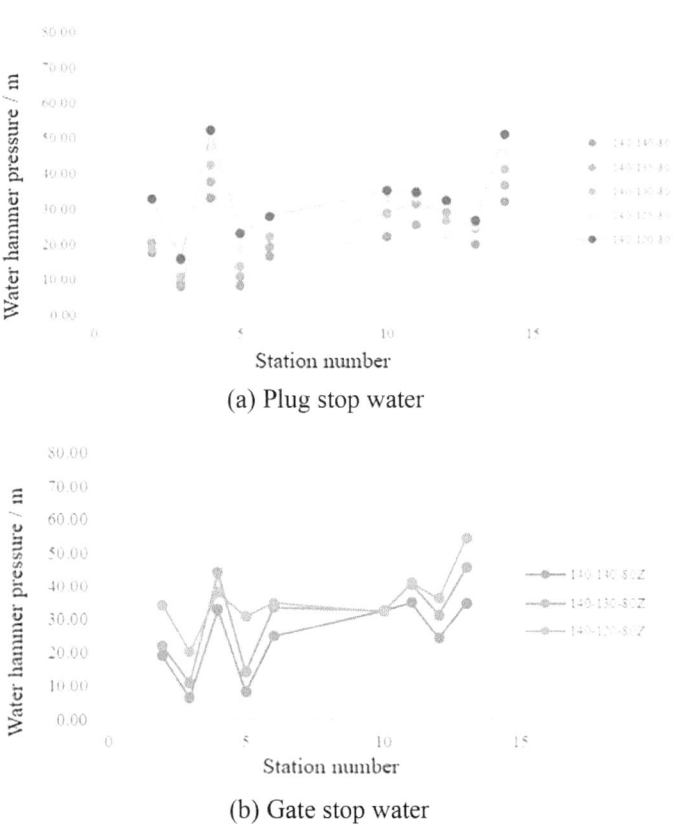

(a) Plug stop water

(b) Gate stop water

Fig. 6 The impact pressure distribution along the reservoir water level at 140 m

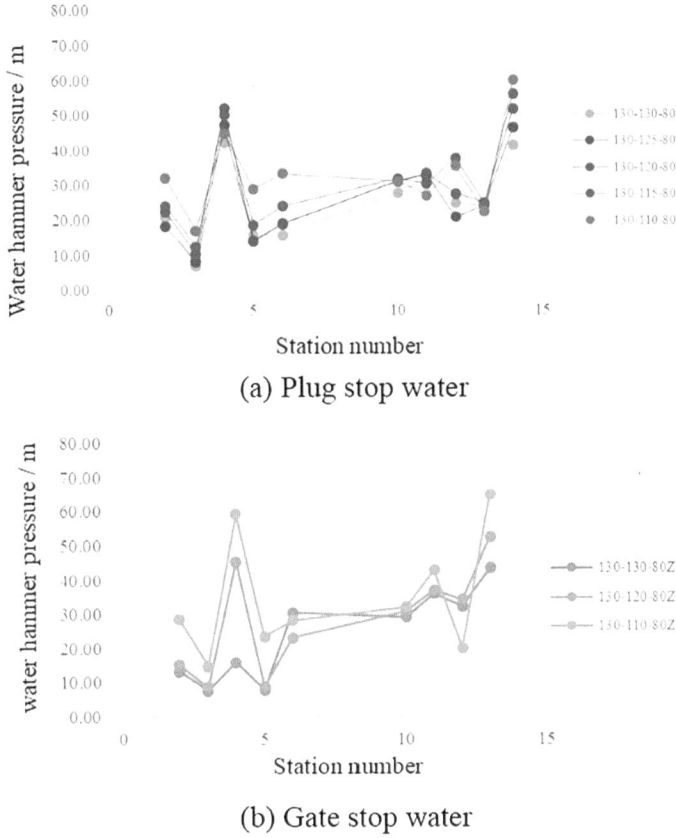

(a) Plug stop water

(b) Gate stop water

Fig. 7 The impact pressure distribution along the reservoir water level at 130 m

5 Conclusion

In this paper, 1:20 integral physical model is used to study the hydraulic characteristics of rock plug blasting at the inlet and outlet of the lower reservoir of a power
station. The results show that:

(1) In the scheme of gate water retaining and plug head water retaining, the distribution law of ballast pile in ballast collecting pit is basically the same, the shape is
still hump-type distribution, and the peak value of ballast is close to the height;
The smaller the water level difference between reservoir and gate well, the closer
the ballast pile is to the rock plug mouth, the larger the water level difference
between reservoir and gate well, the larger the ballast pile moves to the gate
well side, the wider the distribution range, and the top elevation of the ballast
pile decreases slightly. The length and depth of the two schemes are reasonable.

(2) There is a linear relationship between the surge height of gate well and the difference of water level between reservoir and gate well. There is a linear relationship between the swell height of gate well and the inflated air cushion water level of ballast collecting pit. The higher the inflated air cushion water level, the higher the surge water level of gate well.

(3) The impact pressure of the plug increases with the increase of the water level difference between the reservoir and the gate well; Under the condition of water level difference between the same reservoir and gate well, the impact pressure increases with the decrease of reservoir water level. Under the condition of gate water blocking, the maximum pressure of gate is consistent with the water blocking scheme, and the impact pressure is greater than the water blocking scheme under the same condition.

With the rapid development of computer technology, efficient algorithm and high performance computing technology have played an important role in the experimental research of underwater rock plug blasting model. Through efficient algorithms and high-performance computing, researchers can process large amounts of data in a short period of time and quickly get experimental results. The model usually requires a larger site and more input of personnel and materials, and the cost is higher than that of numerical simulation. In the future, numerical simulation and model experiment can be combined to verify the rationality of numerical calculation with model experiment, reduce the model experiment condition and reduce the cost.

References

1. J. Zhongshuai, Analysis on construction technology of underwater rock plug blasting for full row hole of Changdian Hydropower Station. Water Resources Constr. Manage. **40**(12), 53–59 (2020)
2. L. Xilin, Analysis on blasting effect of underwater rock plug in Changdian Hydropower Station reconstruction project. Northeast Water Resources Hydropower **38**(07), 55–58 (2020)
3. Sukarin, Progress of underwater rock plug blasting technology. Water Resour. Hydropower Technol. **50**(08), 110–115 (2019) (in Chinese)
4. L. Feng, Technology research and application of air cushion underwater rock plug blasting. Blasting**2006**(04), 66–70+77 (2006)
5. C. Wen, *Research and Application of New Technology of Underwater Rock Plug Blasting in Xianghongdian Pumped storage Power Station* (China Society of Hydroelectric Power Engineering, 2005), p. 13
6. C. Wen, *Research on New Technology of Underwater Rock Plug Blasting*. Anhui Province, Anhui Water Resources and Hydropower Survey and Design Institute, 1999-08-01

Optimal Arrangement of Pressure Monitoring Points in Water Supply Network Based on Intelligent Optimization Algorithm

Jiangang Fei

Abstract Aiming at the optimal arrangement of pressure monitoring points in urban water supply network, with the goal of maximizing the monitoring range, an optimal arrangement model of monitoring points is constructed, considering the pressure correlation, node flow sensitivity and the effectiveness of the water flow path in the pipe network. Taking an urban water supply pipe network in East China as an example, the model is solved by three kinds of population intelligent algorithms: Krill Harvesting Algorithm (KHA), Bat Algorithm (BA), and Particle Swarm Algorithm (PSO). The optimization results of the three algorithms are compared in various aspects, and it is found that the KHA algorithm shows the most excellent search accuracy and efficiency in the problem of optimal arrangement of monitoring points, because of the strong global optimization ability and not easy to fall into local optimal solutions.

Keywords Pressure monitoring points · Intelligent optimization algorithm · Maximizing the monitoring range · Pressure correlation · Water supply network

1 Introduction

Urban water supply pipeline network pressure monitoring points can not only be used for real-time monitoring of the pressure distribution of the water supply pipeline network [1], but it is also significant for the optimization of the water supply system scheduling, leakage control, intelligent water construction etc. The arrangement of pressure monitoring points should consider the economy and reasonableness, with few numbers of pressure monitoring points to maximize the monitoring range of the water supply system.

In order to optimize the arrangement of pressure monitoring points, scholars in the field have carried out some exploration and made some progress, among which the

J. Fei (✉)
Qingdao Harbour Vocational and Technical College, Qingdao 266404, China
e-mail: feilaoshi2023@163.com

W. Wang et al. (eds.), *Hydraulic Structure and Hydrodynamics*, Lecture Notes in Civil Engineering 608, https://doi.org/10.1007/978-981-97-7251-3_39

451

intelligent optimization algorithm represented by genetic algorithm has been widely used in the research.

Perez et al. [2] proposed a leakage detection method based on water pressure sensitivity analysis. The leakage location was determined by calculating the difference between the monitoring value and the estimated value. At the same time, a pressure sensor placement method based on genetic algorithm was also proposed. Casilllas et al. [3] used the leakage feature space method to optimize the placement of pressure sensors in the pipe network, and used genetic algorithm (GA) and particle swarm optimization (PSO) to solve the problem, so as to improve the accuracy of leakage detection. Sabzkouhi et al. [4] established a new multi-objective model by particle swarm optimization (MO-PSO) in the literature, and coupled it with the hydraulic simulation model to solve the extremal problem of the network pressure effectively. Raei et al. [5] used the K-means clustering method to carry out the preliminary partition of the pipe network. On this basis, an optimal arrangement model of monitoring points was established to determine the optimal position of flow and pressure sensors, and NSGA-II was used to solve the problem. However, previous studies have found that many intelligent optimization algorithms are easy to fall into the local optimal solution in the application process and cannot obtain the global optimal solution, or even far from the optimal solution [6]. According to these problems, this study introduces the krill swarm algorithm (KHA) into the solution of pressure monitoring points optimized arrangement for the first time, which has a great potential of comprehensive performance in the field of population intelligent optimization algorithm. Based on the topology of the pipe network, a mathematical model is established based on the water pressure correlation between nodes and water pressure sensitivity. The effectiveness of the KHA in solving the optimal arrangement model is verified through comparative analysis of three intelligent algorithms, the potential bat algorithm (BA), the traditional particle swarm algorithm (PSO) and the krill swarm algorithm (KHA).

2   The Optimized Arrangement Model Establishment of Pressure Monitoring Points

2.1   Determination of Constraints

Pressure correlation of pipe network

The purpose of installing pressure monitoring points in the pipe network is to reflect the overall condition of the pipe network according to the values and change ranges by monitoring the water pressure at some nodes. Therefore, the selected pressure measurement points should have a strong pressure correlation with other nodes in the pipe network [7], and the quantitative evaluation standard of pressure correlation is the absolute water pressure difference between nodes. That is to meet the conditions:

$$\left| H_i - H_j \right| < h \tag{1}$$

where: H_i, H_j denote the pressure values of nodes i and j;

h is the set minimum pressure difference.

When the water pressure difference between nodes is less than the set value h, it indicates that the pressure between two nodes is related. Equation (1) is used to calculate the absolute pressure difference between two nodes in the pipe network, and father to get the pressure difference matrix $[P]_{n \times n}$ of the nodes of pipe network, which is used as one of the constraints for the optimized arrangement model of pressure monitoring points. At the same time, the water pipe network can be simplified as a fully connected network of nodes, its topological relationship determines that any two nodes have a water flow path. However, there is a lack of distance between nodes or the limitation of the nodes number in the water path, therefore the results cannot effectively represent the pressure changes around the monitoring points [7]. In order to characterize the connectivity between nodes, Dijkstra's algorithm is introduced here to realize the node shortest distance matrix $[L]_{n \times n}$, and this is used as one of the constraints for the optimized arrangement model of pressure monitoring points, meanwhile the effectiveness of the water flow path is demonstrated by the distance limitation d between the pipe network nodes.

Water pressure sensitivity of pipe network

For water supply network operation under certain conditions, the most significant external factor that causes the fluctuation of node water pressure is the change of water consumption. When the water quantity of a node fluctuates, it is bound to cause the water pressure of other nodes in the pipe network to change to varying degrees. Pressure measuring points should be set up at the point where the change of node water pressure is the largest [8]. Therefore, the concept of sensitivity coefficient is introduced here. Assuming that the pipe network contains n nodes, the value of water pressure at each node can be obtained through hydraulic simulation under the given conditions. If the water volume change at node k is ΔQ_k, the corresponding pressure change at node k is ΔH_k, the change pressure at the examined node i is ΔH_i, the pressure change rate of the node i can be expressed as $\frac{\Delta H_i}{\Delta Q_k}$. Due to the poor comparability of pressure and flow, the pressure change rate at node i can be characterized instead by the pressure difference ratio $\frac{\Delta H_i}{\Delta H_k}$, which reflects the sensitivity of water pressure at node i to the change of node flow, and furthermore it is recorded as the sensitivity coefficient. In this study, the basic flow of each node in the pipe network is increased by 10% of the disturbance flow, and the sensitivity coefficient between any two nodes is calculated according to Eq. (2), to form the hydraulic pressure sensitivity matrix $[D]_{n \times n}$.

$$D(i, k) = \frac{H_i' - H_i}{H_k' - H_k} \tag{2}$$

where: H_i, H_k are the water pressures at nodes i and k of the base condition, respectively; H_i', H_k' are the water pressures at nodes i and k after the flow rate at node k is increased by 10%, respectively.

The elements of matrix $[D]_{n \times n}$ are standardized. Firstly, the standard deviation of each column element of matrix $[D]_{n \times n}$ is standardized according to formula (3), and matrix $[D\prime]_{n \times n}$ is obtained.

$$D\prime(i, k) = \frac{D_{(i,k)} - \overline{D}_k}{S_k} \tag{3}$$

where: S_k is the standard deviation of the kth column element, $S_k = \sqrt{\frac{1}{n} \sum_{i=1}^{n} \left[D(i, k) - \overline{D}_k \right]^2}$.

The polar values of each column element of the matrix $[D\prime]_{n \times n}$ are then normalized according to Eq. (4) to obtain the matrix $[D\prime\prime]_{n \times n}$.

$$D\prime\prime_{(i,k)} = \frac{D\prime(i, k) - D'_{k_{\min}}}{D'_{k_{\max}} - D'_{k_{\min}}} \tag{4}$$

The matrix $[D\prime\prime]_{n \times n}$ is analysed by the Euclidean distance method, and the Euclidean distance between two nodes is calculated according to formula (5). Finally, the fuzzy similarity matrix $[R]_{n \times n}$ for pressure sensitive can be obtained as follows.

$$r_{ij} = \sqrt{\frac{1}{n} \sum_{k=1}^{n} \left[D\prime\prime(i, k) - D\prime\prime(j, k) \right]^2} \tag{5}$$

where: r_{ij} is the Euclidean distance between node i and node j, $i, j = 1, 2, \ldots, n$

r_{ij} represents the pressure change sensitivity between node i and node j. The smaller r_{ij} is, the more sensitive node i is to the pressure change of node j. If $r_{ij} < \lambda$ (λ is a set value), it means that the two nodes satisfy the condition of pressure sensitivity, and the fuzzy similarity matrix $[R]_{n \times n}$ can be used as one of the constraints for the optimized arrangement model of pressure monitoring points.

2.2 Objective Function

In this study, the objective function is to maximize the monitoring range of the pressure monitoring points arrangement scheme, which is expressed as follows.

$$maxT = count(A_1 \cup A_2 \cup \ldots \cup A_n) \tag{6}$$

Constraint condition:

$$X = \{X_1, X_2, \ldots, X_n\} \tag{7}$$

$$A_{x_i} = \{j | P(X, j) < h, L(X, j) < d, R(X, j) < \lambda\} \tag{8}$$

where: T is the total number of nodes that satisfy the constraints; X is the set of selected pressure monitoring points; X_i is the node numbered i; A_{x_i} is the set of nodes that satisfy the constraints among the selected monitoring points X_i; $P(X, j)$ denotes the Xth row and jth column elements of the differential pressure matrix; $L(X, j)$ denotes the Xth row and jth column elements of the nodes' shortest distance matrix; $R(X, j)$ denotes the Xth row and jth column elements of the pressure sensitivity fuzzy similarity matrix; and h, d, and λ are the constraints set values.

3 Model Solution

The optimal arrangement of pressure monitoring points is essentially a combinatorial optimization problem, and the decision variable is the combination of monitoring points after the number of monitoring points is determined. Theoretically, the optimal arrangement scheme can be found by exhaustive enumeration, but with the increase of the complexity of the pipe network, the number of feasible solutions is huge, and this kind of problem is very suitable for intelligent optimization algorithms to solve. Compared with other optimization algorithms, the population intelligent optimization algorithm has the advantages of fewer parameter settings, simple algorithm implementation, high search efficiency, etc., and has been widely used in engineering optimization problems.

Krill Herd Algorithm (KHA) is a swarm intelligence optimization algorithm proposed by Gandomi et al. [9] in 2012, which mainly simulates the aggregation behaviours of krill individuals. The movement behaviours of krill individuals in the process of simulation is mainly affected by three aspects, which are the movement caused by the presence of other individuals, the movement caused by foraging activities, and random diffusion. In addition to this, two genetic operators, which are crossover and mutation in the genetic algorithm, are added to the algorithm idea to increase the diversity of the algorithm. Its speed and position update strategy is similar to that of the Bat Algorithm (BA) and Particle Swarm Algorithm (PSO), which is to update the speed and position of the individuals in the population so that the individuals gradually approach the optimal solution. The advantage of KHA is that the crossover and mutation operators are added into the genetic algorithm, which makes it easier to jump out of the local optimal solution compared with the BA and PSO algorithms.

In this study, EPANET2.2 software is used to carry out hydraulic simulation calculation of the arithmetic pipe network. By using the toolbox provided by EPANET, calling the functions in the toolbox to obtain the data such as node flows and pressures of the pipe network, and then calculating to obtain the differential pressure matrix, shortest path matrix, and pressure sensitive fuzzy similarity matrix. Then Python is used to write the optimization program, and KHA, BA and PSO are used to solve the problem separately. KHA is usually applied to optimization problems in continuous domain, while the problem of optimal arrangement of monitoring points belongs to discrete combinatorial optimization. Therefore, the mapping relationship between the solution space and the problem space should be clarified firstly, that is, KHA is suitable for solving the discrete optimization problems by suitable encoding. In this study, the algorithm adopts the real number coding method, which corresponds the monitoring point number to its integer form by rounding.

4 Example Analysis

An urban water supply network in East China was selected as the object of study, and the hydraulic model of the network contains 96 nodes, 99 pipes and one water source, with a daily water consumption of 4980 m^3/d. The maximum pipe diameter is DN400 and the minimum pipe diameter is DN50. Five pressure monitoring points were planned.

4.1 Parameter Setting

The thresholds of the three constraints should be set in combination with the actual situation of the pipe network, and the reasonable setting of the thresholds is crucial to get the accurate monitoring points scheme. The thresholds of constraints are set too large or too small will have impacts on the optimal solution. Therefore, the thresholds of the three constraints are first defined initially, and the accuracy and validity of the results are analysed by repeatedly trial and error calculations. Finally, the thresholds of the three constraints are determined as $\lambda = 0.20$, h = 5 m, and d = 800 m.

The parameters of the optimization algorithm are set as follows: population size in KHA is 30, evolutionary generation is 500, maximum induced speed is 0.01, foraging speed is 0.02. Population size in BA is 30, evolutionary generation is 500, pulse loudness is 0.5, pulse rate is 0.5, frequency range is [0, 2]. Population size is 30 in PSO, evolutionary algebra is 500, individual learning factor is 0.7, global learning factor is 0.5, and maximum flight speed is 1. Twenty calculations are performed for KHA, BA, and PSO, and the optimal fitness values and optimal solutions for each operation are recorded, as shown in Table 1.

Table 1 Optimization results of KHA, BA, and PSO algorithm

Serial no.	Krill Swarm Algorithm (KHA)		Bat Algorithm (BA)		Particle Swarm Algorithm (PSO)	
	Optimal fitness	Encodings	Optimal fitness	Encodings	Optimal fitness	Encodings
1	81	30,33,81,70,5	72	74,51,50,35,80	79	50,82,32,75,20
2	79	9,51,33,88,16	72	24,70,7,38,51	79	9,17,51,33,88
3	80	5,43,16,89,51	71	80,37,2,26,68	77	53,32,47,88,5
4	80	30,32,84,5,70	74	55,84,33,91,10	80	38,90,16,53,83
5	82	32,5,51,80,88	72	10,50,84,77,23	77	10,30,45,70,18
6	80	44,17,51,88,81	70	15,47,86,51,43	76	88,60,21,80,43
7	81	5,91,81,51,32	74	89,10,35,28,85	74	27,8,32,56,74
8	82	32,65,80,50,5	70	16,92,89,59,41	78	32,20,30,5,70
9	81	5,70,32,83,30	69	84,91,17,3,50	75	65,10,52,47,31
10	82	89,32,51,5,80	75	3,72,59,80,15	80	74,40,79,54,16
11	81	8,74,32,81,50	71	44,17,50,18,74	80	75,40,51,83,16
12	82	88,50,81,5,32	65	30,35,2,2,64	79	53,5,81,30,83
13	81	30,81,32,5,70	68	67,29,33,86,47	72	39,93,32,5,54
14	82	51,5,32,74,83	72	62,30,7,2,33	76	68,21,30,78,33
15	80	88,5,53,17,335	76	33,83,71,8,52	76	32,5,96,55,81
16	80	51,17,89,33,80	70	77,81,23,60,21	82	50,33,75,84,6
17	82	32,5,80,74,51	73	92,59,25,39,81	79	16,89,44,81,54
18	82	53,89,5,83,32	73	69,83,35,51,18	80	5,59,89,32,80
19	80	37,74,17,80,53	71	30,36,79,11,92	77	88,68,80,30,19
20	82	5,32,83,53,88	74	79,53,89,42,23	77	35,70,16,83,59

4.2 Analysis of Results

As can be seen from Table 1, among the three population intelligent optimization algorithms, KHA and PSO algorithms can find the optimal monitoring point arrangement scheme, and BA algorithm fails to find the global optimal solution. KHA finds the global optimal value 82 for 8 out of 20 times, and the remaining 12 times fall into the local optimal situation, but the overall fitness value stays above 79. BA algorithm fails to find the global optimum in all 20 runs and the overall fitness value stays above 65. PSO algorithm finds the global optimum in 1 out of 20 runs and the overall fitness value stays above 72. The results show that the KHA algorithm's overall optimization ability is significantly better than the BA and PSO algorithms, in which the BA algorithm has poorer optimization results, this may be related to the coding design of the algorithm own, and the BA algorithm pays more attention to the algorithm's running speed. The 20 operation times of the three algorithms of the KHA, BA, and PSO algorithms are 157.72 s, 18.26 s, and 20.46 s respectively,

Fig. 1 Algorithm operation
result fitness map

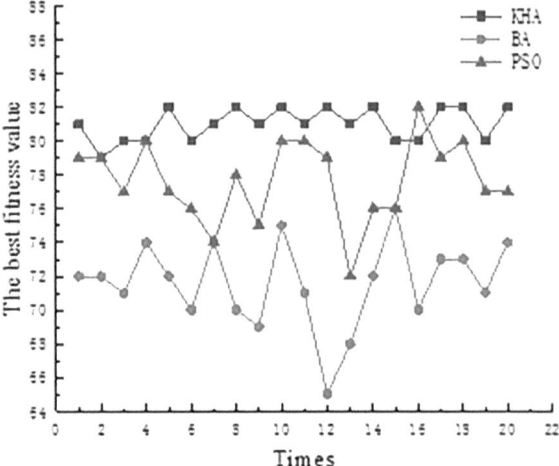

and the BA demonstrates a higher running rate. But the arrangement of the pressure monitoring points of the water supply network pays more attention to the arrangement effect, it is more inclined to the accuracy of the results, so the effect of BA is not good. The PSO algorithm also shows a better performance, and can find the global optimal solution, but the stability is not as good as that of KHA.

Figure 1 shows the adaptability of the three algorithms, which shows that KHA is better than the rest of the algorithms both in terms of optimization stability and optimization accuracy, and the optimal solution obtained by the algorithms is of higher quality, which shows a better application value in the problem of optimal arrangement of monitoring points.

In addition, the optimal solution of the model is not unique, the KHA algorithm in the solution process has resulted in a total of 8 groups of optimal solutions, the number of nodes in its coverage can reach 82. The flow coverages of the monitoring points of different arrangement schemes are compared, as shown in Table 2. It can be concluded that the flow covered by the scheme {32,5,51,80,88} reaches the maximum value. Considering the node coverage and the flow covered by the monitoring points, the scheme {32,5,51,80,88} is selected as the optimum arrangement scheme. The node coverage of this scheme reaches 85.42%, and the node flow coverage reaches 72.16%.

The distribution of monitoring points is shown in Fig. 2. It can be seen that the pressure monitoring points are evenly distributed in each area of the water supply network. The layout position of monitoring points is mostly located at the intersection of large pipe sections and the end of the pipe network, covering the area with dense nodes and large water consumption. The optimization results are representative and in line with the principle of pressure point arrangement, which has certain practical guiding significance. This is because KHA introduces the crossover and mutation operation of genetic algorithm on the basis of the conventional strategy of individual position update, which improves the comprehensive performance of the algorithm

Table 2 Volume of water covered by pressure measuring point arrangement options

Program monitoring point node code	Volume (m³/h)
32,5,51,80,88	149.73
32,65,80,50,5	145.34
89,32,51,5,80	149.23
88,50,81,5,32	149.00
51,5,32,74,83	149.23
32,5,80,74,51	149.23
53,89,5,83,32	149.23
5,32,83,53,88	149.23

Fig. 2 Layout plan of pressure monitoring points

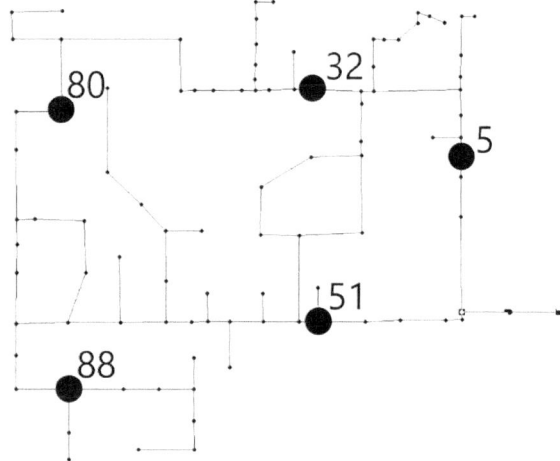

and is easier to jump out of the local optimal solution, but it increases the complexity of the algorithm and prolongs the calculation time of the algorithm.

5 Conclusion

The mathematical model for optimal arrangement of pressure measuring points in water supply network is constructed with the goal of maximizing the monitoring range, which considers the topology of the pipe network, pressure correlation and node flow sensitivity. The KHA algorithm, PSO algorithm and BA algorithm are applied to solve the model, and the optimum arrangement scheme of pressure monitoring points is realized in the example pipe network. After the selection of multiple options, the pressure monitoring points were finally determined, and the maximum monitoring range can cover 85.42% of the total nodes in the pipe network and 72.16% of the total flow in the pipe network. The location of the pressure monitoring points has a high degree of match with the traditional empirical method. Compared with the PSO algorithm and BA algorithm, the KHA algorithm shows better search accuracy and stability in model solving. Based on the accuracy of the calculation results, the KHA algorithm is more advantageous in solving the model for optimal arrangement of pressure monitoring points in water supply network.

The results of this study and the application of the algorithm are universal. The excellent performance of KHA can be widely used in the optimal arrangement of pressure monitoring points of various hydraulic pipe networks. The future research direction will focus on simplifying the operation process of krill herd algorithm, shortening the operation time and improving the efficiency.

Acknowledgements This work was financially supported by the National Natural Science Foundation of China 51778307.

References

1. S. Zhou, S. Xu, Studying optimal locating of pressure monitoring station in urban water distribution system. J. Nanhua Univ. **19**, 59–63 (2005)
2. R. Perez, V. Puig, J. Pascual, Methodology for leakage isolation using pressure sensitivity analysis in water distribution networks. Control. Eng. Pract. **19**, 1157–1167 (2011)
3. M.V. Casillas, L.E. Garza, Optimal sensor placement for leak location in water distribution networks using evolutionary algorithms. Water **7**, 6496–6515 (2015)
4. A.M. Sabzkouhi, A. Haghighi, Uncertainty analysis of pipe-network hydraulics using a many-objective particle swarm optimization. J. Hydraul. Eng. **142**, 1–12 (2016)
5. E. Raei, M.R. Nikoo, Optimal joint deployment of flow and pressure sensors for leak identification in water distribution networks. Urban Water J. **15**, 837–846 (2019)
6. A. Abo-Monasar, M. Al-Zahrani, Framework for water quality monitoring system in water distribution networks based on vulnerability and population sensitivity risks. Water Sci. Technol. **17**, 811–824 (2016)
7. C. Peng, S. Peng, Q. Wu, J.W. Liang, Multi-objective optimization of arrangement of pressure monitoring points in water distribution network based on NSGA-II. China Water Wastewater **35**, 58–62 (2019)
8. S. Liu, H. Wang, P. Xu, S. Xu, Multiobjective genetic algorithms for optimal monitoring station placement in large water distribution systems. J. Tsinghua Univ. **53**, 78–83 (2012)
9. A.H. Gandomi, A.H. Alavi, Krill herd: a new bio-inspired optimization algorithm. Commun. Nonlinear Sci. Numer. Simul. **17**, 4831–4845 (2012)

Spatio-Temporal Characteristics of Waves in Hangzhou Bay Based on ERA5 Wind Field

Guodan Zheng, Taoxiao Chen, Yuan Shi, Ye Liu, and Yuanping Yang

Abstract On the basis of 30-year observed ERA5 wind field analysis data, the wave changes around Hangzhou Bay are generated based on SWAN model. The wave rose diagrams of different locations in Hangzhou Bay have been drawn, the seasonal and interannual variation characteristics of wave height are analyzed. Several results are followed. (1) The direction of dominate wave located in the Hangzhou Bay head are NE and ENE, whereas the strongest wave direction is NE; Meanwhile, in the middle of Bay, the dominate wave direction, which on the northern side are E and ESE, in middle are ENE and E, and on the southern side are NE and ENE. The strongest wave direction from north to south are ESE, E and ENE respectively. In the inlet, the dominate wave direction, which on the northern side are ENE and E, in middle are NE and ENE, and on the southern side is NNE. The strongest wave direction from north to south are ESE, E and NE respectively. (2) The trend of mean significant wave height distribution is consistent all year round. It is diffused from the inlet to the top of Hangzhou Bay. The inlet is larger than the top, and the middle is larger than both sides. The quarterly average significant wave height varies from 0.36 to 0.40 m, while the mean value in autumn is slightly higher than those in other seasons. (3) In Hangzhou Bay, there is no vertical difference on annual mean wind speed of the latest 30 years, and the general trend is relatively stable. The variation tendency of annual mean significant wave height is consistent with that of annual mean wind speed. Also it is found that the correlation between them is good with the correlation coefficient of 0.77.

Keywords Hangzhou Bay · ERA5 wind field · Swan · Wave rose diagrams · Wave seasonal variation

G. Zheng · T. Chen (✉) · Y. Liu · Y. Yang
Zhejiang Institute of Hydraulics & Estuary (Zhejiang Institute of Marine Planning and Design), Hangzhou 310020, Zhejiang, China
e-mail: chentaoxiao2968@dingtalk.com

Key Laboratory of Estuary and Coast of Zhejiang Province, Hangzhou 310020, Zhejiang, China

Y. Shi
Hohai University, Nanjing 210003, Jiangsu, China

W. Wang et al. (eds.), *Hydraulic Structure and Hydrodynamics*, Lecture Notes in Civil Engineering 608, https://doi.org/10.1007/978-981-97-7251-3_40

1 Introduction

Hangzhou Bay is the offshore section of the Qiantang River estuary, with a length of about 98 km. It is dominated by tidal current, the flow path is basically stable, and the riverbed is relatively stable [1]. Both banks are densely populated, the economy is developed, and there are many wading projects in the bay. Waves are one of the main hydrodynamic factors in the bay. Therefore, it is of great significance to study the regional distribution characteristics and interannual variation characteristics of wave elements for disaster prevention and reduction, ocean engineering and energy development.

Long term wave stations in Hangzhou Bay include Zhapu, Tanhu and Zhenhai, which are respectively located in the north of the bay, the middle of the bay and the south of the bay. Location is shown in Fig. 1. The data of these three stations can roughly express the wave characteristics in Hangzhou Bay. Because it is difficult to collect long series of wave station data, it is impossible to systematically analyze the characteristics of long duration waves in Hangzhou Bay. The main researches on waves in Hangzhou Bay are as follows: Xie et al. [2] calculated the typhoon waves at the Qiantang River estuary under the action of super typhoon; Huang et al. [3] studied the wave height process generated by super typhoon landing at different landing points; Bin et al. [4] analyzed the wave spectrum based on the one-year measured wave data of the central station in Hangzhou Bay, improved the standard spectrum one, and statistically analyzed the measured wave height period distribution; Hu et al. [5] calculated the essence of design wave elements in the Qinshan sea area of Hangzhou Bay; Wang et al. [6] calculated the 50 year return period waves in the waters of Hangzhou Bay by numerical simulation. With the gradual publication of reanalysis wind field data set or reanalysis wave data set, the research on long duration waves has been further improved. The literature [7-20] provide scholars with the reanalysis wind field data set to perform repeated analysis on waves in different sea areas or directly use its reanalysis wave data set to conduct wave field characteristics analysis. This paper intends to establish a large-scale wave numerical model to reproduce the wave changes in Hangzhou Bay in the past 30 years through the ERA5 reanalysis wind field data set of the globally recognized European Centre for Medium Range Weather Prediction (ECMWF), and analyze the wave characteristics of Hangzhou Bay in the inter annual and intra annual periods.

2 Description of Wind Field Data

2.1 Basic Situation of Wind Field

The ERA5 reanalysis data set of the European Center for Medium Range Weather Prediction (ECMWF) is used for this wind farm, which is the fifth generation reanalysis data product developed by ECMWF. The wind farm lasted from 1991 to 2020

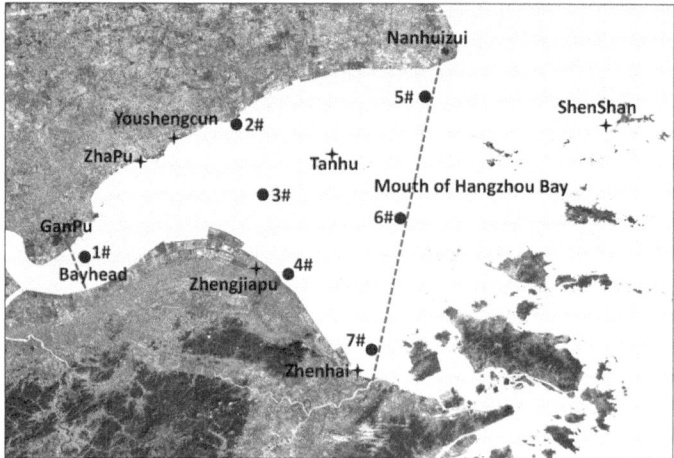

Fig. 1 Schematic diagram of Hangzhou Bay's geographical location, relevant wave wind speed stations and wave representative point

for 30 consecutive years, covering the entire East China Sea with a spatial accuracy of $0.2° \times 0.2°$, and the numerical value is the wind speed at 10 m offshore.

2.2 Quality Analysis of Wind Field in Hangzhou Bay

The measured wind speeds of OBS at Youshengcun Station (Dushangang Town, Pinghu, Jiaxing, 121.21° E, 30.67° N) and Zhengjiapu Station (Fuhai Town, Cixi, Ningbo, 121.45° E, 30.24° N) (The schematic geographic location of the site is shown in Fig. 1) are used to compare the wind speeds of ERA5 data in 2019 and 2020. The comparison results are shown in Figs. 2 and 3. The results show that the ERA5 wind speeds can reflect the seasonal variation of the measured winds; In particular, the strong winds during the summer and autumn typhoon flood season can basically reappear with small errors. The percentage of wind speed error of Youshengcun Station in 2019 (2020) is 26.2% (27.7%), the root mean square error is 1.4 m/s (1.5 m/s), and the average error is small, only 0.4 m/s (0.4 m/s); The percentage of wind speed error of Zhengjiapu Station in 2019 (2020) is 16.5% (18.2%), the root mean square error is 1.3 m/s (1.54 m/s), and the average error is small, only 0.3 m/s (0.3 m/s). It can be considered that the wind field accuracy of ERA5 reanalysis data can basically meet the requirements of the later wave calculation in Hangzhou Bay, Zhejiang Province.

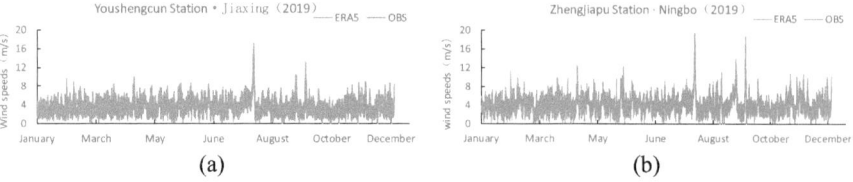

Fig. 2 Comparison of ERA5 wind field and OBS measured values at Yousheng Village and Zhengjiapu station in 2019

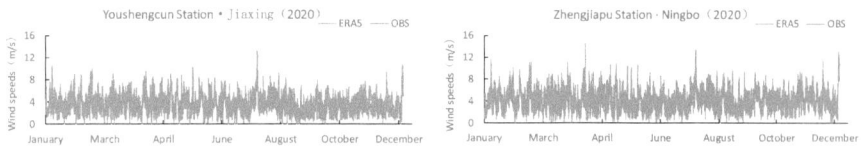

Fig. 3 Comparison of ERA5 wind field and OBS measured values at Yousheng Village and Zhengjiapu station in 2020

3 Wave Mathematical Model

3.1 Basic Principles

The SWAN model was developed and maintained by the Department of Civil Engineering of Delft University of Technology in the Netherlands. From the original version of SWAN 30.51 to the current version of SWAN 41.31, its performance and function have been gradually improved. The dynamic spectrum balance equation is used as the control equation to describe the waves. The control equation of SWAN model in the direct coordinate system is as follows:

$$\frac{\partial}{\partial t}N + \frac{\partial}{\partial x}C_x N + \frac{\partial}{\partial y}C_y N + \frac{\partial}{\partial \sigma}C_\sigma N + \frac{\partial}{\partial \theta}C_\theta N = \frac{S}{\sigma} \tag{1}$$

The first term on the left of the equation represents the rate of change of the dynamic spectral density in time, the second and third terms represent the propagation of the dynamic spectral density in geometric space, the fourth term represents the frequency shift caused by the change of current and water depth, and the fifth term represents the refraction and shallow effect caused by the change of current and water depth. The S on the right of the equation represents the energy source and sink term:

$$S = S_{in} + S_{nl3} + S_{nl4} + S_{ds,w} + S_{ds,b} + S_{ds,br} \tag{2}$$

On the right side of the formula, it is indicated that S_{in} represents the input of wind energy, S_{nl3} represents the redistribution of energy by three-wave interaction, S_{nl4} represents the redistribution of energy by four-wave interaction, $S_{ds,w}$ represents

Fig. 4 Range of wave calculation model

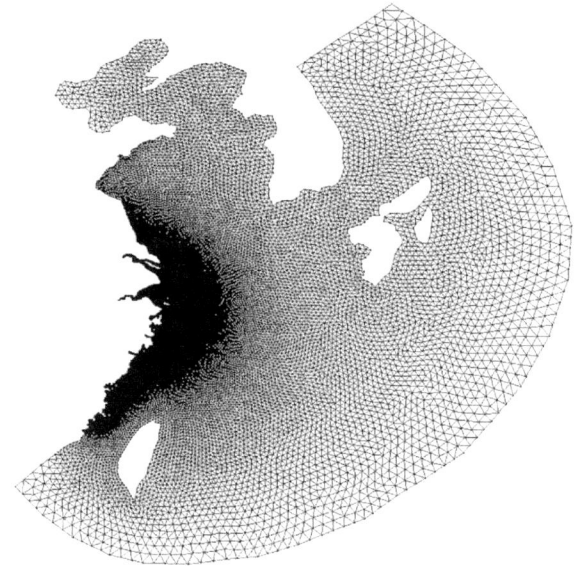

the white hat dissipation, $S_{ds,b}$, $S_{ds,br}$ represents the energy dissipation caused by underwater friction and wave breaking.

3.2 Model Calculation Range

This time, the SWAN unstructured grid is used for wave calculation, and the East China Sea wave model is built, and the Zhejiang sea area and the Hangzhou Bay sea area are densified. The grid in the Hangzhou Bay gradually transits from the shore to the middle of the bay. The grid scale is between 100 and 500 m. The shoreline and terrain are pieced together using the latest measured terrain, which can reflect the actual topography of the project sea area. The calculation range of the model is shown in Figs. 4 and 5.

3.3 Model Validation

The wave mathematical model used the collected observation data during the typhoon to effectively verify the Zhejiang waters and the Hangzhou Bay, and the results are good. It can be seen from the literature2. This paper further verified the waves in the Hangzhou Bay during the 2019 "Lichma" typhoon on the basis of the predecessors, taking Shengshan outside the bay and Zhapu in the bay as the verification point, and the verification diagram is shown in Fig. 6. It is shown that the model can well

Fig. 5 Local grid of
Hangzhou Bay

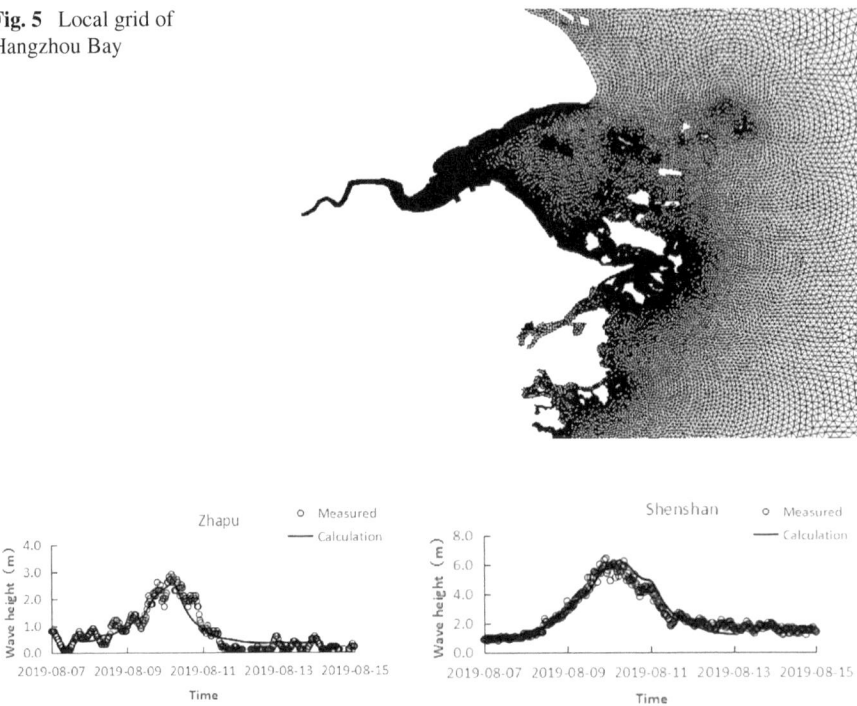

Fig. 6 Comparison between calculated and measured waves at Zhapu station and Shengshan station
during Liqima Typhoon

reproduce the wave change process in the waters of Hangzhou Bay and can be used
for the next stage of calculation.

4 Wave Feature Analysis

4.1 Analysis of Constant Wave Direction and Strong Wave Direction

In order to study the wave characteristics of different spatial locations in the
Hangzhou Bay, 1 point is arranged at the top of the bay, 3 points are arranged at
the middle of the bay and the mouth of the bay, a total of 7 points. According to the
calculation results of the past 30 years, the wave rose diagram of 7 representative
points is drawn, as shown in Fig. 7, which can basically represent the characteristics
of constant and strong waves in the entire Hangzhou Bay. From the figure, we can
see:

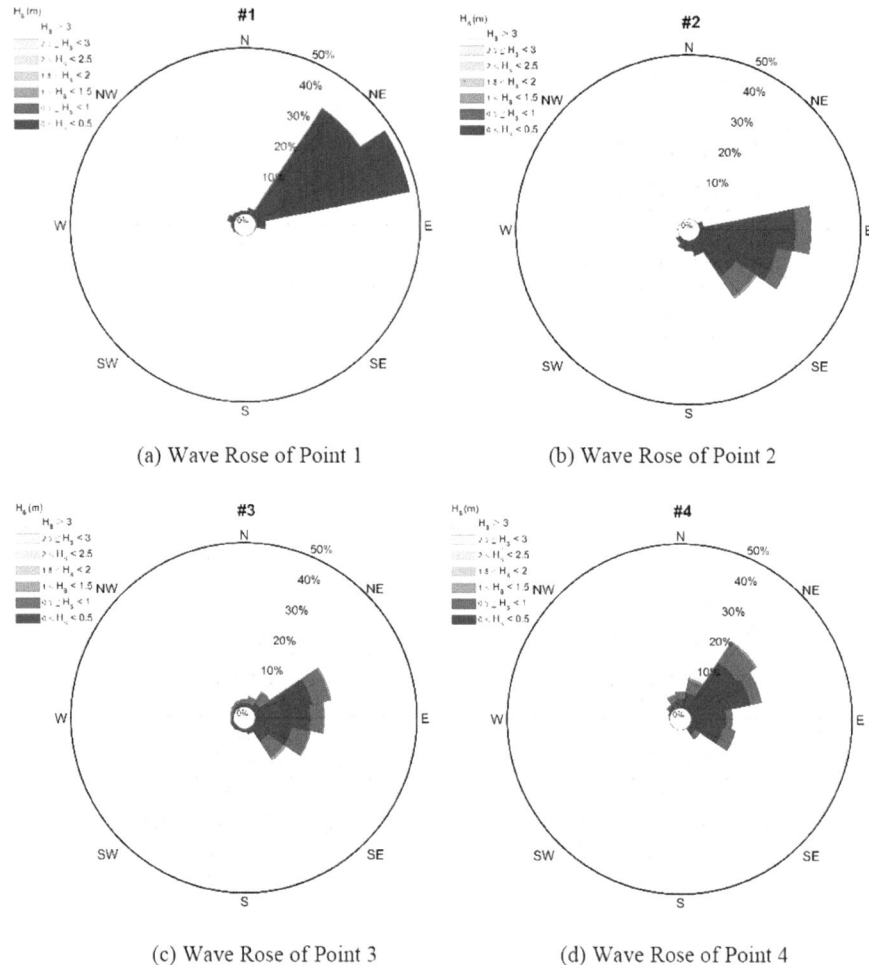

(a) Wave Rose of Point 1 (b) Wave Rose of Point 2

(c) Wave Rose of Point 3 (d) Wave Rose of Point 4

Fig. 7 Rose diagram of each wave representing point

(1) At the top of the Hangzhou Bay, the shoreline trend is southwest-northeast, and its normal wave direction is NE-ENE, accounting for about 86.9%, followed by E-direction, accounting for about 2.97%. Restricted by the shorelines on both sides, the length of the wind zone is relatively small. The wave height is mostly 0–0.5 m, accounting for about 98.9%. The strong wave direction is NE, and the maximum effective wave height is not more than 2 m.

(2) For the middle of Hangzhou Bay, the normal wave direction of No. 2 point on the north side is E~ESE, accounting for about 64.2%, followed by SE, accounting for about 21.9%; The normal wave direction of No. 3 point near the center is ENE~E direction, accounting for about 45.9%, followed by ESE direction, accounting for about 17.7%; The normal wave direction at No. 4 point

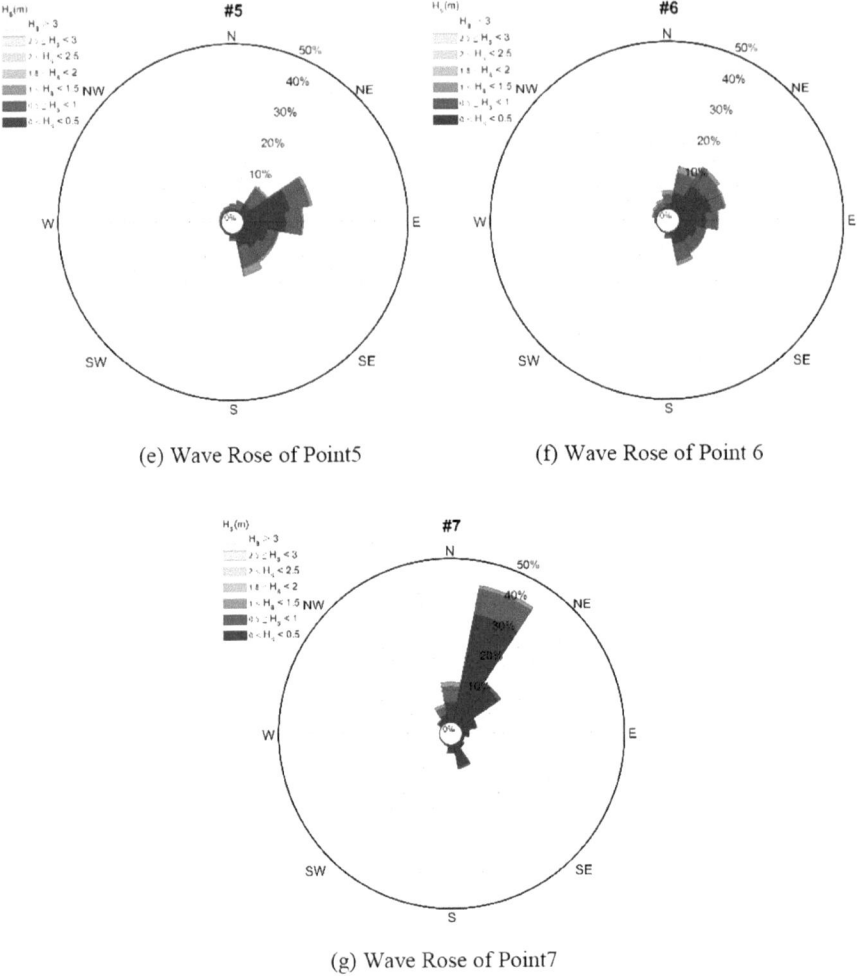

(e) Wave Rose of Point5   (f) Wave Rose of Point 6

(g) Wave Rose of Point7

Fig. 7 (continued)

on the south side is NE~ENE, accounting for about 47.5%, followed by ESE, accounting for 14.0%. The normal wave direction changes from north to south, from east to northeast.

The strong wave direction from north to south are ESE, E and ENE respectively, which are close to the regular wave direction change rule, and change from north to south from southeast to northeast. The maximum effective wave height at points 2 and 3 is 3–3.5 m, and the maximum effective wave height at point 4 is not more than 3 m.

(1) For the location of Hangzhou Bay mouth, the normal wave direction of No. 5 point on the north side is ENE~E direction, accounting for about 39.8%,

followed by SSE direction, accounting for about 13.3%, and the ESE and SE directions are also relatively close, accounting for about 11.2%; The normal wave direction of No. 6 point near the center is NE~ENE direction, accounting for about 30.5%, followed by NNE direction and E direction, accounting for 13.1% and 12.1% respectively, and ESE and SE directions are also relatively close, accounting for about 8.7%; The normal wave direction of Point 7 to the south is NNE, accounting for about 42.7%, followed by NE and N, accounting for 15.3% and 12.4% respectively, and the other directions are relatively small. For the bay mouth, except that the south side is directly covered by the Zhoushan Islands, it is difficult to form easterly and southeasterly waves, and the directions of waves coming from other places are not as concentrated as those in the bay and in the bay, with different proportions between NNE and SSE directions.

The strong wave direction from north to south is ESE, E and NE respectively. The maximum effective wave height at points 5 and 6 is 3–3.5 m, and point 7 is not more than 3 m. The change trend of strong wave direction from north to south is close to that in the bay.

4.2 Analysis of Seasonal Variation of Wave Height

Influenced by the southeast monsoon, Hangzhou Bay belongs to the subtropical monsoon area, with obvious seasonal changes. In winter, it is located in the southeast of the strong and stable Mongolian high, with northerly winds predominating. In summer, under the control of the subtropical high, southeasterly winds prevail. Spring and autumn are the transition period. In order to study the seasonal variation characteristics of the effective wave height in Hangzhou Bay, the data calculated hourly for 30 consecutive years are averaged for many years. Figure 8 shows the distribution of the average effective wave height in spring (March to May), summer (June to August), autumn (September to November), and winter (December to February of the next year) in Hangzhou Bay. It can be seen from the figure:

Influenced by the southeast monsoon, Hangzhou Bay belongs to the subtropical monsoon area, with obvious seasonal changes. In winter, it is located in the southeast of the strong and stable Mongolian high, with northerly winds predominating. In summer, under the control of the subtropical high, southeasterly winds prevail. Spring and autumn are the transition period. In order to study the seasonal variation characteristics of the effective wave height in Hangzhou Bay, the data calculated hourly for 30 consecutive years are averaged for many years. Figure 8 shows the distribution of the average effective wave height in spring (March to May), summer (June to August), autumn (September to November), and winter (December to February of the next year) in Hangzhou Bay. It can be seen from the figure:

The trend of the multi-year mean effective wave height distribution map of the Hangzhou Bay in spring, summer, autumn and winter is basically the same. The wave height is diffused from the mouth to the top of the bay. The mouth of the bay is

Fig. 8 Distribution of
perennial mean effective
wave heights in Hangzhou
Bay in Spring, Summer,
Autumn and Winter

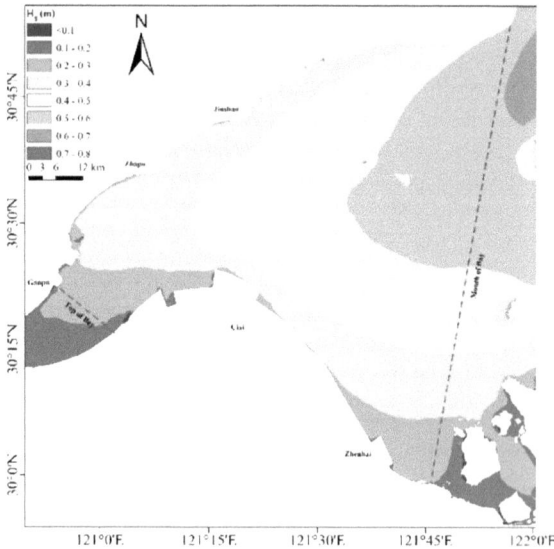

(a) Distribution of perennial mean effective
wave heights in Hangzhou Bay in spring

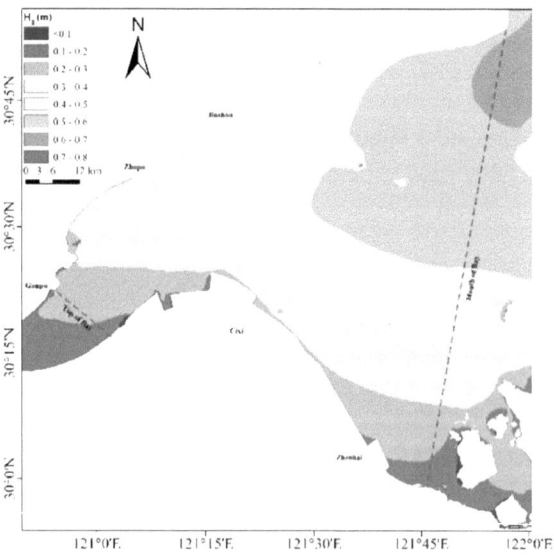

(b) Distribution of perennial mean effective
wave heights in Hangzhou Bay in summer

Fig. 8 (continued)

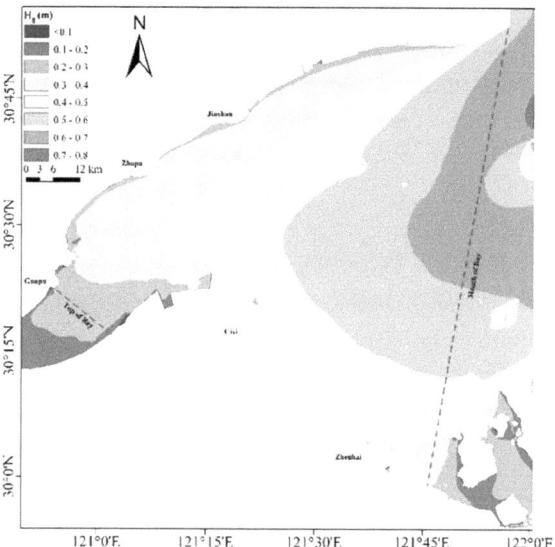

(c) Distribution of perennial mean effective wave heights in Hangzhou Bay in autumn

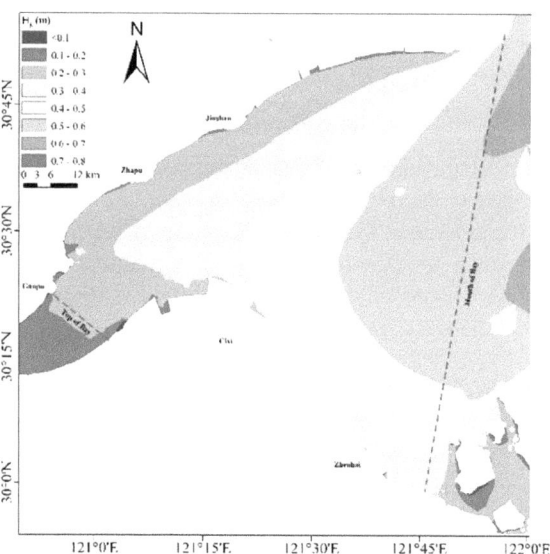

(d) Distribution of perennial mean effective wave heights in Hangzhou Bay in winter

larger than the top of the bay, and the middle of the bay is larger than both sides. The average effective wave height at the mouth of the bay can reach about 0.6–0.7 m, and it is basically below 0.3 m at the top of the bay. From the numerical point of view, the multi-year average effective wave heights in spring, summer, autumn and winter in Hangzhou Bay are 0.36 m, 0.37 m, 0.39 m and 0.35 m respectively. The average wave height changes little in four seasons, but slightly in autumn.

4.3   Analysis of Interannual Variation of Wave Height

Average the annual effective wave height data of Hangzhou Bay from 1991 to 2020 in the bay mouth and the bay top area, and analyze the trend change of the annual average effective wave height and the average wind speed in the sea area of Hangzhou Bay for 30 consecutive years, as well as the relevant situation. See Fig. 9 for the variation of the annual average effective wave height and the average wind speed in Hangzhou Bay, and see Fig. 10 for the relationship between the annual average wind speed and the annual average effective wave height.

It can be seen from Fig. 9 that the annual average wind speed in the Hangzhou Bay area has a small difference in the past 30 years, and the average wind speed in the previous year has changed between 4.74 and 5.25 m/s, with no obvious increase or decrease trend, and the trend is relatively stable. The annual mean wave height is restricted by the wind field on the surface of the Hangzhou Bay. The change trend of the annual mean wave height is basically consistent with the change trend of the annual mean wind speed, and the interannual difference is also small. The annual mean effective wave height changes between 0.34 and 0.40 m, and there is no obvious increase or decrease trend.

It can be seen from Fig. 10 that the annual average wind speed in the Hangzhou Bay has a good correlation with the annual average effective wave height, and its correlation coefficient has reached 0.77 in the past 30 years, which further shows

Fig. 9 Variation of annual mean effective wave height and annual mean wind speed in Hangzhou Bay

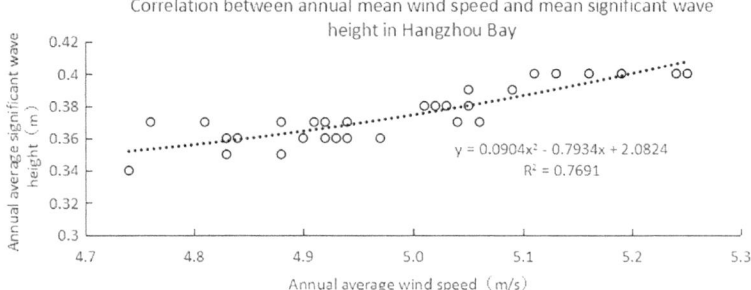

Fig. 10 Correlation between annual mean wind speed and annual mean effective wave height in Hangzhou Bay

that the waves in the waters of the Hangzhou Bay are basically dominated by wind waves, and the wave growth is affected by the wind speed, which is consistent with the conclusion that the Hangzhou Bay is dominated by wind waves observed by the measured wave stations in the Hangzhou Bay.

5 Conclusion

Based on the ERA5 wind field data for 30 consecutive years from 1991 to 2020, the long-duration waves in the Hangzhou Bay have been re-calculated by SWAN model, and the wave height data set has been constructed. The regular and strong wave directions at different spatial locations in the Hangzhou Bay have been systematically analyzed, and the seasonal and annual changes of the effective wave height have been analyzed. The following conclusions have been obtained:

(1) The normal wave direction at the top of Hangzhou Bay is NE~ENE, and the strong wave direction is NE; In the bay, the normal wave direction is E~ESE in the north, ENE~E in the middle, NE~ENE in the south, and the strong wave direction is ESE, E and ENE from north to south; At the mouth of the bay, the normal wave direction is ENE~E on the north side, NE~ENE on the middle, NNE on the south, and ESE, E and NE on the north to south. For the mouth of the bay, except that the south side is directly covered by the Zhoushan Islands, the east and southeast waves are difficult to form, and the wave directions from other locations are not as concentrated as those in the bay and in the bay. The NNE~SSE directions have different proportions of waves.

(2) The distribution trend of multi-year mean effective wave height in spring, summer, autumn and winter in Hangzhou Bay is basically the same. The wave height is diffused from the mouth to the top of the bay. The mouth of the bay is larger than the top of the bay, and the middle of the bay is larger than both sides. The average effective wave height at the mouth of the bay can reach about 0.6– 0.7 m, and it is basically below 0.3 m at the top of the bay. The average

value of the four seasons is between 0.36 and 0.40 m, and it is slightly larger in autumn.

(3) The wind speed difference in Hangzhou Bay in recent 30 years is small. The average wind speed in the previous year changes between 4.74 and 5.25 m/s, and the trend is relatively stable. The change trend of the annual average effective wave height is consistent with the change trend of the annual average wind speed. The average effective wave height in the previous year changes between 0.34 and 0.40 m, and there is no obvious increase or decrease trend. The correlation between wind speed and effective wave height is relatively good, up to 0.77, which further shows that the waves in the waters of Hangzhou Bay are basically dominated by wind waves, and the wave growth is subject to the influence of wind speed.

Acknowledgements This work was financially supported by Presidential Foundation of Zhejiang Institute of Hydraulics & Estuary (Zhejiang Institute of Marine Planning and Design) (ZIHE21Y001); Zhejiang Province Water Resources Department Science and Technology Planning Project (RC2020).

References

1. C. Pan, Z. Han, et al., *Research on Conservation and Regulation of Qiantang Estuary* (Water Power Press, Beijing, China, 2018)
2. Y. Xie, S. Huang, et al., Calculation of typhoon-generated wave due to a super typhoon in Qiantang Estuary. Acta Oceanologica Sinica **35**(1), 38–43 (2013)
3. S. Huang, X. Zhao, et al., Typhoon-generated wave height due to the super typhoon in the coastal region of Zhejiang Province. Marine Sci. **4**, 369–375 (2012)
4. B. Yang, J. Zhang, et al., Analysis of measured spectral characteristics of central Hangzhou Bay, in *Proceedings of the 18th China Offshore Engineering Symposium* (2017), pp. 295–299
5. J. Hu, Q. Yang, et al., Numerical simulation of design waves in Qinshan Water area of Hangzhou Bay. Zhejiang Hydrotechnics **3**, 15–16 (2011)
6. W. Wang, Q. He, et al., Numerical simulation research of wave with a return period of 50 years in the Hangzhou Bay. J. Marine Sci. **4**, 44–48 (2013)
7. S. Hang, Y. Dong, et al., Study of the temporal and spatial variations of wave in South China Sea. Trans. Oceanol. Limnol. **2**, 1–9 (2020)
8. B. Yang, Q. Ye, et al., Statistical wave distribution of central Hangzhou Bay. J. Waterway Harbor **39**(1), 38–43 (2018)
9. H. Tan, Z. Shao, et al., A comparative study on the applicability of EAR5 wind and NCEP wind for wave simulation in the Huanghai Sea and East China Sea. Marine Sci. **5**, 524–540 (2021)
10. D. Xie, Y. Chen et al., On wave distribution of the East China Sea. Port Waterway Eng. **22**, 14–21 (2012)
11. C. Zheng, L. Zhou, Wave climate and wave energy analysis of the South China Sea in recent 10 years. Acta Energiae Solaris Sinica **8**, 1349–1356 (2012)
12. F. Yi, W. Feng, et al., Wave analysis based on ERA-Interim reanalysis data in the South China Sea. Marine Forecasts **1**, 44–51 (2018)
13. S. Qiao, J. Sun, et al., Spatial and temporal characteristics of wave energy resources in the Yellow Sea and Bohai Sea using ERA5 datasets. Oceanologia et Limnologia Sinica **1**, 44–51 (2018)

14. C.B. Gramcianinvo, R.M. Campos, R. de Camargo et al., Analysis of Atlantic extratropical storm tracks characteristics in 41 years of ERA5 and CFSR/CFSv2 databases. Ocean Eng. **216**, 108–111 (2020)
15. T.M. Naseef, V.S. Kumar, Climatology and trends of the Indian Ocean surface waves based on 39-year long ERA5 reanalysis data. Int. J. Climatol. **40**(2), 979–1006 (2020)
16. P. Kumar, S.K. Min, E. Weller et al., Influence of climate variability on extreme ocean surface wave heights assessed from ERA-Interim and ERA-20C. J. Clim. **29**(11), 4031–4046 (2016)
17. Z. Kai, C. Xi et al., Comparison between two sea surface wind fields and their influence on the wave simulations. Marine Forecasts **3**, 9–14 (2012)
18. L. Min, D. Zhao, Comprehensive studies on wind and wave climates in China Seas based on ERA-20C reanalysis data. Periodical Ocean Univ. China **7**, 1–10 (2019)
19. L. Wang, B. Liang, et al., Simulation error of wave models forced by reanalysis wind data in the South China Sea. Periodical Ocean Univ. China (11), 96–104 (2020)
20. C.W. Zheng, L. Zhou, C.F. Huang et al., The long-term trend of the sea surface wind speed and wave height (wind wace, swell, mixed wave) in global ocean during the last 44a. Acta Oceanol. Sin. **32**(10), 1–4 (2013)

Development and Application of an Intelligent Analysis System for Three-Dimensional Geological Modeling in Dredging Engineering

Zheng Lu, Hui Sun, Yuchi Hao, Guoquan Zhao, and He Bai

Abstract In this manuscript, a methodology for three-dimensional geological modeling and visualization in dredging engineering was proposed to understand and analyze the spatial distribution of soil conditions in complex underwater geological settings for dredging engineering comprehensive and enhance the operational efficiency of dredgers. A three-dimensional digitized platform for dredging engineering was established by using computer technology, database management, visualization techniques and 3D geological modeling. The intelligent system of dredging engineering geological database is mainly developed using the .NET as the program development platform and C# and C++ as language. This system realizes the functions of engineering geological data storage and management, 3D geological modeling, dredging engineering analysis, and so on. The practical application showed that the system boasts not only fast 3D modeling speed, strong operability and high conformity with the actual geological situation, but also the dredging engineering analysis module, which greatly improves the quality and efficiency of construction. It can meet the actual needs of dredging engineering well and has broad promotion and application prospects.

Keywords Dredging engineering · Exploration borehole · 3D geological modeling · Digitability analysis

Z. Lu · H. Sun · Y. Hao (✉) · H. Bai
CCCC National Engineering Research Center of Dredging Technology and Equipment Co., Ltd., Shanghai 201208, China
e-mail: haoyuchi@ccccltd.cn

Key Laboratory of Dredging Technology, CCCC, Shanghai 201208, China

G. Zhao
CCCC Shanghai Dredging Co., Ltd., Shanghai 201208, China

© The Author(s) 2025
W. Wang et al. (eds.), *Hydraulic Structure and Hydrodynamics*, Lecture Notes in Civil Engineering 608, https://doi.org/10.1007/978-981-97-7251-3_41

1 Introduction

Dredging engineering refers to the process of excavating underwater soil and rock materials using dredgers or other machinery, followed by their transportation or placement through either blowing or filling operations [1]. In dredging engineering, underwater geological conditions often tend to be complex, characterized by multi-layered distribution, anisotropy, and dynamic variations [2]. Traditional geological data in dredging engineering are typically presented in a two-dimensional, static format through borehole diagrams. This method represents the spatial structure of soil conditions and lacks the capability of dynamic spatial geological displays inadequately and parameterized design analysis [3]. Its low utilization of diagrams impacts both the construction efficiency and design quality of projects significantly, making it challenging to meet the current design requirements of dredging engineering. Presently, there is a limited amount of research focused on three-dimensional geological modeling and visualization analysis specifically for dredging engineering. Much of the research conducted by foreign scholars tends to concentrate on theoretical aspects, with minimal practical application in engineering [4, 5]. Domestic scholars in China have predominantly conducted research that combines theoretical frameworks with practical applications in soil modeling and visualization analysis. They have achieved notable research outcomes in various fields such as mining engineering, tunnel engineering, slope engineering, among others. However, these achievements are confined to specific professional domains and lack general applicability [6–12].

Based on borehole data and utilizing computer technology, database management, 3D modeling, and visualization techniques, this paper developed a three-dimensional geological model and intelligent dredging analysis system suitable for dredging engineering. This system employs .NET as the primary programming platform, it enables comprehensive analysis and utilization of soil information within construction mining areas. Design and construction personnel can gain a more intuitive understanding of underwater soil distribution, thereby enhancing the efficiency of dredging engineering surveys and design processes while elevating the level of construction intelligence. This system can serve as a reference and a source of inspiration for subsequent related engineering projects.

2 Overall System Structure

This system utilizes SQL Server database technology and employs.NET as the primary programming platform, utilizing languages such as C# and C++. It also incorporates the ArcGIS suite as the geographic information platform, using ArcGIS Server as the management platform for map services. Operating in a Client/Server (C/S) architecture, the system adopts plugin technology to enhance system stability and scalability. The overall framework of the system consists of a network support layer,

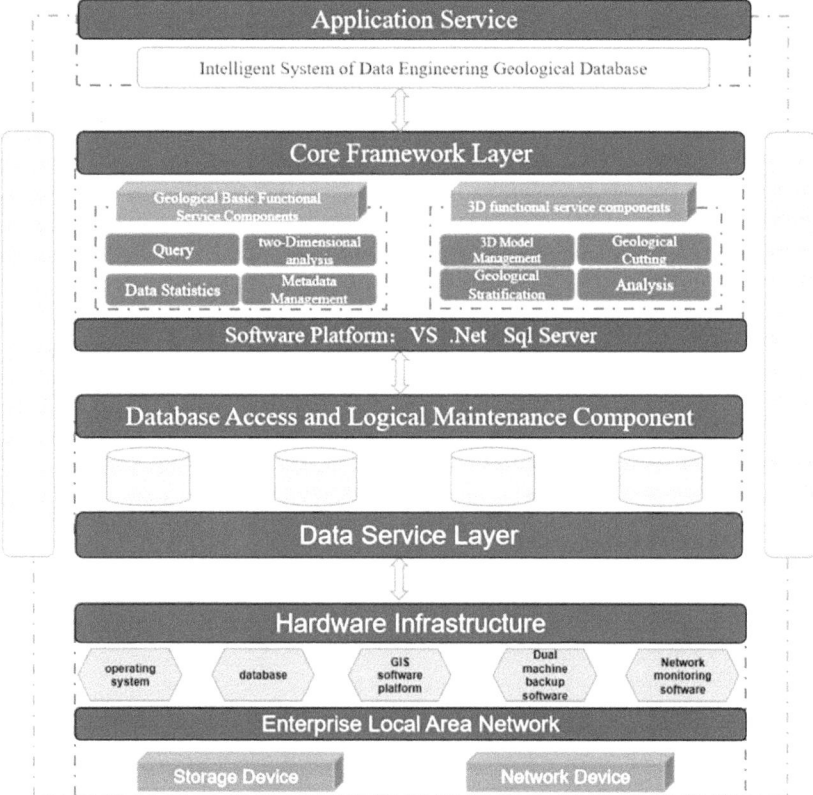

Fig. 1 System framework diagram

data service layer, core framework layer, and hardware infrastructure, as depicted in Fig. 1.

Figure 1 illustrates the application service layer, which constitutes the primary system functional tier directly engaging with users. It is dedicated to crafting interactive interfaces tailored to user requirements and is compatible across prevalent operating systems. The core framework layer serves as the foundation, offering developmental frameworks, fundamental functional components, and an upper-tier application-oriented functional component library. This tier plays a pivotal role in furnishing technical support for data management, system application functionalities, and ensuring stable operations. The data service layer assumes the principal responsibility for storing and managing project data, survey databases, and spatial data. Its function involves the meticulous classification and integration of intricate data, furnishing a crucial groundwork for subsequent modules such as 3D modeling and dredging analysis. The hardware infrastructure primarily encompasses foundational software and hardware platforms. This encompasses an array of components: the software platform includes operating system software, database management system software, GIS software platforms, dual-machine backup software, and network monitoring software. Meanwhile, the hardware platform encompasses servers, storage devices, security equipment, and similar components.

3 System Key Technology

The dredging engineering three-dimensional geological model and intelligent analysis system primarily comprise three modules. The first module is the foundational database module, responsible for managing and storing fundamental survey data as well as GIS data. This module forms the basis for subsequent three-dimensional geological modeling. The technology for constructing this database is well-established and will not be extensively discussed in this paper. The second module is the three-dimensional geological modeling module, primarily tasked with creating three-dimensional models based on the borehole information stored in the system. Lastly, the dredging analysis module encompasses functionalities such as dredging engineering quantity statistics, rock and soil classification, dredging feasibility analysis of rock and soil, dredging output, and more. Due to space limitations, this paper will focus on elaborating on the system's three-dimensional stratigraphic modeling techniques and the analysis of dredgeable rock and soil.

3.1 Three-Dimensional Stratigraphic Modeling Techniques

The creation of three-dimensional geological models using borehole information has been a focal point of research both domestically and internationally. Existing modeling methods have high demands on borehole data, requiring strict adherence to a topological structure where the stratigraphic information provided by boreholes follows a continuous top-down sequence. However, in practical engineering scenarios, the geological conditions are often intricate and complex. Single boreholes can reach depths of hundreds of meters, and their distribution spans a wide area. Various strata often intersect or even exhibit discontinuities within boreholes. Determining the stratigraphic sequence poses a significant challenge in creating three-dimensional geological models. Building upon prior research, this system has improved traditional three-dimensional modeling algorithms. It reevaluates the stratigraphic sequence based on existing borehole data to enhance the algorithm's adaptability to situations where strata may intersect or be discontinuous.

Based on geological structural theories, the system inputs all borehole information into a database and then sorts it from top to bottom based on the frequency of occurrence of each stratum. The strata are arranged in descending order according to their frequency of occurrence. Below is a detailed explanation of the algorithm, illustrated with an example.

Fig. 2 Drilling schematic

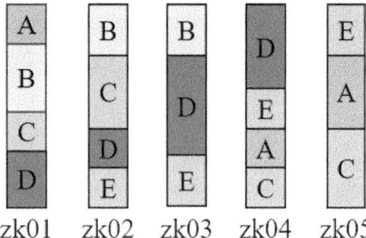

$$zk01 \quad zk02 \quad zk03 \quad zk04 \quad zk05$$

As shown in Fig. 2, there are five boreholes labeled as zk01, zk02, zk03, zk04, and zk05 respectively. Different letters within each borehole represent various soil strata. Based on the existing borehole data analysis, the stratigraphic sequence of each stratum in the geological model has been determined. This serves as the basis for subsequent three-dimensional geological modeling. The steps for determining the stratigraphic sequence are as follows:

Step 1: Traverse through the different soil stratum types within the boreholes. As shown in Fig. 2, there are five boreholes displaying five distinct stratum types, labeled as A, B, C, D, and E respectively.

Step 2: Taking zk01 as an example: determine the stratigraphic sequence by organizing and analyzing the relative positional information of each stratum within the boreholes systematically. There are four distinct strata identified in this borehole.

For instance, we take β_{AB}^{01} as the relative positional relationship between stratum A and stratum B within borehole zk01. $\beta_{AB}^{01} = -1$ indicates that stratum A is below stratum B; $\beta_{AB}^{01} = 1$ signifies that stratum A is above stratum B; when $\beta_{AB}^{01} = 0$, it implies that at least one stratum between A and B is missing.

In this specific case, $\beta_{AB}^{01} = 1$. S_{AB} represents the relative relationships between different strata across all boreholes. As shown in Table 1, if $S_{AB} = \sum_{i=01}^{n} \beta_{AB}^{i}$ is defined, the relative positional relationship values for each stratum in this particular case can be calculated.

Step 3: The stratigraphic sequence for each stratum can be preliminarily determined based on Table 1. However, there might be contradictions in some of the strata sequences. For example, if $S_{AB} = 1$ and $S_{BE} = 2$ imply a sequence of A, B, E from top to bottom, but $S_{AE} = -2$ deduces results that contradict the aforementioned conclusion. A secondary adjustment should be made to the Table 1 results, following the principle that prioritizes the frequency of occurrence of the same stratigraphic sequence. The higher the absolute value of S_{mn}, the higher the priority for recognizing the stratigraphic sequence. If there are contradictions between other stratigraphic sequences and the one with higher frequency, they should be excluded; otherwise,

Table 1 Table of relative positions of different strata

S_{AB}	S_{AC}	S_{AD}	S_{AE}	S_{BC}	S_{BD}	S_{BE}	S_{CD}	S_{CE}	S_{DE}
1	3	0	-2	2	3	2	1	−1	3

Table 2 Adjusted table of relative positions among different strata after modification

S_AC	S_BD	S_DE	S_BC	S_BE	S_AB	S_CD
3	3	3	2	2	1	1

they are included. Adjustments to the Table 1 results following this approach are presented in Table 2.

Step 4: Based on the information from Table 2, inferring the stratigraphic sequence reveals that the sequence from top to bottom is uniquely determined as A, B, C, D, and E, with no contradictions present.

Once the stratigraphic sequence is determined, the construction of the geological model proceeds layer by layer from bottom to top, adhering to the established sequence. In instances where there's disorder in the stratigraphic sequence, the boreholes are segmented into multiple subsections, each having an orderly stratigraphic sequence. Subsequently, interpolation modeling is performed for each subsection based on the determined sequence. Commonly employed techniques include Kriging or inverse distance weighting methods for rapid modeling, although this paper won't delve into detailed explanations of these methods. These individual layer models are then assembled from the bottom upwards, resulting in the final three-dimensional geological model. This approach ensures comprehensive utilization of borehole data, prevents loss of individual borehole information, and accurately applies it to stratigraphic modeling. Moreover, it adapts well to complex geological structures such as stratigraphic truncation, overlying layers, and lens bodies.

3.2 Analysis of Excavatability of Dredged Rock and Soil

Currently, there is existing literature that grades the excavatability difficulty of homogeneous dredged rock and soil. However, in practical construction, due to the complex soil conditions in dredging projects, encountering a single and clear soil condition during construction is challenging. Dredged soil excavated often consists of a combination of various soil types, presenting an ever-changing combination. Without quantitative excavatability indicators beforehand, it becomes disadvantageous for the selection of dredging equipment, thereby impacting construction efficiency.

Based on the aforementioned issues, this system can assess the excavatability of dredged rock and soil within the dredging area. It automatically categorizes different soil types and calculates suitable dredging volumes for different dredging equipment, serving as a reference for dredging equipment selection. Based on various standards considering different types of dredgers, categories of dredged rock and soil, and engineering characteristics, the excavatability of dredged rock and soil is classified into seven levels: easy, relatively easy, acceptable, relatively difficult, difficult, very difficult and unsuitable. Within the system, these seven levels are represented by different representative values, where a higher value signifies a greater excavation difficulty. The results are shown in Table 3.

Based on the representative values defined in Table 3 and the three-dimensional modeling data, the system can conduct an analysis of excavatability for dredged rock and soil. The steps to achieve this are as follows:

Step 1: Firstly, within the system, create a weight vector $\vec{R}$ where $\vec{R} = (Q_1, Q_2...Q_n)^T$, and '$n$' represents the number of layers of strata in the dredging area. Calculate the weights of each soil layer based on the excavation volume for different soil grades.

$$Q_i = \frac{W_i}{\sum_{j=1}^{n} W_j} \tag{1}$$

Q_i represents the weight value of the excavation volume for the i-th layer of rock and soil, while W_i denotes the excavation volume of the i-th layer of rock and soil.

Step 2: Calculate the excavation representational values of each layer of excavated rock and soil under different dredging equipment. Within the system, create several representative value vectors such as '$\vec{a}$', '$\vec{b}$', etc. The representational value of the excavation of each layer of excavated soil under different types of dredging equipment is denoted as M.

$$\vec{a} = (a_1, a_2...a_n)^T$$
$$\vec{b} = (b_1, b_2...b_n)^T$$
$$M_a = \vec{a} \cdot \vec{R} = a_1Q_1 + a_2Q_2 + ... + a_nQ_n \tag{2}$$

a_i represents the representational value of the excavation for the i-type dredging equipment on the a-th layer of rock and soil, b_i represents the representational value of the excavation for the i-type dredging equipment on the b-th layer of rock and soil, and so on. M_a represents the representational value of excavating soil under the 'a' type dredging equipment.

Step 3: The system recommends the optimal dredging equipment based on the calculation results. The preferred equipment is denoted by 'Min$\{M_a, M_b, M_c, ...\}$'. In cases where multiple devices have the lowest representational values, the system displays the model of the top-ranking equipment.

Table 3 The representative values indicating the degree of difficulty in excavating dredged rock and soil

Excavation difficulty	Easy	Moderately easy	Moderate	Relatively difficult	Difficult	Very difficult	Unsuitable
Representative value	1	2	3	4	5	6	7

4   Engineering Application

Based on the aforementioned research findings, an intelligent three-dimensional geological modeling and analysis system applicable to dredging engineering has been developed. This system has been applied to the Daxiaodeng Land Reclamation Project in Xiamen successfully. Situated in the Xiang'an District of Xiamen City, the project involves the reclamation of land at Daxiaodeng. For this project, a dredger is required to travel to the southeast sea area at the entrance of Xiamen Bay, near Dongding Island's marine mining area, for sand extraction and filling. According to the regional geological survey report, the overlying strata in the extraction area comprise marine sand deposits for backfilling, with intermediate sediments varying in thickness and including silt and sandy soils. Below these lie bedrock formations consisting of altered rocks, mixed rocks, and granite. The geological conditions for the project are complex, presenting significant challenges in both preliminary design and construction phases. Exploration borehole data, totaling 46 boreholes within the area, containing up to 8 different soil strata, were input into the system and visualized (as shown in Fig. 3). The topological relationships between these soil strata layers are complex.

The three-dimensional geological model generated based on borehole data is shown in Fig. 4. The system automatically displays different colors on the 3D model according to the different soil qualities, with labels in the legend. This method provides a clearer and more intuitive representation of the soil distribution in the sand mining area compared to traditional geological survey data, offering valuable references for design and construction. Figure 5 illustrates the geological cross-section of the created 3D geological model from any angle. It allows cutting the model along the dredging trajectory of the trailing suction hopper dredger, enabling construction personnel to visually understand the soil characteristics at various digging depths along different dredging trajectories. This aids in selecting suitable construction techniques and provides an intuitive, visual platform for three-dimensional soil analysis, further enhancing the efficiency of sand extraction operations and shortening the construction period.

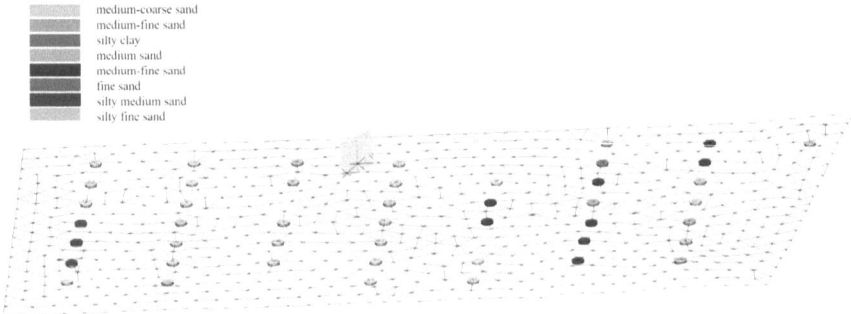

Fig. 3 Mining site borehole data

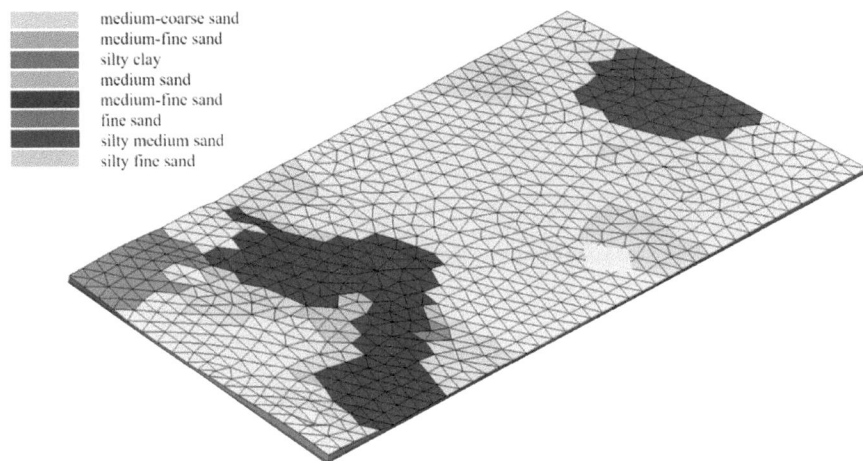

Fig. 4 The three-dimensional geological model of the mining area

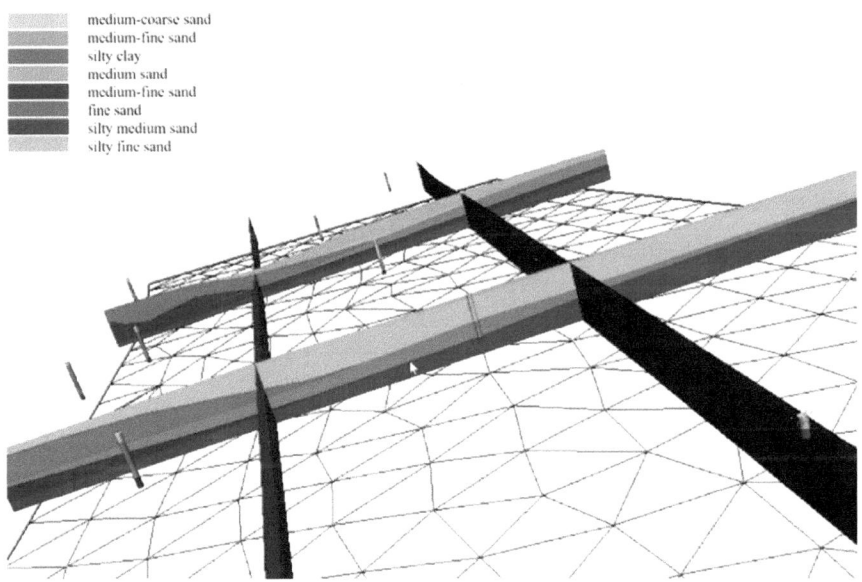

Fig. 5 Diagram illustrating the geological cross-section of the mining area

Figure 6 presents the system's excavatable layer analysis within the dredging area and recommends the optimal dredging equipment based on a comprehensive assessment of the three-dimensional geological model and survey data. According to the three-dimensional geological model, the soil distribution in the mining area predominantly comprises medium-coarse sand, muddy medium sand, and medium-fine sand.

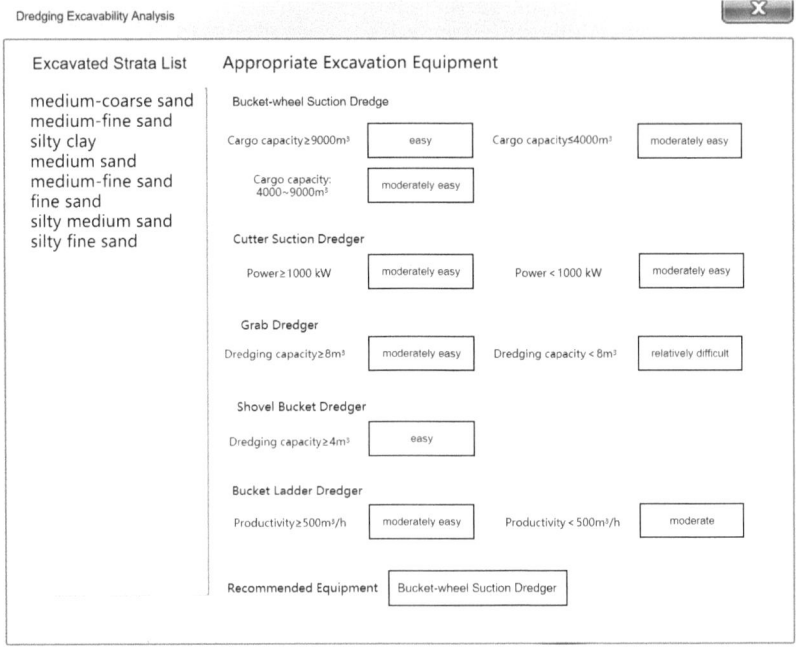

Fig. 6 Excavation suitability analysis display chart for mining area

The soil distribution is uneven and diverse, with multiple soil types. Choosing inappropriate dredging equipment could significantly impact sand extraction efficiency. As depicted in Fig. 6, the system computes the excavatability of various types of dredging equipment based on soil distribution weights. It indicates that cutter suction dredgers, grab dredgers, and bucket chain dredgers have moderate suitability within this dredging area. Based on equipment priority, the system recommends the use of a trailing suction hopper dredger for sand extraction in the mining area. This recommendation provides a basis for equipment selection and ensures the engineering quality of sand extraction operations.

5 Conclusion

In order to address the technical challenge of complex underwater geological conditions in dredging projects and the inability of construction personnel to accurately analyze the engineering properties of the soil in the construction area, a three-dimensional geological modeling and engineering analysis study for dredging projects was conducted based on computer technology, database technology, visualization technology, and three-dimensional geological modeling technology. An intelligent system for dredging project geological database was developed. The system can generate a three-dimensional geological model based on survey data and perform dredging project analysis based on this model, providing a convenient and fast visualization platform for dredging project personnel. This greatly improves the work efficiency of engineering personnel, reduces design costs, and improves the accuracy of design and engineering quantity calculation. At the same time, during the modeling process of this system, users only need to import drilling data and set the grid size of the stratum, and the system can automatically perform three-dimensional geological modeling. It can adapt well to complex geological structures such as stratum extinction, stratum overburden, and lens bodies. The modeling module is also applicable to other professional fields and has broad prospects for promotion and application.

Due to constraints on the system's development timeframe, certain functionalities are still in need of refinement. Subsequent efforts will focus on integrating the ArcGIS platform into the dredging analysis module, utilizing the latitude and longitude coordinates of borehole data. This integration will involve incorporating real-time vessel positioning data alongside water depth information. The primary objective is to enable intelligent analysis of sediment quality parameters by correlating the positions of dredging vessels and their dredging depths. This enhancement aims to further elevate both the efficiency of dredging operations conducted by suction dredgers in offshore settings and the overall standard of construction management practices.

References

1. Q. Li, *Dredging Engineering* (China Communication Press, Beijing, 2000)
2. J. Tian, M. Gu, S. Ding et al., Computer-aided dredging monitoring and decision-making system of cutter suction dredger. Port Waterway Eng. **03**, 20–23 (2005)
3. X. Wu, Y. Sun, J. Ding et al., Field and experimental investigations on three reclaimed yards of dredged soil in Eastern Route of South-to-North Water Diversion project. Adv. Sci. Technol. Water Resources **32**(04), 47–50 (2012)
4. K. Olivier, M. Thierry, Reprint of"3D geological modelling from boreholes, cross-sections and geological maps ,application over former natural gas storages in coal mines. Comput. Geosci. **35**(1), 70–82 (2009)
5. M. Michael, R. Philippe, C. Gabriel et al., A workflow to facilitate three-dimensional geometrical modelling of complex poly-deformed geological units. Comput. Geosci. **35**(3), 644–658 (2009)
6. B. Wang, B. Shi, Z. Song, A simple approach to 3D geological modeling and visualization. Bull. Eng. Geol. Environ. **68**(4), 559–565 (2009)
7. J. Ming, M. Pan, H. Qu et al., GSIS: a 3D geological multi-body modeling system from netty cross-sections with topology. Comput. Geosci. **36**(6), 756–767 (2010)
8. L. Zhu, C. Zhang, M. Li et al., Building 3D solid models of sedimentary stratigraphic systems from borehole data: an automatic method and case studies. Eng. Geol. **127**, 1–13 (2012)
9. D. Zhong, M. Li, J. Yang, 3D visual construction of complex engineering rock mass structure and its application. Chinese J. Rock Mech. Eng. **2005**(03):575–580+2 (2005)
10. D. Zhong, M. Li, L. Song et al., Enhanced NURBS modelling and visualization for large 3D geoengineering applications: an example from the Jinping first-level hydropower engineering project, China. Comput. Geosci. **32**(9), 1270–1282 (2006)
11. D. Zhong, M. Li, J. Liu, 3D integrated modeling approach to geo-engineering objects of hydraulic and hydroelectric projects. Sci. China Ser. E-Technol. Sci. **50**(3), 329–342 (2007)

12. D. Zhong, J. Liu, M. Li et al., NURBS reconstruction of digital terrain for hydropower engineering based on TIN model. Progr. Nat. Sci. **18**(11), 1409–1415 (2008)